Trigonometry With Applications

An **appropriate emphasis** on theory gives students excellent preparation for college mathematics, pp. 101–103.

Worked-out examples provide valuable reference for completing the exercises, pp. 87–89.

Applications smoothly integrated into the text and exercises show students how trigonometry is used to solve problems in everyday life and in the sciences, pp. 1, 15, 32; 2, 18.

An **abundance of exercises** reinforce theory, skills, and applications at each of three levels of difficulty. Extra practice for each chapter is also provided, pp. 103–106, 312.

Compatibility with calculators allows students to use the book successfully with or without calculators, pp. 2–3, 352, 360.

"Computer Investigations" extend mathematical concepts and generate enthusiasm, pp. 120–121.

Algebra and Geometry Review Exercises brush-up skills specifically needed for the course, p. 329.

Comprehensive Tests (two per chapter) not only review concepts and skills but also prepare students for the type of questions that appear on college entrance examinations, pp. 299, 303.

"Challenge" exercises provide extra motivation for very capable students, pp. 106, 132.

Illustrated essays on history and applications expand students' knowledge of many mathematical topics, p. 64.

TRIGONOMETRY
WITH APPLICATIONS

John A. Graham
Robert H. Sorgenfrey

EDITORIAL ADVISERS
Edwin F. Beckenbach
William Wooton

HOUGHTON MIFFLIN COMPANY/BOSTON

Atlanta   Dallas   Geneva, Ill.   Lawrenceville, N.J.   Palo Alto   Toronto

AUTHORS

John A. Graham, Mathematics Teacher, Buckingham Browne and Nichols School, Cambridge, Massachusetts. Mr. Graham formerly taught mathematics at Wellesley College in Wellesley, Massachusetts. He is a member of the Association of Advanced Placement Mathematics Teachers.

Robert H. Sorgenfrey, Professor of Mathematics, University of California, Los Angeles. Dr. Sorgenfrey has won the Distinguished Teaching Award at U.C.L.A.

EDITORIAL ADVISERS

Edwin F. Beckenbach, Professor of Mathematics, Emeritus, University of California, Los Angeles. Dr. Beckenbach served as Chairman of the Committee on Publications of the Mathematical Association of America.

William Wooton, former Professor of Mathematics, Los Angeles Pierce College.

ABOUT THE TEXT

Trigonometry With Applications is a comprehensive text offering a variety of course options. The first seven chapters may be considered as the core of the book. The materials in these chapters are necessary for a basic understanding of trigonometry. The remaining chapters provide extensions of the core topics. These chapters may be presented in any order after the main part of the text is covered.

ISBN 0-395-38539-3
CDEFGHIJ-RM-943210/898

CONTENTS

6 *Vectors in the Plane*   **147**

VECTOR OPERATIONS AND APPLICATIONS

THE DOT PRODUCT AND ITS APPLICATIONS

7 *Complex Numbers*   **175**

POLAR COORDINATES

REPRESENTING COMPLEX NUMBERS

USING THE POLAR FORM OF COMPLEX NUMBERS

Transformations   **213**

Vectors in Space   **233**

Spherical Trigonometry   **259**

11 *Infinite Series*   **285**

ACKNOWLEDGEMENTS

Illustrations by ANCO/Boston

Photographs provided by the following sources:
Cover, © Pete Turner/The Image Bank
Opposite page 1, The Picture Cube/Frank J. Staub
page 33, Mt. Wilson & Palomar Observatories
page 36, NASA
page 65, Honeywell Inc.
page 68, Leo DeWys Inc./Barton Silverman
page 97, The Picture Cube/Andrew Brilliant
page 100, The Port Authority of New York and New Jersey
page 118, Photo Researchers/Bruce Roberts
page 122, SUNY/Stony Brook Photos
page 143, Photo Researchers/Tom McHugh
page 146, Editorial Photocolor Archives/Thomas Sheehan
page 172, Monkmeyer Press Photo Service/E. Meerkamper
page 174, H. S. Rice/courtesy of The American Museum of
   Natural History
page 212, Rainbow/Dan McCoy
page 232, Department of the Interior/Bureau of Reclamation
page 258, NOAA
page 284, Pittsburgh-Des Moines Corporation

LIST OF SYMBOLS

THE GREEK ALPHABET

A	α	Alpha	B	β	Beta	Γ	γ	Gamma	Δ	δ	Delta
E	ϵ	Epsilon	Z	ζ	Zeta	H	η	Eta	Θ	θ	Theta
I	ι	Iota	K	κ	Kappa	Λ	λ	Lambda	M	μ	Mu
N	ν	Nu	Ξ	ξ	Xi	O	o	Omicron	Π	π	Pi
P	ρ	Rho	Σ	σ	Sigma	T	τ	Tau	Υ	υ	Upsilon
Φ	ϕ	Phi	X	χ	Chi	Ψ	ψ	Psi	Ω	ω	Omega

METRIC UNITS OF MEASURE

Length: cm centimeter
        m meter
        km kilometer

Speed: m/s meters per second
      km/h kilometers per hour
      rpm revolutions per minute

Force: N newton

Area: cm^2 square centimeter
     m^2 square meter
     km^2 square kilometer

Mass: g gram
     kg kilogram

Work and Energy: J joule
               kW·h kilowatt hour

The distance between Earth and the moon can be estimated by using trigonometric functions. Such functions can also be used to determine the distance from Earth to nearby stars.

TRIGONOMETRIC FUNCTIONS

OBJECTIVES

1. To find values of the trigonometric functions.
2. To apply the trigonometric functions to solve right-triangle problems.

Basic Concepts

1-1 *Angles and Degree Measure*

One of the first accomplishments of trigonometry was comparing the distance between Earth and the moon with the distance between Earth and the sun. The discovery that made this accomplishment possible was that if

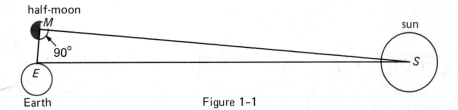

Figure 1–1

the measure of angle *E* in Figure 1-1 is known, then the ratio of the distances *EM* and *ES* can be found, regardless of the size of right triangle *EMS*. In fact, for any given acute angle of a right triangle, the ratio of the lengths of the sides corresponding to $\overline{EM}$ and $\overline{ES}$ can be calculated. The writing down of this ratio by Hipparchus of Nicaea, about 140 B.C., constituted the formal beginning of trigonometry.

The basic ideas used by Hipparchus are used today in many applications.

Example 1   An awning is needed to completely shade a glass door at noon. Given the height of the door and the maximum angle that the sun's rays make with the door, we can find the minimum width x that the awning can be.

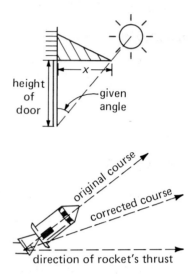

Example 2   A directional rocket attached to a space vehicle is used to correct the course of the vehicle. Given the specifications of the original and corrected courses, we can compute the angle, force, and duration of the rocket's thrust.

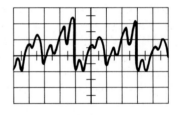

There are other applications of trigonometry that have no obvious connection with angles.

Example 3   The pattern produced by a vibrating violin string as shown on an electronic device called an oscilloscope can be described by a combination of trigonometric functions.

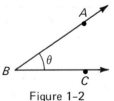

After studying trigonometry you will have a better understanding of these applications. To begin this study, we will first review a few essential facts about angles and their measure.

Recall from geometry that an **angle** consists of two rays with a common endpoint, called the **vertex.** The angle shown in Figure 1-2 can be named $\angle ABC$, $\angle B$, or simply θ. The Greek letters, such as θ, listed on page ix are often used to name angles.

We denote the measure of an angle θ by m(θ). When the context is clear, we will use θ to stand for m(θ). Angles are often measured in **degrees.** Each degree can be subdivided either into decimal parts, as is done by most calculators, or, more traditionally, into 60 equal parts called **minutes.** Each minute can be subdivided further into 60 equal parts called **seconds.** Degrees are denoted by the symbol °, minutes by the symbol ′, and seconds by the symbol ″.

Figure 1–2

To convert from decimal degrees to degrees and minutes, keep the whole number of degrees and multiply the decimal part by 60. A calculator display of 21.8°, for instance, can be written as $21°(0.8 \times 60)' = 21°48'$.

Example 4 Express 37.0417° in the degree-minute-second system.

Solution $37.0417° = 37°(0.0417 \times 60)' = 37°2.502' = 37°2'(0.502 \times 60)'' = 37°2'\,30.12''$, which can be rounded off to $37°2'30''$.

To convert from the degree-minute-second system to decimal degrees, begin by dividing the seconds by 60.

Example 5 Express $37°2'30''$ in decimal degrees.

Solution $37°2'30'' = 37°\left(2 + \dfrac{30}{60}\right)' = 37°2.5' = \left(37 + \dfrac{2.5}{60}\right)° = 37.041\overline{6}°$, which can be rounded off to 37.0417°.
(The bar over the 6 indicates a repeating decimal.)

In geometry, you probably used angles whose degree measures were between 0° and 180°. In trigonometry, we shall consider angles of any degree measure, positive and negative, and we shall extend the concept of an angle to include the idea of *rotation*.

In applied problems, an angle often represents a rotation about a point. Thus, in trigonometry, we consider an angle as being formed by the rotation of one side of the angle about its vertex to the position of the other side. To describe this rotation, we distinguish between the two sides of the angle. One side is called the **initial side** and the other is called the **terminal side.** We draw a curved arrow from the initial side to the terminal side to indicate the direction, or sense, of rotation about the vertex. An angle whose rotation has been specified in this way is called a **directed angle.**

Figure 1-3   Figure 1-4

If the arrow from the initial side to the terminal side of an angle goes in a *counterclockwise* direction, we take the measure to be positive. For example, if θ is half a right angle with a counterclockwise direction, then $m(\theta) = 45°$ or, more simply, $\theta = 45°$ (Figure 1-3). If the arrow goes in a *clockwise* direction, we take the measure to be negative (Figure 1-4). One full revolution, or rotation, about a point in a counterclockwise direction has measure 360°. Thus, when speaking of directed angles, one degree can be defined as $\dfrac{1}{360}$ of a full counterclockwise revolution.

An angle is in standard position in a Cartesian coordinate system if its initial side lies along the positive *x*-axis and its vertex is at the origin. Some examples are shown in Figure 1-5.

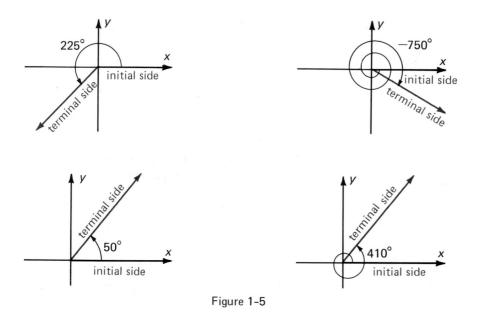

Figure 1-5

Notice the similarity between the 50° and the 410° angle. Both angles begin and end in the same place. Two angles of different measure that have coincident terminal sides when in standard position are called **coterminal angles.** It is important to remember that, in spite of their appearance, coterminal angles are *not equal.* Every angle in standard position is either coterminal with *or* equal to some angle θ such that $0° \le \theta < 360°$.

Exercises 1-1

Draw an angle in standard position having the given measure.

A 1. 90°   2. 270°   3. 330°   4. 170°

   5. −45°   6. −210°   7. −280°   8. −570°

   9. 405°   10. 1200°   11. −900°   12. 1040°

Express the following in the degree-minute-second system.

   13. 53.4°   14. 102.15°   15. −22.13°   16. −212.54°

Express the following in decimal degrees.

17. $-61°9'$ 18. $-430°45'$ 19. $12°4'30''$ 20. $99°45''$

Give the degree measure of an angle between $0°$ and $360°$ that is coterminal with the given angle.

21. $-280°$ 22. $-570°$ 23. $405°$ 24. $1200°$

25. $-900°$ 26. $1040°$ 27. $670°30'$ 28. $-29°30'$

B 29. $-49.9°$ 30. $440.72°$ 31. $-141°42'$ 32. $-535°21'$

33. Two positive angles are complementary if the sum of their measures is $90°$. What is the degree-minute-second measure of the complement of an angle of $52°14'50''$? (Hint: $90° = 89°59'60''$)

34. Two positive angles are supplementary if the sum of their measures is $180°$. What is the degree-minute-second measure of the supplement of an angle of $105°48'7''$?

35. Let k be any positive or negative integer. Give an expression in terms of k for the degree measure of all angles coterminal with an angle of $65°$.

36. Let k be any positive or negative integer. Give an expression in terms of k for the degree measure of all angles coterminal with an angle of $-320°$.

C 37. Let θ be the degree measure of an angle and let k be any positive or negative integer. Give an expression in terms of k for the degree measure of all angles coterminal with angle θ.

38. Let n be a fixed positive integer greater than 2 and let $\theta = k\left(\dfrac{360}{n}\right)°$.

    Describe the pattern formed by the locations of a point on the terminal side of θ as k takes on the values $1, 2, 3, \ldots, n$.

In Exercises 39 and 40, if θ is any angle, give a general method of finding an angle γ coterminal with θ such that $0° \leq \gamma < 360°$, for the given values of θ.

39. $\theta > 360°$ 40. $\theta < 0°$

41. A machine part rotates through an angle of $102°$ every minute. How many revolutions does it make in an hour?

42. A long-playing record makes $33\dfrac{1}{3}$ revolutions per minute. Through what angle does it rotate in one second?

The Trigonometric Functions

1-2 *Sine and Cosine*

Let θ be an acute angle: $0° < \theta < 90°$. Place θ in standard position. Choose any point $P(x, y)$, other than the vertex O, on the terminal side of θ, and let r be the distance OP (Figure 1-6).

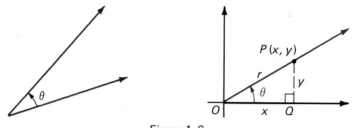

Figure 1-6

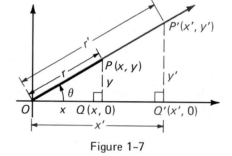

Figure 1-7

Although the numbers x, y, and r certainly depend on the choice of P, the *ratios* $\dfrac{y}{r}$ and $\dfrac{x}{r}$ do not. They depend *only on the angle* θ. To see this, let $P'(x', y')$ be any other point, besides the vertex, on the terminal side of θ, and let $r' = OP'$ (Figure 1-7). Recall from geometry that if two angles of one triangle are equal to two angles of another triangle, the triangles are similar and their corresponding sides are in proportion. The right triangles OQP and $OQ'P'$ are therefore similar and

$$\frac{y}{y'} = \frac{r}{r'} \quad \text{and} \quad \frac{x}{x'} = \frac{r}{r'}.$$

From these equations it follows that $\dfrac{y}{r} = \dfrac{y'}{r'}$ and $\dfrac{x}{r} = \dfrac{x'}{r'}$.

Look again at Figure 1-6. We can now conclude that given angle θ, the ratio $\dfrac{QP}{OP}$ will always be the same, regardless of the size of right triangle OQP.

We can summarize the foregoing by saying that $\dfrac{y}{r}$ in Figure 1-6 is a function of the angle θ. We call this function the **sine function** and we write

$$\sin \theta = \frac{y}{r}.$$

Likewise, we define another function, the **cosine function,** as $\frac{x}{r}$. We write

$$\cos \theta = \frac{x}{r}.$$

We can restate the definitions just given to apply to any acute angle of a right triangle as follows (Figure 1-8).

$$\sin \theta = \frac{\text{length of the side } \textbf{opposite } \theta}{\text{length of the } \textbf{hypotenuse}}$$

$$\cos \theta = \frac{\text{length of the side } \textbf{adjacent to } \theta}{\text{length of the } \textbf{hypotenuse}}$$

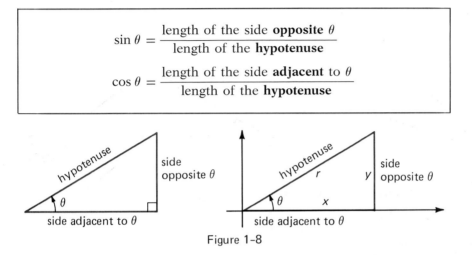

Figure 1-8

(Note that in trigonometric definitions involving right triangles "the side adjacent to θ," where θ is an acute angle, means the leg adjacent to θ and not the hypotenuse.)

Example 1   Find $\sin \theta$ and $\cos \theta$ to four decimal places for an angle θ in standard position whose terminal side passes through the point (12, 5).

Solution   By the Pythagorean theorem, $r^2 = 5^2 + 12^2$. Hence, $r^2 = 169$, and $r = 13$. Therefore,

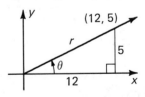

$$\sin \theta = \frac{5}{13} \approx 0.3846$$

and

$$\cos \theta = \frac{12}{13} \approx 0.9231.$$

In order to illustrate how a table of values of the sine and cosine functions can be used to solve problems, some of the values of these functions are listed in Table 1-1. These values have been rounded to four decimal places.

θ	$\sin \theta$	$\cos \theta$
5°	.0872	.9962
10°	.1736	.9848
15°	.2588	.9659
20°	.3420	.9397
25°	.4226	.9063
30°	.5000	.8660

θ	$\sin \theta$	$\cos \theta$
35°	.5736	.8192
40°	.6428	.7660
45°	.7071	.7071
50°	.7660	.6428
55°	.8192	.5736
60°	.8660	.5000

θ	$\sin \theta$	$\cos \theta$
65°	.9063	.4226
70°	.9397	.3420
75°	.9659	.2588
80°	.9848	.1736
85°	.9962	.0872

TABLE 1-1

Example 2   A ladder 4.0 m long is placed against a wall so that it makes an angle of 75° with the ground. How far up the wall will the ladder reach?

Solution   From the adjoining diagram we have

$$\sin 75° = \frac{x}{4.0}.$$

From Table 1-1,

$$\sin 75° \approx 0.9659.$$

Therefore,

$$0.9659 \approx \frac{x}{4.0}$$

$$x \approx 3.9.$$

The ladder will reach 3.9 m up the wall.

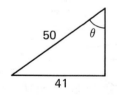

Example 3   In the diagram at the right, find θ to the nearest 5°.

Solution   From the diagram, $\sin \theta = \frac{41}{50} = 0.82$. From Table 1-1 we see that the angle whose sine is closest to 0.82 is 55°. Therefore, to the nearest 5°, $\theta = 55°$.

Significant Digits

Whenever we use measurements we must be careful to remember that measurements are only approximations. Notice that in Example 2 we approximated the value of x to two significant digits. A **significant digit** is any nonzero digit or any zero that has a purpose other than indicating the position of the decimal point. Which of the zeros in a number like 45,000 is significant, however, is unclear. For our purposes in this text we will consider all zeros that appear to the right of any nonzero digit to be significant.

Examples	Number of Significant Digits
4	1
40.01	4
4.0	2
0.040	2

The significant digits of a number can be shown clearly by using **scientific notation.** To write a number in scientific notation, express it in the form $a \times 10^n$, where a is a terminating decimal such that $1 < |a| < 10$ and n is an integer. The digits in a are the significant digits of the number.

Examples	Scientific Notation	Number of Significant Digits
400	4.00×10^2	3
0.000040	4.0×10^{-5}	2

When adding or subtracting measurements, round off the answer to the least number of decimal places in any of the numbers.

Example   0.001 cm + 10.9 cm = 10.9 cm

We will use "=" in place of "≈" since the approximation is understood.
When multiplying or dividing measurements, round the answer to the least number of significant digits in any of the numbers.

Example   0.2588 × 4.0 m = 1.0 m

When working with angles in the degree-minute system, use the following guide.

If lengths are rounded to	angles should be rounded to the nearest
1 significant digit	$10°$
2 significant digits	$1°$
3 significant digits	$10'$
4 significant digits	$1'$

Exercises 1-2

Use Table 1-1 on page 8 and the square root table on page 369 (Table 4) as needed. Let θ be an angle in standard position whose terminal side passes through the given point. Find (a) sin θ and cos θ to four decimal places and (b) θ to the nearest 5°.

A   1. (3, 4)   2. (15, 8)   3. (8, 6)   4. (20, 21)

5. (7, 24)   6. $(1, 2\sqrt{2})$ 7. $(3, \sqrt{7})$ 8. $(1, \sqrt{3})$

9. $(\sqrt{5}, 2)$ 10. (2, 1)   11. $(4\sqrt{3}, 1)$ 12. $(\sqrt{2}, \sqrt{2})$

Find *x* or θ in each right triangle. Give angle measures to the nearest 5°.

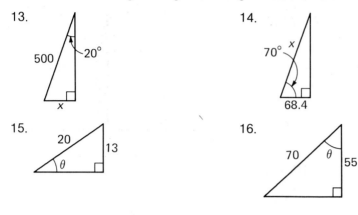

13.

500 20° *x*

14.

70° *x* 68.4

15.

20 13 θ

16.

70 θ 55

In Exercises 17 and 18, let θ be any acute angle of a right triangle. Use the given diagram to show the following:

B 17. $\sin \theta = \cos (90° - \theta)$

18. $\cos \theta = \sin (90° - \theta)$

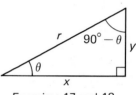

r 90° − θ *y* θ *x*

Exercises 17 and 18

19. Each leg of an isosceles triangle is 260 cm long and each base angle measures 35°. Find the length of the base.

20. The base of an isosceles triangle is 143.4 cm long and each base angle measures 55°. Find the length of each leg.

21. The sine bar is an instrument used by toolmakers to measure angles. In the diagram, if $AB = 12.7$ cm, how high must the stack of gauge blocks be to set the sine bar at an angle of 20°?

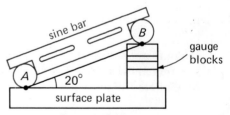

sine bar *B* gauge blocks *A* 20° surface plate

Exercise 21

22. The Archimedean Snail, used in ancient times for irrigation, consisted of a wooden spiral encased in a cylinder. The snail would be slanted into water at an angle of about 25° with the horizontal. Turning the cylinder caused the water to climb the spiral. About how long should a cylinder be to raise water to a height of 2.0 m?

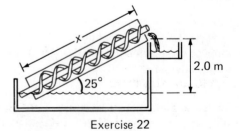

x 25° 2.0 m

Exercise 22

Exercises 23–26 refer to the given diagram.

23. A guy wire attached to the transmitter tower at point D is to be anchored into the ground at point B. If point B is 11.4 m from the base of the tower and $\angle DBC = 50°$, how long should the wire be?

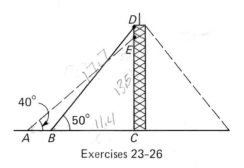

24. If the guy wire in Exercise 23 was attached at a point E below point D, and anchored into the ground at point A, how far away from point B would point A be if $\angle EAC = 40°$?

Exercises 23–26

C 25. Given that $DB = EA$, use Exercise 17 or 18 to show that $EC = BC$.

26. Given that $DB = EA$, use Exercise 17 or 18 to show that $DE = AB$.

27. Let α and β be complementary angles. Given that $\cos 90° = 0$, show that $\cos(\alpha + \beta) = \cos \alpha \cos \beta - \sin \alpha \sin \beta$. (Use the result of Exercise 17 or 18.)

28. A windshield wiper arm 30.5 cm long pivots at point C in the given diagram and sweeps out an angle of 110°. Find AB.

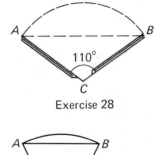

Exercise 28

29. In the given diagram, $\overline{AB}$ is a chord of a circle with center O. The first known trigonometry table, constructed by Hipparchus of Nicaea, essentially gave values for the ratio $\dfrac{AB}{AO}$ as a function of the angle θ. Give an equivalent expression for this ratio using the definition of sine given in this section.

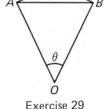

Exercise 29

1-3 *Values of the Sine and Cosine Functions*

Special Values of θ

Three angles that appear often in the study of trigonometry are the 30°, 45°, and 60° angles. Exact values for the sine and cosine of these angles can be determined very easily by using geometry.

Recall that each of the acute angles of an isosceles right triangle measures 45°, and if each leg has length 1, the hypotenuse has length $\sqrt{2}$ (by the Pythagorean theorem) [Figure 1-9]. Therefore,

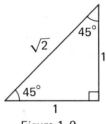

Figure 1-9

$$\sin 45° = \frac{1}{\sqrt{2}} = \frac{\sqrt{2}}{2}$$

$$\cos 45° = \frac{1}{\sqrt{2}} = \frac{\sqrt{2}}{2}.$$

Also recall that the bisector of an angle of an equilateral triangle cuts off a right triangle having acute angles 30° and 60° (Figure 1-10). If each side of the original equilateral triangle has length 2, the sides of the 30-60 right triangle will have lengths 2, 1, and $\sqrt{3}$. Therefore,

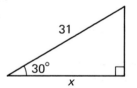

Figure 1-10

$$\sin 30° = \frac{1}{2} \qquad \sin 60° = \frac{\sqrt{3}}{2}$$

$$\cos 30° = \frac{\sqrt{3}}{2} \qquad \cos 60° = \frac{1}{2}$$

Example   Find the exact value of *x*. Leave your answer in simplest radical form.

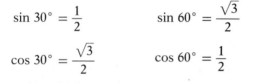

Solution $\cos 30° = \dfrac{x}{31}$

$$\frac{\sqrt{3}}{2} = \frac{x}{31}$$

$$2x = 31\sqrt{3}$$

$$x = \frac{31\sqrt{3}}{2}$$

Values of the sine and cosine functions for other angles are in general more difficult to compute. You can find the principal ones, rounded to four significant digits, in Table 1 on page 352 or Table 2 on page 360. If you have access to a calculator with keys that give the values of trigonometric functions, you may not need these tables. Nevertheless, it is a good idea to learn how to use them. Note that unless specified otherwise, we will use values of trigonometric functions rounded to four significant digits, to correspond with the tables.

The Use of Trigonometric Tables (Optional)

In order to find the sine or cosine of an angle θ to four significant digits, use the following steps. (Refer to Tables 1 and 2 as you read.)

1. Find the degree measure of θ in either the extreme left-hand column or the extreme right-hand column. If $0° \leq \theta \leq 45°$, the listing for θ will appear in the left-hand column. If $45° \leq \theta \leq 90°$, this listing will appear in the right-hand column. (The tables are arranged to take advantage of the fact that $\sin \theta = \cos (90° - \theta)$. (See Exercise 17 on page 10.)

2. If the degree measure of θ appears in the left-hand column, read the value of sine or cosine in the column with the appropriate label at the *top* of the column. If the degree measure appears in the right-hand column, read the value in the column with the appropriate label at the *bottom* of the column.

3. Note that the angle measures increase as you read *down* the left-hand column and as you read *up* the right-hand column. For example, the value for 37°30′ will be below that for 37°, and the value for 55°30′ will be above that for 55°.

Example   Find the sine of 66°30′ in Table 2.

Solution   Find the angle in the right-hand column. Read the value in the column labeled "sin θ" at the bottom of the column. (Remember to find the 30′ designation above 66°.)
sin 66°30′ = 0.9171 to four significant digits.

If you do not have a calculator and you wish to find the value of the sine or cosine function for an angle that is *between* two angles listed in a table, you can use a method called **linear interpolation.** This method is outlined on page 350.

The other trigonometric functions listed in the tables, namely tan θ, sec θ, csc θ, and cot θ, will be discussed in the following section. You can find values for these functions using the same procedures as for sine and cosine.

Exercises 1-3

Use the square root values in Table 4 on page 369 as needed.

Let θ be an acute angle in standard position whose terminal side passes through the given point. Find θ without using a calculator or a table.

A   1. $(5, 5)$ 2. $(4, 4\sqrt{3})$ 3. $(2\sqrt{3}, 2\sqrt{3})$ 4. $(2\sqrt{3}, 2)$

5. $\left(\dfrac{\sqrt{3}}{2}, \dfrac{1}{2}\right)$ 6. $(\sqrt{2}, \sqrt{6})$ 7. $\left(1, \dfrac{1}{\sqrt{3}}\right)$ 8. $\left(a, \dfrac{a}{\sqrt{3}}\right)$

Find the exact value of *x* in each triangle. Leave your answers in simplest radical form.

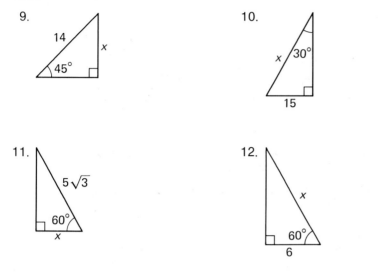

9.

14

x

45°

10.

x / 30°

15

11.

$5\sqrt{3}$

60°

x

12.

x

60°

6

For the given angle θ, find $\sin \theta$ and $\cos \theta$ to four significant digits.

13. 12°50′   14. 38°20′   15. 56°10′   16. 85°40′

Find a value of θ that satisfies the given equation. Round your answer to the nearest minute or to the nearest hundredth of a degree.

17. $\sin \theta = 0.6626$ 18. $\cos \theta = 0.9964$ 19. $\sin \theta = 0.9827$

20. $\sin \theta = \dfrac{1}{4}$ 21. $\cos \theta = \dfrac{2}{3}$ 22. $\cos \theta = \dfrac{\sqrt{5}}{9}$

In Exercises 23 and 24, leave your answer in simplest radical form.

B 23. Given that $\sin (\alpha - \beta) = \sin \alpha \cos \beta - \cos \alpha \sin \beta$, find the exact value of $\sin 15°$. (Use $\alpha = 45°$ and $\beta = 30°$.)

24. Given that $\cos (\alpha - \beta) = \cos \alpha \cos \beta + \sin \alpha \sin \beta$, find the exact value of $\cos 15°$. (Use $\alpha = 60°$ and $\beta = 45°$.) Compare your answer with the value of $\cos 15°$ given either in Table 1, Table 2, or by a calculator.

25. Find *DC* if $\angle ABC = 90°$, $\angle A = 50°20′$, and $AB = 125$.

26. Find *AD* if $\angle ABC = 90°$, $\angle C = 42.5°$, and $BC = 130$.

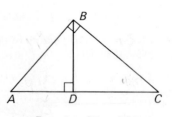

Exercises 25 and 26

Exercises 27 and 28 refer to Figure 1-1 from page 1. Assume that $\angle E$ is about 89°50′.

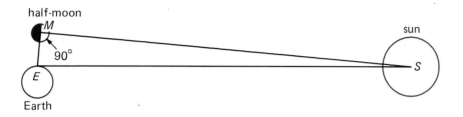

27. Give a rough comparison of the distance between Earth and the moon with the distance between Earth and the sun.

28. If the moon is roughly 4×10^5 km away from Earth, about how far away from Earth is the sun?

29. The given diagram illustrates the plans for a roof. Given the specifications shown, find the height allowed for the skylight.

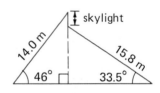

Exercise 29

30. In the given diagram, rods BC and AC are joined at C so that as C moves in a clockwise circle, the piston attached at A moves back and forth inside the cylinder. Suppose that $AC = 40$ cm. If C moves from the position shown, where $\angle ABC = 90°$ and $\angle C = 50°$, to a position in line with $\overline{AB}$, how far forward must the piston move?

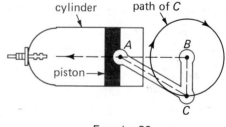

Exercise 30

C 31. (a) Use similar triangles to show that the value of x in the given diagram is $\dfrac{\sqrt{5} - 1}{2}$.

   (b) Use (a) to find cos 72° in simplest radical form.

32. Use Exercise 31 to find cos 36° in simplest radical form.

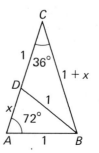

Exercise 31 and 32

The diagram shows the path of a light ray as it passes from one medium to another. Let the constants v_1 and v_2 indicate the speed of light in each medium, respectively. The angles α and β that the ray makes with the normal line (a line perpendicular to the common boundary) are related by the equation

$$\frac{\sin \alpha}{\sin \beta} = \frac{v_1}{v_2}$$

which is a form of Snell's Law of Optics.

33. Suppose a light ray passes from air, where $v_1 = 2.999 \times 10^8$ m/sec, into water. If $\alpha = 53°$ and $\beta = 37°$, find the approximate speed of light in water.

34. Suppose a light ray passes from ice, where $v_1 = 2.290 \times 10^8$ m/sec, into air, where $v_2 = 2.999 \times 10^8$ m/sec. If $\alpha = 21°$, find β to the nearest degree.

35. Use Snell's Law of Optics to show that if a light ray travels along the normal line in one medium, it continues to do so in the second medium.

36. Use the given diagram and Snell's Law of Optics to show how light can pass through a flat pane of glass surrounded by air without changing direction. Assume that $\alpha = 50°$, the speed of light in air $= 2.999 \times 10^8$ m/sec, and the speed of light in the glass $= 1.974 \times 10^8$ m/sec.

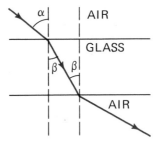

1-4 *Other Trigonometric Functions*

In describing some situations mathematically it is convenient to use ratios other than the sine or cosine. For example, in describing the grade, or slope, of a road up the side of a hill, we use the ratio $\dfrac{y}{x}$ in Figure 1-11, since $\dfrac{y}{x}$ tells by how much the road rises for each unit of horizontal distance. (Recall from algebra that the slope of a line is defined to be $\dfrac{\text{vertical change}}{\text{horizontal change}}$ or $\dfrac{\text{rise}}{\text{run}}$.) In fact, the ratio $\dfrac{y}{x}$ written as a percent is called the percent of grade of the road.

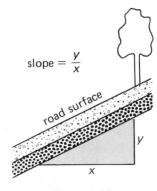

Figure 1-11

With these facts in mind, we define the **tangent** (tan) ratio for any acute angle θ in standard position as follows (Figure 1-12).

$$\tan \theta = \frac{y}{x}$$

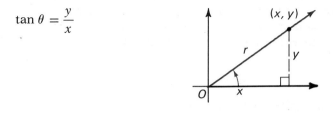

Figure 1-12

For any right triangle containing an acute angle θ, the tangent function can be defined by:

$$\tan \theta = \frac{\text{length of side } \textbf{opposite } \theta}{\text{length of side } \textbf{adjacent to } \theta}$$

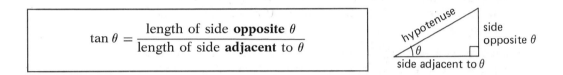

Three other functions traditionally used in trigonometry are the reciprocals of the sine, cosine, and tangent functions. These were originally introduced to simplify calculations (as Example 3 will demonstrate). The functions, **secant** (sec), **cosecant** (csc), and **cotangent** (cot), are defined as follows for any acute angle θ in standard position (Figure 1-13).

$$\sec \theta = \frac{1}{\cos \theta} = \frac{r}{x}$$

$$\csc \theta = \frac{1}{\sin \theta} = \frac{r}{y}$$

$$\cot \theta = \frac{1}{\tan \theta} = \frac{x}{y}$$

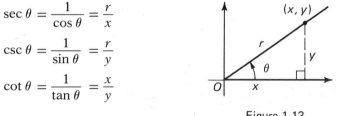

Figure 1-13

Example 1   Find the values of the six trigonometric functions for an angle θ in standard position whose terminal side passes through the point $(3, 4)$. Leave your answers in ratio form.

Solution   By the Pythagorean theorem, $r^2 = x^2 + y^2 = 3^2 + 4^2 = 25$. Therefore, $r = 5$ and

$$\sin \theta = \frac{y}{r} = \frac{4}{5} \qquad \cos \theta = \frac{x}{r} = \frac{3}{5}$$

$$\tan \theta = \frac{y}{x} = \frac{4}{3} \qquad \cot \theta = \frac{x}{y} = \frac{3}{4}$$

$$\sec \theta = \frac{r}{x} = \frac{5}{3} \qquad \csc \theta = \frac{r}{y} = \frac{5}{4}$$

Values for the tangent, secant, cosecant, and cotangent functions are given in Tables 1 and 2 at the back of the book. If you are using a calculator that has keys for the cosine, sine, and tangent functions you can take the reciprocals of these to obtain values of the secant, cosecant, and cotangent functions, respectively.

Example 2   Suppose that in Example 1 on page 2 the height of the door is 215 cm and the maximum angle that the sun's rays make with the door is 40°. Find the minimum width of the awning.

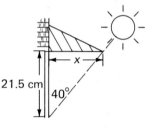

Solution   From the adjoining diagram we have

$$\tan 40° = \frac{x}{215}$$

$$0.8391 = \frac{x}{215}$$

$$x = 180$$

The minimum width that the awning can be is 180 cm.

Example 3   Find the lengths of the sides of an isosceles triangle if the base has length 580 and each base angle has a measure of 43°30′.

Solution   First draw a diagram. Since ABC is an isosceles triangle we know that $AD = \frac{1}{2}(580) = 290$. Either of two possible equations can be used to find AB:

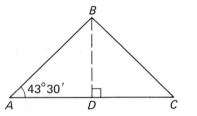

(a) $\cos 43°30′ = \dfrac{290}{AB}$

(b) $\sec 43°30′ = \dfrac{AB}{290}$

Using equation (a), $AB \cos 43°30′ = 290$ so

$$AB = \frac{290}{\cos 43°30′} = \frac{290}{0.7254} = 400.$$

Using equation (b), $AB = 290 \sec 43°30′ = (290)(1.379) = 400$.

Note that method (b) is preferable if a calculator is not available because it is generally easier to multiply than to divide.

Exercises 1-4

Find the values of the six trigonometric functions for an angle θ in standard position whose terminal side passes through the given point. Simplify radicals and leave your answers in ratio form.

A 1. $(5, 12)$ 2. $(15, 8)$ 3. $(3, \sqrt{7})$ 4. $(4, 2\sqrt{2})$

5. $(2, 3)$ 6. $(7, 1)$ 7. $\left(\dfrac{\sqrt{5}}{3}, \dfrac{2}{3}\right)$ 8. $\left(\dfrac{5}{7}, \dfrac{2\sqrt{6}}{7}\right)$

Find x or θ in each right triangle. Give angle measures to the nearest 10' or to the nearest tenth of a degree.

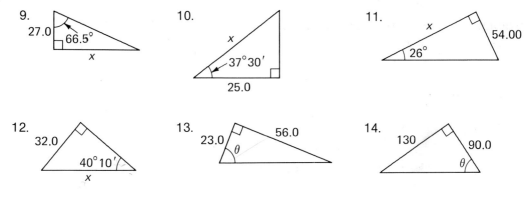

9. 27.0 66.5° x

10. x 37°30' 25.0

11. x 26° 54.00

12. 32.0 40°10' x

13. 23.0 θ 56.0

14. 130 90.0 θ

Find the exact value of each function. Leave your answers in simplest radical form.

15. $\tan 45°$ 16. $\csc 30°$ 17. $\sec 30°$ 18. $\cot 60°$

19. $\sec 45°$ 20. $\cot 30°$ 21. $\csc 45°$ 22. $\csc 60°$

B 23. An isosceles triangle has a 70 cm base. The altitude drawn to the base is 75 cm long. Find the measure of each base angle to the nearest degree.

24. Each base angle of an isosceles triangle measures 35° and the altitude drawn to the base is 25 cm long. Find the length of each side.

25. In order for prevailing summer breezes to cool a house, there should be an angle θ of 20° to 70° between the wind direction and the side of the house. In the given diagram, the winds move in the direction of ray AB. Suppose that $AC = 1.7$ m and $BC = 5.2$ m. Is the house situated in an ideal position?

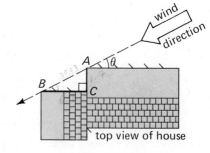

Exercise 25

26. For safety reasons, a stovepipe should be kept away from combustible materials at a distance equal to three times the diameter of the pipe. Suppose a pipe 15 cm in diameter and 60 cm long connects a wood-burning stove and a chimney as shown in the diagram. If $\theta = 32°$, is the pipe in a safe location? (Hint: Find x in the given diagram.)

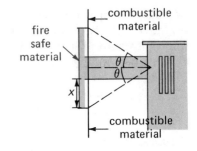

The diagram for Exercises 27 and 28 illustrates the plans for a 50° bay window unit consisting of a window seat in the shape of an isosceles trapezoid and four separate windows all of the same size.

27. The window seat shown measures 45 cm at its widest point. Find the width x of each window.

28. Use Exercise 27 to find the length y of the edge of the window seat.

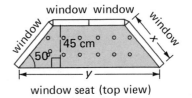

window seat (top view)

Exercises 27 and 28

29. Show that $\sec \theta = \csc (90° - \theta)$.

30. Show that $\tan \theta = \cot (90° - \theta)$.

C 31. Use the given diagram to show that
$$\frac{\tan A}{BD} = \frac{\tan B}{AD}.$$

32. Use the given diagram to show that
$$\frac{\cot ACD}{BD} = \frac{\cot BCD}{AD}.$$

Exercises 31 and 32

In Exercises 33 and 34, let O be a circle of radius 1 and let $\overleftrightarrow{AB}$ be a tangent to O at A.

33. Show that (a) $\tan AOB = AB$
       (b) $\cot AOB = \tan ABO$.

34. Show that (a) $\sec AOB = OB$
       (b) $\csc AOB = \sec ABO$.

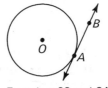

Exercises 33 and 34

1-5   *Solving Right Triangles*

Finding values for all the sides and angles of a triangle is often referred to as **solving the triangle**. As you may have discovered in the previous sections, we can find the value of a specific side or angle of a right triangle if we are given certain information about the triangle. In fact, we can solve any right triangle given only the length of one side, together with either the length of another side or the measure of one of the acute angles. The following examples offer a few suggestions for solving right triangles and for finding particular sides and angles.

CASE 1 Given the length of a leg and the length of the hypotenuse of a right triangle.

Example Solve $\triangle ABC$ in which $\angle C = 90°$, $c = 49.7$, and $b = 25.0$.
(Note: It is customary to associate angles with capital letters and the sides opposite the angles with the corresponding lower-case letters.)

Solution Draw a diagram and label all known parts. Drawing your diagram to scale when possible will be helpful in checking your answers.

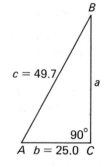

　　To find $\angle A$, we can use the equation $\cos A = \dfrac{25.0}{49.7}$. Since

$\cos A = 0.503$, $\angle A = 59.8°$ or $59°50'$.
　　To find $\angle B$ we can use the fact that the acute angles of a right triangle are complementary. Thus, $\angle B = 90° - \angle A$ and $\angle B = 30.2°$ or $30°10'$.

　　To find side a, we can use the Pythagorean theorem $a^2 = c^2 - b^2$, or the equation $\tan A = \dfrac{a}{25.0}$, to obtain $a = 43.0$. Since the former method involves only given values, using it will help avoid errors.

CASE 2 Given the measure of an acute angle and the length of the hypotenuse of a right triangle.

Example Triangle ABC is inscribed in a circle of radius 7.20 cm. If $\angle C = 90°$ and $\angle A = 55°$, find the lengths of sides a and b.

Solution Draw a diagram. Since $\angle C = 90°$, we know from geometry that AB is the diameter of the circle. Then $c = (7.20)(2) = 14.4$. To find a, we can use the equation $\sin 55° = \dfrac{a}{14.4}$, so

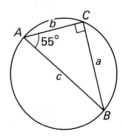

that
$a = 14.4 \sin 55° = (14.4)(0.8192) = 11.8$ cm.
To find b, we can use the equation
$\cos 55° = \dfrac{b}{14.4}$, so that
$b = 14.4 \cos 55° = (14.4)(0.5736) = 8.26$ cm.

CASE 3   Given the length of a leg and the measure of an acute angle of a right triangle.

Example   An angle between a horizontal ray and another ray above the horizontal is called an **angle of elevation.** A section of the plan for a proposed cantilever bridge is shown in the diagram. The angle of elevation from point A to point B' is specified as 50.0°. The angle of elevation from C to B' is 44.5°. If the distance between supports AA' and BB' is given as 7.5 m, what should the distance be between supports BB' and CC'?

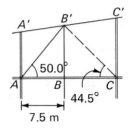

Solution   First find BB' and use this measurement to find BC. To find BB', we can use the equation $\tan 50.0° = \dfrac{BB'}{7.5}$, so $BB' = 7.5 \tan 50.0° = (7.5)(1.192) = 8.9$ m. To find BC, we can use the equation $\cot 44.5° = \dfrac{BC}{BB'}$, so $BC = 8.9 \cot 44.5 = (8.9)(1.018) = 9.1$ m.

CASE 4   Given the lengths of the legs of a right triangle.

Example   An angle made by a horizontal ray and a ray below the horizontal is called an **angle of depression.** A closed-circuit camera in a bank is mounted at point A so that the lens is 2.5 m above and 3.9 m to the left of a teller's window at point B. If the camera is to be aimed at B, what angle of depression θ should the lens make with the horizontal?

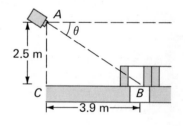

Solution   Recall from geometry that if two parallel lines are cut by a transversal, alternate interior angles are equal, so $\angle ABC = \theta$. We can use the equation $\tan \theta = \dfrac{2.5}{3.9} = 0.64$. Therefore, $\theta = 33°$.

Exercises 1-5

In Exercises 1-12, solve $\triangle ABC$, in which $\angle C = 90°$.

A   1. $c = 240$; $\angle A = 22°40'$
2. $c = 80.0$; $\angle A = 65°10'$

3. $a = 65.0$; $\angle B = 56°40'$
4. $a = 125$; $\angle B = 20.5°$

5. $a = 80.0$; $\angle A = 64°$
6. $a = 36.0$; $\angle A = 25°$

7. $a = 65.0; b = 72.0$

8. $a = 48.0; b = 55.0$

9. $b = 28.0; c = 53.0$

10. $b = 105; c = 137$

In Exercises 11 and 12, leave lengths in simplest radical form.

11. $a = 5\sqrt{3}; c = 5\sqrt{6}$

12. $a = 7\sqrt{6}; c = 14\sqrt{2}$

13. Find $\angle B$ in the diagram if $\angle A = 50°40'$, $AD = 25.0$, and $BC = 40.0$.

14. Find $\angle A$ in the diagram if $\angle B = 14°50'$, $BC = 150$, and $AD = 46.0$.

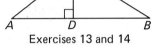

Exercises 13 and 14

B 15. Triangle ABC is inscribed in a circle of radius 8.24 cm. If $\angle C = 90°$ and $\angle B = 63°$, find the lengths of sides a and b.

16. An end of a cylindrical piece of wood 2.5 cm in diameter is cut down to form a cube. Find the length of an edge of the largest cube that can be made.

17. A spotlight is mounted 7.3 m high on a pole to illuminate the center of a parking area at point A. If A is 10.2 m from the base of the pole, at what angle of depression θ should the spotlight be aimed?

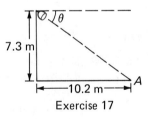

Exercise 17

18. In one minute an airplane descending at a constant angle of depression of $10°$ travels 1600 m along its line of flight. How much altitude has the plane lost?

Exercises 19 and 20 refer to the given diagram of a crane. The boom (or arm) of the crane is 65 m long and can be raised or lowered by varying its angle of elevation β.
The distance from the center of the load to the center of the tower frame is called the radius.

19. Suppose the radius for a certain load should be 63 m. At what angle β should the boom be set?

20. Suppose the largest angle β at which the boom can be set is $70°$. Find the minimum possible radius.

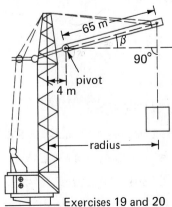

Exercises 19 and 20

21. In its original form, the Great Pyramid at Giza had a square base of about 230 m on each side. The angle of elevation represented by $\angle PRO$ in the given diagram measured about $51°50'$. Find the original height of the Great Pyramid.

22. Use the information given in Exercise 21 to find the angle of elevation represented by $\angle PAO$.

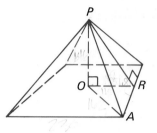

Exercises 21 and 22

C 23. Use the given diagram to show that

$$CA \sin A = CB \sin B.$$

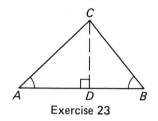

Exercise 23

Exercise 24 demonstrates a method of surveying known as triangulation.

24. A forest fire at point C is spotted by fire station A and fire station B. The distance from A to B is known to be 2.42 km. At station A, $\angle CAB$ is measured to be $44.5°$. At station B, $\angle CBA$ is measured to be $50.5°$. How far from the fire are stations A and B? Use Exercise 23.

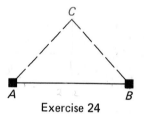

Exercise 24

25. In the given $\triangle ABC$, $\overline{CD}$ is the altitude to side AB. Show that

$$CD = (AB) \left(\frac{\tan A \tan B}{\tan A + \tan B} \right).$$

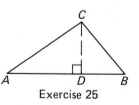

Exercise 25

26. Point P lies in the horizontal line through the bases of two telephone poles of equal height. The angles of elevation from point P to the tops of the poles are α and β, as shown. If P is c meters from the nearest pole and the poles are x meters apart, show that

$$x = c \left(\frac{\tan \alpha}{\tan \beta} - 1 \right).$$

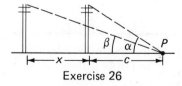

Exercise 26

Challenge

A weight is suspended from a string that is attached to two vertical poles of unequal height. The weight can slide along the string and is allowed to come to a rest position at point B as shown in the diagram. A law of physics tells us that the two angles (labeled θ) between the string and a vertical line through B are equal. If the poles are b meters apart and if the string is c meters long, show that $\sin \theta = \dfrac{b}{c}$ (and hence the angle θ is independent of the relative heights of the poles).

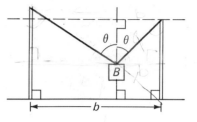

1-6 *Trigonometric Functions of Arbitrary Angles*

Thus far we have defined the trigonometric functions for acute angles. We now define them for angles of any measure.

Given an angle θ, place it in standard position. Choose a point $P(x, y)$ on the terminal side of θ and let r be the distance OP (Figure 1-14).

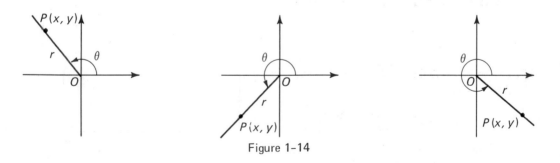

Figure 1–14

In general, we can define the following.

$$\sin \theta = \frac{y}{r} \qquad \cos \theta = \frac{x}{r} \qquad \tan \theta = \frac{y}{x} \text{ if } x \neq 0$$

$$\csc \theta = \frac{r}{y} \text{ if } y \neq 0 \qquad \sec \theta = \frac{r}{x} \text{ if } x \neq 0 \qquad \cot \theta = \frac{x}{y} \text{ if } y \neq 0$$

Example 1 Let θ be an angle in standard position whose terminal side passes through $P(12, -5)$. Find the six trigonometric functions of θ.

Solution   Let $r = OP$. Then $r^2 = 12^2 + (-5)^2 = 169 = 13^2$.
Thus $r = 13$, $x = 12$, $y = -5$.

$$\sin \theta = \frac{-5}{13} = -\frac{5}{13} \qquad \cos \theta = \frac{12}{13}$$

$$\tan \theta = \frac{-5}{12} = -\frac{5}{12} \qquad \cot \theta = -\frac{12}{5}$$

$$\csc \theta = -\frac{13}{5} \qquad \sec \theta = \frac{13}{12}$$

Recall that the coordinate axes divide the plane into four regions called **quadrants,** numbered as indicated in Figure 1-15. An angle θ is **in a** particular **quadrant** if its terminal side lies in that quadrant when θ is in standard position. Thus a 240° angle is in Quadrant III, or is a third-quadrant angle. Angles such as those measuring 90°, 180°, and 270°, whose terminal sides lie along an axis, are called **quadrantal angles.**

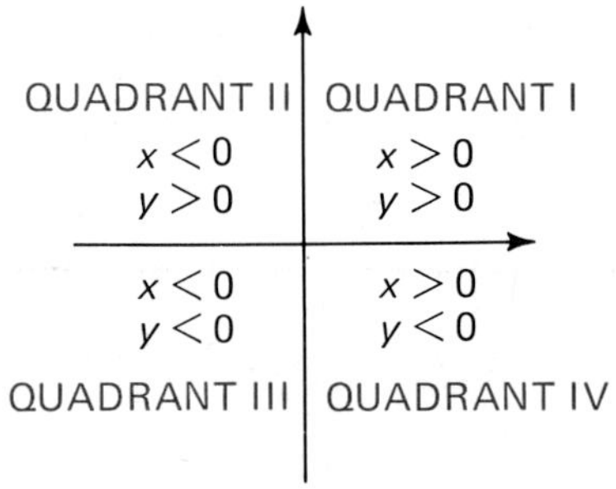

QUADRANT II	QUADRANT I
$x < 0$	$x > 0$
$y > 0$	$y > 0$
$x < 0$	$x > 0$
$y < 0$	$y < 0$
QUADRANT III	QUADRANT IV

Figure 1-15

The following table gives the signs of the trigonometric functions of angles in the various quadrants. (See Exercise 25.)

θ in Quadrant	$\sin \theta$ $\csc \theta$	$\cos \theta$ $\sec \theta$	$\tan \theta$ $\cot \theta$
I	+	+	+
II	+	−	−
III	−	−	+
IV	−	+	−

TABLE 1-2

Although the general definition of the trigonometric functions does not use right triangles, our work will be simplified if we introduce the so-called reference triangle. Refer to Figure 1-14 and let Q be the foot of the perpendicular from P to the x-axis (Figure 1-16).

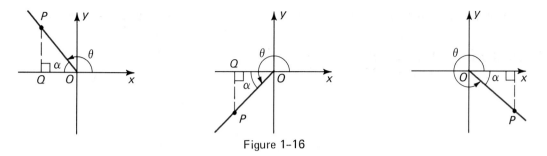

Figure 1-16

Right triangle *POQ* is the **reference triangle** of θ, and the acute angle $\alpha = \angle POQ$ is the **reference angle** of θ. It is not hard to see that if α is the reference angle of an arbitrary angle θ, then

$$\sin \theta = \sin \alpha \text{ or } -\sin \alpha \qquad \cos \theta = \cos \alpha \text{ or } -\cos \alpha$$

$$\tan \theta = \tan \alpha \text{ or } -\tan \alpha \qquad \cot \theta = \cot \alpha \text{ or } -\cot \alpha$$

$$\csc \theta = \csc \alpha \text{ or } -\csc \alpha \qquad \sec \theta = \sec \alpha \text{ or } -\sec \alpha$$

The proper sign is determined by the quadrant of θ.

Example 2 Find $\sin 321°$ and $\cos 321°$.

Solution   A sketch may be helpful. In this case,

$$\alpha = 360° - 321° = 39°.$$

Hence, since $321°$ is a fourth-quadrant angle, we have

$$\sin 321° = -\sin 39° = -0.6293,$$

$$\cos 321° = \cos 39° = 0.7771.$$

Example 3 If $180° < \theta < 360°$ and $\tan \theta = 2.175$, find θ to the nearest 10'.

Solution   Since $180° < \theta < 360°$, θ is in Quadrant III or Quadrant IV; because $\tan \theta > 0$, θ is in Quadrant III. Thus θ and its reference angle α are related as shown in the sketch. Moreover,

$$\tan \alpha = \tan \theta = 2.175$$

Therefore, $\alpha = 65°20'$ and $\theta = 180° + 65°20' = 245°20'$

In the case of quadrantal angles, certain of the trigonometric functions are not defined.

Example 4   Find the values of the trigonometric functions of $-90°$ which are defined.

Solution   We use the definition directly. $P(0, -1)$ is on the terminal side of $-90°$, and $OP = 1$. Therefore,

$$\sin(-90°) = \frac{-1}{1} = -1$$

$$\cos(-90°) = \frac{0}{1} = 0$$

$$\tan(-90°) = \frac{-1}{0}, \text{ undefined}$$

$$\cot(-90°) = \frac{0}{-1} = 0$$

$$\sec(-90°) = \frac{1}{0}, \text{ undefined}$$

$$\csc(-90°) = \frac{1}{-1} = -1$$

Exercises 1-6

Find exact values of the six trigonometric functions for each angle. Leave your answers in simplest form.

A 1. $210°$ 2. $120°$ 3. $315°$ 4. $225°$

  5. $330°$ 6. $240°$ 7. $90°$ 8. $180°$

Use a reference triangle to find each of the following.

  9. $\sin 261°20'$ 10. $\cos 173°30'$ 11. $\tan 137°$ 12. $\cot 310°$

  13. $\sec 325.5°$ 14. $\csc 125.5°$ 15. $\sin -65°$ 16. $\cos -152.5°$

In Exercises 17–24, (a) find $\sin\theta$ and $\cos\theta$ for an angle θ in standard position whose terminal side passes through the given point, and (b) find θ to the nearest 10' or nearest tenth of a degree. Assume that $0° \le \theta < 360°$.

  17. $(-8, 6)$ 18. $(-15, -12)$ 19. $(15, -8)$ 20. $(-20, 21)$

B 21. $(-2, -\sqrt{5})$ 22. $(-1, 2\sqrt{2})$ 23. $(-\sqrt{7}, 3)$ 24. $(-\sqrt{11}, -5)$

25. Use the general definition of the trigonometric functions and the information given in Figure 1-15 to explain the entries in Table 1-2.

Exercises 26 and 27 refer to the given diagram.

26. Use the Pythagorean theorem to show that $(\sin \theta)^2 + (\cos \theta)^2 = 1$. (Hint: Divide each side of the equation $x^2 + y^2 = r^2$ by r^2.)

27. Use the Pythagorean theorem to show that $1 + (\tan \theta)^2 = (\sec \theta)^2$. (Hint: Divide each side of the equation $x^2 + y^2 = r^2$ by x^2.)

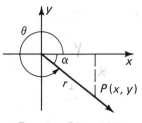

Exercises 26 and 27

Find $\tan \theta$ for θ in the given quadrant and satisfying the given equation.

Example $\sin \theta = \dfrac{2}{3}$; Quadrant II

Solution   Draw a reference triangle for θ. By the Pythagorean theorem, side $a = \sqrt{3^2 - 2^2} = \sqrt{5}$. But θ is in Quadrant II so $(x, y) = (-\sqrt{5}, 2)$ and
$$\tan \theta = \frac{2}{-\sqrt{5}} = -\frac{2\sqrt{5}}{5}.$$

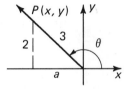

28. $\sin \theta = -\dfrac{4}{5}$; Quadrant III

29. $\cos \theta = -\dfrac{8}{17}$; Quadrant III

30. $\sec \theta = \dfrac{41}{9}$; Quadrant I

31. $\csc \theta = \dfrac{25}{24}$; Quadrant II

32. $\cos \theta = \dfrac{5}{7}$; Quadrant IV

33. $\sin \theta = -\dfrac{5}{6}$; Quadrant IV

In Exercises 34–39, let θ be any angle and let $P(x, y)$ be any point on the terminal side of θ. Use congruent triangles to show each of the following.

C 34. $\sin (-\theta) = -\sin \theta$

35. $\cos (-\theta) = \cos \theta$

36. $\sin \theta = \sin (180° - \theta)$

37. $\cos \theta = -\cos (180° - \theta)$

38. $\tan \theta = -\tan (180° - \theta)$

39. $\cot \theta = -\cot (180° - \theta)$

40. Let α and β be supplementary angles. Use Exercises 36 and 37 to show that $\sin (\alpha + \beta) = \sin \alpha \cos \beta + \cos \alpha \sin \beta$.

Challenge

The diagram is an illustration of one of the first steam engines. Steam pressure forced the beam to pivot up and down at point X. The beam, in turn, moved the shaft. Gear **A** then revolved in a path around a Gear **B** of equal diameter, causing Gear **B** to rotate and turn the attached wheel. Suppose the beam moved from Y to Z, sweeping out an angle of $40°$. If YZ equals the diameter of the path of Gear **A** around Gear **B**, and if $XY = 203$ cm, find the diameter of Gear **B**.

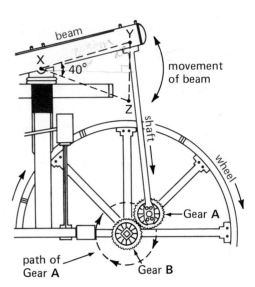

CHAPTER SUMMARY

1. An angle consists of two rays with a common endpoint called the vertex. Angles are measured in degrees, minutes, and seconds, or in decimal degrees. A directed angle has an initial side and a terminal side. Two angles of different measure with coincident terminal sides when in standard position are called coterminal angles.

2. The sine function and the cosine function of an acute angle θ of a right triangle are defined as follows:

$$\sin \theta = \frac{\text{length of side opposite } \theta}{\text{length of hypotenuse}}$$

$$\cos \theta = \frac{\text{length of side adjacent to } \theta}{\text{length of hypotenuse}}$$

3. Significant digits are used to express approximations. A significant digit is any nonzero digit or any zero that has a purpose other than showing the position of the decimal point.

4. The exact values for the sines and cosines of $30°$, $45°$, and $60°$ angles can be found by using isosceles right triangles and equilateral triangles. Approximations for the trigonometric functions of all acute angles can be found by using tables or a calculator.

5. The tangent function of an acute angle θ of a right triangle is defined as $\tan \theta = \dfrac{\text{length of side opposite } \theta}{\text{length of side adjacent to } \theta}$. The three other trigonometric functions of θ are $\sec \theta = \dfrac{1}{\cos \theta}$, $\csc \theta = \dfrac{1}{\sin \theta}$, and $\cot \theta = \dfrac{1}{\tan \theta}$.

6. The six trigonometric functions may be used to solve right triangles.

7. The trigonometric functions may be defined for angles of any measure.

CHAPTER TEST

1-1   1. Express $37°46'32''$ in decimal degrees.

   2. Express $251.42°$ in the degree-minute-second system.

1-2   3. Find θ to the nearest 5°.

1-3   4. Find the exact value of x.

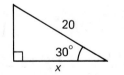

1-4   5. Find the values of the six trigonometric functions for θ.

1-5   6. Solve $\triangle XYZ$ in which $\angle X = 90°$, $y = 17$, and $x = 25$.

1-6   7. Find the six trigonometric functions for an angle θ whose terminal side passes through $(-6, 4)$. Give the measure of θ to the nearest $10'$.

ASTRONOMICAL DISTANCES

As we saw on page 1, the distances from Earth to the sun and the moon occupied a central place in the investigations of the early astronomers. Aristarchus of Samos (310–230 B.C.) was one of the first astronomers to make extensive use of mathematics to estimate these distances.

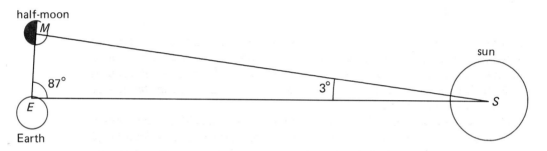

By observing that a half-moon appeared in the sky 87° from the position of the sun, he concluded that (1) lines *EM* and *MS* in the given diagram are perpendicular, and (2) angle *ESM* is 3°. Unfortunately, Aristarchus' observation was inaccurate. (The half-moon actually appears in the sky 89°51′ from the position of the sun.) As a result, by finding what is now called the sine of 3°, he concluded that the sun is about 19 times as far from Earth as is the moon, when it is actually about 382 times as far. Aristarchus' method, however, was quite sound.

A century later, Hipparchus of Nicaea was able to compute the distance to the moon with a remarkable degree of accuracy.

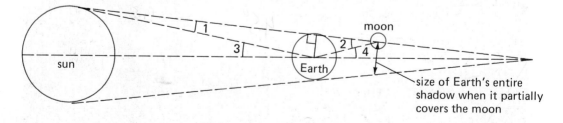

size of Earth's entire shadow when it partially covers the moon

By observing Earth's shadow on the moon during a partial eclipse of the moon, Hipparchus estimated that the diameter of the shadow was twice the diameter of the moon. He noted that $\angle 1 + \angle 2 = \angle 3 + \angle 4$ and that $\angle 1$ is so small, compared to the other angles, as to be negligible. Thus, he was able to find $\angle 2$. Because the radius of Earth was known at the time of Hipparchus, he was able to use the sine function to find the distance from Earth to the moon. In fact, his figure of 59 Earth radii is very close to the actual value of 60.3 Earth radii.

It was not until the nineteenth century that a method known as trigonometrical parallax was used to determine the distance from Earth to nearby stars other than the sun.

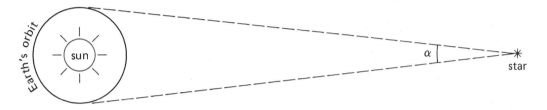

As Earth revolves around the sun, the position of a nearby star relative to a distant star will appear to alter by a small angle α. One half of α is called the parallax of the nearby star. Given the parallax of a star and the radius of Earth's orbit around the sun, the distance from Earth to the star can be found using the sine function.

Knowledge about the distances of nearby stars provided astronomers with a ruler for measuring the distances to remote stars. The discoveries of Henrietta Leavitt (1868–1921), for example, made it possible to find the distance to a remote star by comparing its brightness with the brightness of a star whose distance is known.

COMPUTING VALUES OF TRIGONOMETRIC FUNCTIONS

Because you can find values of the trigonometric functions directly from a scientific calculator or from a table, for all practical purposes you will not need to actually compute them. It is interesting, however, to see how the values can be computed.

For instance, in advanced mathematics it is shown that

$$\sin x = x - \frac{x^3}{3!} + \frac{x^5}{5!} - \frac{x^7}{7!} + \cdots$$

(The notation 3!, read "three factorial," means $1 \cdot 2 \cdot 3$. Similarly, $5! = 1 \cdot 2 \cdot 3 \cdot 4 \cdot 5 = 3!(4 \cdot 5)$.) By adding successive terms of the expression, the value of $\sin x$ can be computed to any desired number of decimal places. This can be done by using the following computer program.

```
10   PRINT "TO FIND THE VALUE OF THE SINE"
20   PRINT "OF AN ANGLE IN DEGREES:"
30   PRINT
40   PRINT "WHAT IS THE DEGREE MEASURE";
50   INPUT A
60   LET P=3.141593
70   LET X=A*P/180
80   LET E=1
90   LET T=X
100  LET S=0
110  PRINT
120  PRINT " T"," S"
130  PRINT
140  PRINT T,
150  IF ABS(T)<5.E-07 THEN 210
160  LET S=S+T
170  PRINT S
180  LET E=E+2
190  T=T*X↑2/(-(E-1)*E)
200  GOTO 130
210  PRINT
220  PRINT "SIN (";A;" DEG.) =";S
230  PRINT
240  PRINT "COMPUTER VALUE:";SIN(X)
250  END
```

The "GOTO loop" in the program reflects how each term after the first can be derived from the preceding term. The following display shows how the terms are related. "*E*" stands for the exponent of *x* and "*D*" stands for the denominator.

$$\sin x = \frac{x}{1} \qquad\qquad\qquad E = 1$$
$$D = 1$$

$$+ \left(\frac{x}{1}\right)\left(\frac{x^2}{-2\cdot 3}\right) \qquad\qquad E = 1 + 2 = 3$$
$$D = -1\cdot 2\cdot 3$$

$$+ \left(\frac{x^3}{-1\cdot 2\cdot 3}\right)\left(\frac{x^2}{-4\cdot 5}\right) \qquad E = 3 + 2 = 5$$
$$D = 1\cdot 2\cdot 3\cdot 4\cdot 5$$

$$+ \left(\frac{x^5}{1\cdot 2\cdot 3\cdot 4\cdot 5}\right)\left(\frac{x^2}{-6\cdot 7}\right) \qquad E = 5 + 2 = 7$$
$$D = -1\cdot 2\cdot 3\cdot 4\cdot 5\cdot 6\cdot 7$$

$$+ \cdots$$

The general expression for any term is X↑E/D. Then for each succeeding term,

$$E = E + 2 \text{ and } D = -D * (E - 1) * E.$$

These expressions are incorporated in line 190 of the program. (In line 150, 5.E−07 means 0.0000005.)

Note that in any computer calculation, answers are rounded off to a certain number of digits. Thus, when a value is arrived at by different methods, the last digits may not agree.

Exercises

1. Copy and RUN the given program for the following values.

   a. 0°   b. 20°   c. 30°   d. 45°

   e. 50°   f. 60°   g. 70°   h. 80°

2. To find the cosine of an acute angle, make the following changes in the program and RUN it for the values listed in Exercise 1.

```
250   PRINT
260   LET C=SQR(1−S*S)
270   PRINT "COS (";A;" DEG.) =";C
280   PRINT
290   PRINT "COMPUTER VALUE:";COS(X)
300   END
```

3. Extend the program to find the tangent of an acute angle. RUN the revised program for the values listed in Exercise 1.

A satellite moving in uniform circular motion has an
angular speed as well as a linear speed. The location of
the satellite can be determined by using circular functions.

CIRCULAR FUNCTIONS AND THEIR GRAPHS

OBJECTIVES

1. To define the circular functions.
2. To develop the graphs of the circular functions.
3. To define and graph the inverse circular functions.

Circular Functions

2-1 *Radian Measure*

The degree is not the only unit used for measuring angles. In advanced mathematics, for example, a unit called the *radian* is generally used. To define the *radian measure* of an angle θ, let r be the radius of any circle with center at the vertex of θ (Figure 2-1). Let s be the length of the arc intercepted by θ. The **radian measure** of θ is $\frac{s}{r}$ or $-\frac{s}{r}$, depending upon the direction of θ. One **radian,** or 1^R, then, is defined to be the measure of θ when θ is a counterclockwise angle with $s = r$. (The raised symbol ^{R} denotes radians.) The size of the circle used does not matter because of the properties of similar figures.

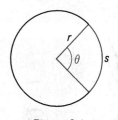

Figure 2-1

An angle θ of 180° intercepts a semicircular arc. Since πr is the length of a semicircle of radius r, the radian measure of $\theta = 180°$ is $\dfrac{\pi r}{r}$, or π. We can write this as:

$$180° = \pi^R$$

We can use this equation to convert from degree measure to radian measure or from radian measure to degree measure.

Example 1   Convert the following to radian measure.
(a) 30°   (b) 45°   (c) 60°   (d) 90°

Solution   (a) $\dfrac{30°}{180°} = \dfrac{\theta^R}{\pi^R}; \; \theta^R = \dfrac{30\pi}{180} = \dfrac{\pi}{6}$

(b) $\dfrac{45°}{180°} = \dfrac{\theta^R}{\pi^R}; \; \theta^R = \dfrac{45\pi}{180} = \dfrac{\pi}{4}$

(c) $\dfrac{60°}{180°} = \dfrac{\theta^R}{\pi^R}; \; \theta^R = \dfrac{60\pi}{180} = \dfrac{\pi}{3}$

(d) $\dfrac{90°}{180°} = \dfrac{\theta^R}{\pi^R}; \; \theta^R = \dfrac{90\pi}{180} = \dfrac{\pi}{2}$

Example 2   Convert $\dfrac{3}{4}\pi^R$ to degree measure.

Solution   Since $\pi^R = 180°, \; \dfrac{3}{4}\pi^R = \dfrac{3}{4}(180°) = 135°$

Example 3   (a) Find the length of an arc on a circle of radius 45 cm intercepted by a central angle of 216°. Leave your answer in terms of π.
(b) Find the radius of a circle in which a central angle of 135° intercepts an arc of length 6π cm.

Solution   (a) $\dfrac{216°}{180°} = \dfrac{\theta^R}{\pi^R}; \; \theta^R = \dfrac{216\pi}{180} = \dfrac{6}{5}\pi$

Let $s = $ the length of the arc. Then $\dfrac{6}{5}\pi = \dfrac{s}{45}$ and

$s = \dfrac{6}{5}\pi(45) = 54\pi$ cm.

(b) $\dfrac{135°}{180°} = \dfrac{\theta^R}{\pi^R}; \; \theta^R = \dfrac{135\pi}{180} = \dfrac{3}{4}\pi$

Let $r = $ the radius. Then $\dfrac{3}{4}\pi = \dfrac{6\pi}{r}$ and $r = 8$ cm.

To avoid confusion between radians and degrees, we will always use the symbol ° when discussing degrees. When the context is clear, we will drop the symbol ^{R}.

Exercises 2-1

Convert the following to radian measure. Leave your answers in terms of π.

A 1. $-90°$ 2. $270°$ 3. $315°$ 4. $-135°$ 5. $120°$ 6. $300°$

7. $72°$ 8. $324°$ 9. $22.5°$ 10. $-7.5°$ 11. $720°$ 12. $-660°$

Convert the following to degree measure.

13. $\dfrac{\pi^R}{3}$ 14. $\dfrac{\pi^R}{4}$ 15. $-\dfrac{2\pi^R}{3}$ 16. $\dfrac{5\pi^R}{6}$ 17. $-\dfrac{11\pi^R}{6}$ 18. $\dfrac{5\pi^R}{4}$

19. $\dfrac{7\pi^R}{12}$ 20. $-\dfrac{11\pi^R}{12}$ 21. $\dfrac{5\pi^R}{8}$ 22. $\dfrac{15\pi^R}{8}$ 23. $5\pi^R$ 24. $-\dfrac{7\pi^R}{2}$

Complete the following table where r is the radius of a circle with center at the vertex of θ and s is the length of the arc intercepted by θ. Give values of r or s to the nearest tenth.

	$\theta°$	θ^R	r	s
25.		$\dfrac{7\pi}{6}$	30	
26.		$\dfrac{7\pi}{4}$	640	
27.	$135°$		28	
28.	$240°$		9.6	
29.	$150°$			6π
30.			27	$\dfrac{9\pi}{2}$

B 31. Use the definition of radian measure to derive a formula for the length s of an arc of a circle of radius r intercepted by a central angle of radian measure θ.

32. Show that the formula derived in Exercise 31 is equivalent to the formula for arc length from geometry, $\dfrac{\theta°}{360°} \cdot 2\pi r$.

33. Find the length of an arc on a circle of radius 15 cm intercepted by a central angle of $140°$. Leave your answer in terms of π.

34. Find the length of an arc on a circle of radius 412 cm intercepted by a central angle of $25°$. Leave your answer in terms of π.

35. Find the radius of a circle in which a central angle of 20° intercepts an arc of length 4π cm.

36. Find the radius of a circle in which a central angle of 282° intercepts an arc of length 2π cm.

37. In the given diagram, the small gear has a radius of 48 cm and the large gear has a radius of 56 cm. Through what angle will the large gear rotate when the small gear rotates $\dfrac{7\pi}{4}$ radians?

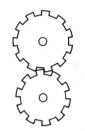

38. If the radius of the small gear in Exercise 37 is 75 mm, and the large gear rotates $\dfrac{3\pi^R}{4}$ when the small gear rotates $\dfrac{12\pi^R}{5}$, what is the radius of the large gear?

Exercises 37 and 38

C 39. About 200 B.C., a Greek mathematician named Eratosthenes found that when the sun's rays (assumed to be parallel) struck Earth at a certain time, they made an angle of about 7.2° with a vertical pole in the city of Alexandria. At the same time, the rays were exactly vertical at the city of Syene, 800 km to the south, since the bottom of a deep well there was illuminated. From these data, estimate the circumference of Earth.

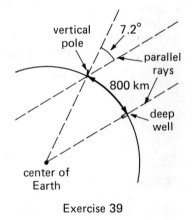

Exercise 39

40. A belt passes tightly over two wheels, A and B, of equal diameter. The wheels are in line, with centers 75 cm apart. If the diameter of each wheel is 10 cm, find the length of (a) the part of the belt in contact with wheel A and (b) the entire belt.

41. Suppose the wheels in Exercise 40 are in contact. Find the length of the entire belt.

2-2   Circular Functions

Many applications of trigonometry (including those using calculus) require that the functions sine, cosine, tangent, and so forth, be defined for *real numbers* as well as for angles. The resulting functions are often called **circular functions** because a circle is used in defining them.

Let C be a circle of radius r with center at the origin. Given any real number t, there is precisely one angle in standard position, θ_t, that has radian measure t. Let $P(x, y)$ be the point where the terminal side of θ_t intersects the circle C (Figure 2-2). We define

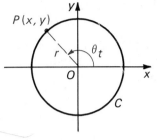

$$\sin t = \frac{y}{r} \quad \text{and} \quad \cos t = \frac{x}{r}.$$

Referring to the definition on page 25, we see that

$$\sin t = \sin \theta_t \quad \text{and} \quad \cos t = \cos \theta_t.$$

Figure 2-2

Note carefully, however, that the newly defined sine and cosine are functions of real numbers, not of angles.

If C is a **unit circle**, that is, if $r = 1$, then

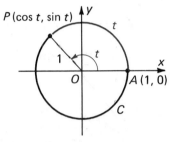

$$\sin t = y \quad \text{and} \quad \cos t = x.$$

Thus, the coordinates of $P(x, y)$ are $(\cos t, \sin t)$. Moreover, the length of the arc from $A(1, 0)$ to P is t (if $t > 0$). The radian measure of $\angle AOP$ is, of course, t also. These facts are illustrated in the **unit-circle diagram** in Figure 2-3.

We define the other circular functions as follows.

Figure 2-3

$$\tan t = \frac{\sin t}{\cos t}, \ \cos t \neq 0 \qquad \cot t = \frac{\cos t}{\sin t}, \ \sin t \neq 0$$

$$\sec t = \frac{1}{\cos t}, \ \cos t \neq 0 \qquad \csc t = \frac{1}{\sin t}, \ \sin t \neq 0$$

We can find the exact values of circular functions for numbers that are multiples of $\frac{\pi}{6}, \frac{\pi}{4}, \frac{\pi}{3}$, and $\frac{\pi}{2}$ just as we could for angles whose measures were multiples of $30°$, $45°$, $60°$, and $90°$. For example, to find $\cos \frac{\pi}{3}$, we can use the fact (from Section 2-1) that $\frac{\pi}{3} = 60°$, so that $\cos \frac{\pi}{3} = \cos 60° = \frac{1}{2}$. You should learn the frequently-occurring values in the following table.

$t \rightarrow$	0	$\dfrac{\pi}{6}$	$\dfrac{\pi}{4}$	$\dfrac{\pi}{3}$	$\dfrac{\pi}{2}$	π
$\sin t$	0	$\dfrac{1}{2}$	$\dfrac{\sqrt{2}}{2}$	$\dfrac{\sqrt{3}}{2}$	1	0
$\cos t$	1	$\dfrac{\sqrt{3}}{2}$	$\dfrac{\sqrt{2}}{2}$	$\dfrac{1}{2}$	0	-1
$\tan t$	0	$\dfrac{\sqrt{3}}{3}$	1	$\sqrt{3}$	—	

TABLE 2-1

Example 1   Find the exact values of $\sin \dfrac{5\pi}{6}$, $\cos \dfrac{5\pi}{6}$, and $\tan \dfrac{5\pi}{6}$.

Solution   The angle having radian measure $\dfrac{5\pi}{6}$ is in the second quadrant, and its reference angle measures $\pi - \dfrac{5\pi}{6} = \dfrac{\pi}{6}$ radians. (See the accompanying unit-circle diagram.) Thus,

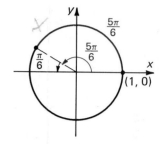

$$\sin \dfrac{5\pi}{6} = \sin \dfrac{\pi}{6} = \dfrac{1}{2}$$

$$\cos \dfrac{5\pi}{6} = -\cos \dfrac{\pi}{6} = -\dfrac{\sqrt{3}}{2}$$

$$\tan \dfrac{5\pi}{6} = -\tan \dfrac{\pi}{6} = -\dfrac{\sqrt{3}}{3}.$$

For most numbers t the values of the circular functions of t can only be approximated. This can be done with the help of a scientific calculator or a table.

Example 2   Use Table 3 to find $\cos 3.7$ to two significant digits.

Solution   The angle having radian measure 3.7 is in the third quadrant, and its reference angle has radian measure

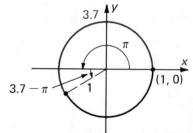

$3.7 - \pi \approx 3.7 - 3.14 = 0.56$.

From Table 3, $\cos 0.56 = 0.8473$. Therefore, $\cos 3.7 = -\cos 0.56 = 0.85$ to two significant digits.

Example 3   A point P moves counterclockwise a distance of 3π on a circle of radius $\dfrac{9}{4}$ with center at the origin. If P starts at the point $\left(\dfrac{9}{4}, 0\right)$, what are the coordinates of its final position $P(x, y)$?

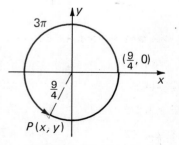

Solution   Let t be the radian measure of an angle that intercepts an arc of length 3π on the given circle. Then

$$t = \frac{3\pi}{\dfrac{9}{4}} = \frac{12\pi}{9} = \frac{4\pi}{3}.$$

The coordinates of $P(x, y)$ are $\left(\dfrac{9}{4}\cos\dfrac{4\pi}{3}, \dfrac{9}{4}\sin\dfrac{4\pi}{3}\right)$. Since $\dfrac{4\pi}{3}$ is a multiple of $\dfrac{\pi}{3}$, we can use Table 2-1 to find that $\cos\dfrac{4\pi}{3} = -\dfrac{1}{2}$ and $\sin\dfrac{4\pi}{3} = -\dfrac{\sqrt{3}}{2}$. Therefore

$$\frac{9}{4}\cos\frac{4\pi}{3} = \frac{9}{4}\left(-\frac{1}{2}\right) = -\frac{9}{8} \quad \text{and}$$

$$\frac{9}{4}\sin\frac{4\pi}{3} = \frac{9}{4}\left(-\frac{\sqrt{3}}{2}\right) = -\frac{9\sqrt{3}}{8}.$$

We then have $x = -\dfrac{9}{8}$ and $y = -\dfrac{9\sqrt{3}}{8}$.

Exercises 2-2

Find the exact values of sin t, cos t, and tan t for the given values of t.

A 1. $\dfrac{7\pi}{6}$ 2. $\dfrac{5\pi}{3}$ 3. $\dfrac{3\pi}{4}$ 4. $\dfrac{3\pi}{2}$

5. $-\dfrac{\pi}{2}$ 6. $-\dfrac{3\pi}{4}$ 7. $\dfrac{8\pi}{3}$ 8. $-\dfrac{17\pi}{6}$

Find the exact values of sec t, csc t, and cot t for the given values of t.

9. $-\dfrac{7\pi}{4}$ 10. $\dfrac{11\pi}{6}$ 11. $\dfrac{5\pi}{3}$ 12. $\dfrac{4\pi}{3}$

13. $-\dfrac{4\pi}{3}$ 14. $-\dfrac{3\pi}{4}$ 15. $-\dfrac{11\pi}{6}$ 16. π

Use Table 3 on page 365 to find each value to two significant digits.

17. sin 4.28   18. cos 3.67   19. cos 2.01

20. sin 3   21. cos (-4) 22. sin $(--5)$

23. sin 1.5   24. cos 5   25. sin 12

26. cos 12   27. tan 12   28. tan 4.1

Suppose that a point P moves counterclockwise a distance s on a circle of radius r with center at the origin. If P starts at the point $(r, 0)$, find the coordinates of its final position $P(x, y)$ for the given values of s and r.

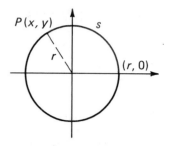

Exercises 29-34

B 29. $s = \dfrac{7\pi}{4}; r = 1$ 30. $s = \dfrac{5\pi}{4}; r = 1$

31. $s = \dfrac{8\pi}{3}; r = 2$ 32. $s = \dfrac{10\pi}{3}; r = 5$

33. $s = \dfrac{11\pi}{4}; r = \dfrac{3}{2}$ 34. $s = \dfrac{25\pi}{9}; r = \dfrac{5}{3}$

In Exercises 35 and 36, let t be the radian measure of an angle θ.

35. Show how the definition of $\tan t$ coincides with the definition of $\tan \theta$.

36. Show how the definition of $\sec t$ coincides with the definition of $\sec \theta$.

C 37. Imagine the unit circle with center at the origin, rolling to the left along the line $y = -1$. Imagine also a point P starting at position $(1, 0)$ and ending at position P' after the circle has rolled a distance t. Find the coordinates of P' in terms of t. (Hint: arc AP' has length t.)

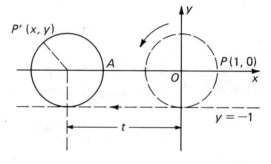

Exercise 37

38. Suppose the unit circle described in Exercise 37 rolls to the right a distance s along the line $y = -1$. Find the coordinates in terms of s of the final position P' of a point P on the circle that starts at position $(1, 0)$. (Hint: arc AP' has length s.)

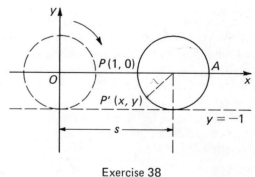

Exercise 38

39. Explain why for very small values of θ, the radian measure of θ approximates the value of $\sin \theta$.

2-3 *Uniform Circular Motion*

When a point P moves with a constant speed in a circular path, we can speak of its *angular speed,* as well as its *linear speed.* We call the motion of P **uniform circular motion.**

Imagine radius $\overline{OA}$ in Figure 2-4 as fixed and radius $\overline{OP}$ as moving uniformly in a counterclockwise direction. The **linear speed** v of P is simply the distance P moves along an arc s in one unit of time t. The **angular speed** ω of P is the rate with respect to time t at which the angle θ generated by $\overline{OP}$ changes. We write

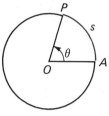

Figure 2-4

$$v = \frac{s}{t} \quad \text{or} \quad s = vt \qquad \omega = \frac{\theta}{t} \quad \text{or} \quad \theta = \omega t.$$

Angular speeds can be given in various units; for example, revolutions per minute (rpm), degrees per minute, or radians per minute. If ω represents the angular speed of a point in radians per unit of time, the linear speed v of the point is given by the formula

$$v = r\omega$$

where r is the radius of the point's circular path. (See Exercise 25 on p. 48.)

Example 1 A centrifuge is a machine that can be rotated at high speeds in order to separate fluid particles of different densities. Suppose a centrifuge is turning with an angular speed of 1.5×10^3 radians per second. If the radius is 35 cm, what is the linear speed of the centrifuge?

Solution Use the equation $v = r\omega$. Then
$v = 35(1.5 \times 10^3) = 53 \times 10^3$ or 5.3×10^4 cm/s.

The location of a point moving in uniform circular motion can be specified by using coordinates. Suppose that P is a point moving uniformly in a counterclockwise direction around the origin and suppose that, at time $t = 0$, P is at the position P_0 (Figure 2-5). During a given interval of time t, P moves from P_0 to P_1 so that radius $\overline{OP}$ forms an angle $\theta = \theta_0 + \theta_1$ with the x-axis. Since $\theta_1 = \omega t$, we have

$$\theta = \omega t + \theta_0.$$

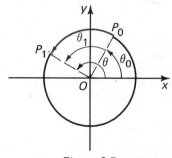

Figure 2-5

If $OP_1 = a$, the coordinates of P at position P_1 can be written as

$$x = a \cos (\omega t + \theta_0), \quad y = a \sin (\omega t + \theta_0).$$

Example 2   A point P travels uniformly at 4 rpm in a counterclockwise direction around a circular path of radius 6 cm with center at the origin. Find the coordinates of P and the distance s that P has traveled, 5 seconds after P moves from the location $(3\sqrt{3}, 3)$.

Solution   Draw the path of P. The angular speed of P is 4 rpm or

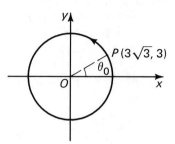

$\omega = 4 \times 2\pi = 8\pi$ radians/min.
The linear speed of P is
$v = 8\pi \times 6 = 48\pi$ cm/min. Thus in 5 seconds, or $\dfrac{5}{60} = \dfrac{1}{12}$ min, P moves a distance $s = 48\pi \times \dfrac{1}{12} = 4\pi$ cm.

To find the coordinates of P, let θ_0 be the angle that $\overline{OP}$ makes with the x-axis when P is at $(3\sqrt{3}, 3)$. Then
$\sin \theta_0 = \dfrac{3}{6} = \dfrac{1}{2}$ and $\cos \theta_0 = \dfrac{3\sqrt{3}}{6} = \dfrac{\sqrt{3}}{2}$. From Table 2-1 on page 41 and the fact that $(3\sqrt{3}, 3)$ is in Quadrant I, $\theta_0 = \dfrac{\pi}{6}$. Thus, after 5 seconds the coordinates of P are

$$x = 6 \cos \left(8\pi \times \frac{1}{12} + \frac{\pi}{6} \right) = 6 \cos \frac{5\pi}{6} = -3\sqrt{3}$$

$$y = 6 \sin \left(8\pi \times \frac{1}{12} + \frac{\pi}{6} \right) = 6 \sin \frac{5\pi}{6} = 3.$$

Exercises 2-3

A   1. The crankshaft pulley of a car has a radius of 10.5 cm and turns at 6π radians/s. What is the linear speed of the pulley?

2. Find to the nearest cm/s the linear speed of a point on the rim of a wheel of radius 24 cm turning at an angular speed of $\dfrac{17\pi}{12}$ radians.

3. The linear speed of a point 15.3 cm from the center of a phonograph record is 17π cm/s. What is the angular speed of the record in radians/s?

4. Find the linear speed of a point 5.2 cm from the center of the phonograph record in Exercise 3.

5. Find to the nearest cm/s the linear speed of a point on the rim of a wheel of radius 36 cm turning at an angular speed of 324 degrees/s.

6. The fan pulley of a car has a radius of 10.8 cm and turns at 390 rpm. What is the linear speed of the pulley?

In Exercises 7–16, suppose that a point P travels uniformly at an angular speed of ω in a counterclockwise direction around a circular path of radius a with center at the origin. Find the coordinates of the final position of P if P starts at the point $(a, 0)$ and moves for t units of time.

7. $a = 2$ cm; $\omega = \dfrac{3\pi}{2}$ radians/s; $t = 4$ s

8. $a = 3$ cm; $\omega = \dfrac{\pi}{6}$ radians/s; $t = 18$ s

9. $a = 6$ cm; $\omega = 5$ rpm; $t = 32$ min

10. $a = 4$ cm; $\omega = 4$ rpm; $t = 2.5$ min

11. $a = 10$ cm; $\omega = 10$ rpm; $t = 9$ s

12. $a = 4$ cm; $\omega = 8$ rpm; $t = 5$ s

13. $a = 6$ cm; $\omega = \dfrac{5\pi}{3}$ radians/s; $t = 5$ s

14. $a = 8$ cm; $\omega = \dfrac{4\pi}{3}$ radians/s; $t = 15$ s

15. $a = 4$ cm; $\omega = 1.5$ rpm; $t = 2$ min

16. $a = 6$ cm; $\omega = 2.5$ rpm; $t = 2$ min

In Exercises 17–22 suppose that a point P travels uniformly at an angular speed of ω in a counterclockwise direction around a circular path of radius a cm with center at the origin. Find the coordinates of P and the distance s that P has traveled, t seconds after P moves from the location P_0.

B 17. $a = 1$; $P_0 = \left(-\dfrac{\sqrt{3}}{2}, \dfrac{1}{2}\right)$; $\omega = \dfrac{4\pi}{3}$ radians/s; $t = 5$ s

18. $a = 1$; $P_0 = \left(\dfrac{1}{2}, \dfrac{\sqrt{3}}{2}\right)$; $\omega = \dfrac{3\pi}{4}$ radians/s; $t = 4$ s

19. $a = 4$; $P_0 = (-2, 2\sqrt{3})$; $\omega = \dfrac{3\pi}{2}$ radians/s; $t = 3$ s

20. $a = 8$; $P_0 = (-8, 0)$; $\omega = 2.5$ rpm; $t = 30$ s

21. $a = 6$; $P_0 = (-3\sqrt{2}, -3\sqrt{2})$; $\omega = \dfrac{2\pi}{3}$ radians/s; $t = 2$ s

22. $a = 2$; $P_0 = (-2, 0)$; $\omega = 1.2$ rpm; $t = 45$ s

23. Find the angular speed of Earth's rotation on its axis in radians per hour. What is the linear speed, to the nearest km/h, of a point on the equator if Earth's diameter at the equator is assumed to be 12,700 km?

24. If Earth's orbit around the sun is assumed to be a circle and a year is assumed to have 365 days, what is Earth's angular speed in its orbit in radians per hour? If the radius of Earth's orbit is 149,000,000 km, what is Earth's linear speed in its orbit to the nearest km/h?

25. Derive the formula $v = \omega \cdot r$, where v is the linear speed and ω the angular speed in radians per unit of time (the same unit of time as the linear speed) of a point moving around a circle of radius r. (Hint: Use the fact that $s = \theta r$.)

26. Derive the formula $v = 2\pi\omega r$, where v is the linear speed and ω the angular speed in revolutions per unit of time (the same unit of time for both) of a point moving around a circle of radius r.

C 27. As Earth rotates on its axis, what is the linear speed of a point P on the surface of Earth at latitude 60°N, if Earth's diameter is 12,700 km? (Hint: See Exercise 23 and note that if P is at latitude 60°N, the angle between Earth's radius drawn to P and the plane of the equator is 60°.)

28. A circle of radius a begins with its center at the point $(0, a)$ and rolls to the right along the x-axis with angular velocity ω. A point on the circumference of this circle starts at the origin and is at position P at time t. Find equations for the coordinates of P at time t. (Hint: OA = the length of arc $PA = a\theta$ (θ in radians).)

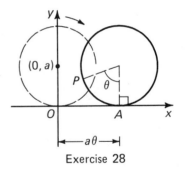

Exercise 28

Use the following information to solve Exercises 29 and 30. Particles moving in uniform circular motion are held in place by a centripetal, or "center seeking," force, directed toward the center of rotation. The centripetal force F_c acting on a particle can be described by the equation $F_c = \dfrac{mv^2}{r}$ where m is the mass of the particle, r is the radius of rotation, and v is the linear speed of the particle.

29. If the mass and radius are held constant, how will the linear speed change when the centripetal force is doubled?

30. Write an equation for centripetal force in terms of angular speed ω.

Graphs

2-4 *Graphing the Circular Functions*

Properties Used in Graphing Functions

As we have seen, the circular functions are defined for real numbers. Before we graph the circular functions in particular, then, we will first discuss some of the general properties used to graph functions defined for real numbers.

Even Functions

A function f is called an **even function** if whenever x is in the domain of f, $-x$ is also in the domain, and $f(-x) = f(x)$ for all x in the domain. For example, the functions $f(x) = \cos x$ and $f(x) = \sec x$ are even functions. The graph of any even function is **symmetric with respect to the y-axis.** In other words, by thinking of the y-axis as a mirror, we can draw the entire graph by reflecting the right side of the graph in the y-axis to produce the left side of the graph.

Examples $y = f(x) = x^2$

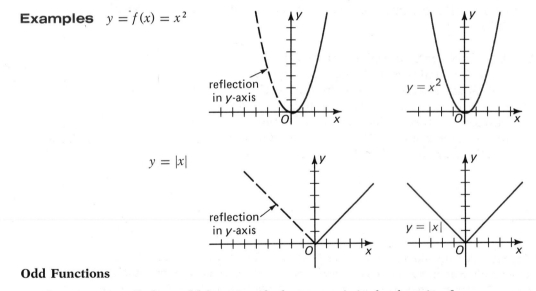

$y = |x|$

Odd Functions

A function f is called an **odd function** if whenever x is in the domain of f, $-x$ is also in the domain, and $f(-x) = -f(x)$ for all x in the domain. The functions $\sin x$, $\tan x$, $\cot x$, and $\csc x$, for example, are odd functions. The graph of any odd function is **symmetric with respect to the origin.** This means that by thinking of the origin as a mirror, we can reflect any point on the graph in the origin to produce another point on the graph. We can also draw the entire graph by reflecting the right side of the graph in the y-axis and then in the x-axis to produce the left side.

Example $y = x^3$

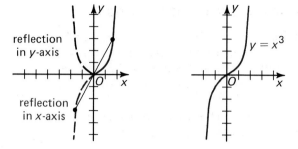

Asymptotes

Roughly speaking, an **asymptote** is a line that a curve approaches more and more closely. The line $y = a$ is a **horizontal asymptote** of a function f if, as we take progressively larger positive or negative values of x, $f(x)$ approaches closer and closer toward a. The line $x = a$ is a **vertical asymptote** of f if, as we take values of x closer and closer to a, $|f(x)|$ gets larger and larger without bound. The function $\tan x$, for example, has a vertical asymptote at $x = \frac{\pi}{2}$, and $\cot x$ has a vertical asymptote at $x = 0$. The functions $\sec x$ and $\csc x$ also have vertical asymptotes.

Example $y = \dfrac{1}{x}$

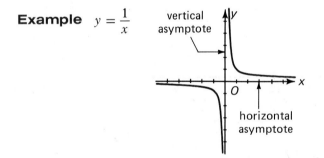

Periodicity

A function f is said to be **periodic** if there is some fixed positive number p such that whenever x is in the domain of f, both $x - p$ and $x + p$ are also in the domain, and $f(x - p) = f(x) = f(x + p)$ for all x in the domain. If there is a least such positive number p, then p is called the **fundamental period** of f. Most functions, for example the polynomial functions, are not periodic. All the circular functions, however, are periodic.

Example

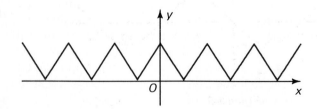

The Graphs of y = sin x and y = cos x

The domain of the functions sin t and cos t is the entire set of real numbers. Their range is the set of numbers between and including -1 and 1, since all of the values for these two functions are coordinates of points on the unit circle.

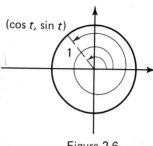

Figure 2-6

For any real number t, the angles corresponding to t and $t + 2\pi$ are coterminal (Figure 2-6). Therefore,

$$\sin (t + 2\pi) = \sin t \text{ and } \cos (t + 2\pi) = \cos t,$$

or equivalently in terms of the variable x,

$$\sin (x + 2\pi) = \sin x \text{ and } \cos (x + 2\pi) = \cos x.$$

To draw the graph of $y = \sin x$, then, we begin by choosing values for x in the interval $0 \le x \le 2\pi$. Since we know the values of sin x for multiples of $\dfrac{\pi}{6}$ and $\dfrac{\pi}{4}$, we can conveniently plot the points shown in Figure 2-7.

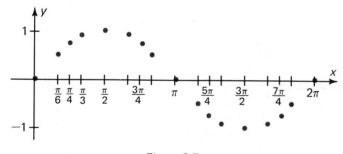

Figure 2-7

We next join these points by a smooth, continuous curve. By periodicity, we can draw the graph for all positive values of x. We can then use the fact that sin x is an odd function to reflect the graph in the y-axis and then in the x-axis. The final graph is shown in Figure 2-8.

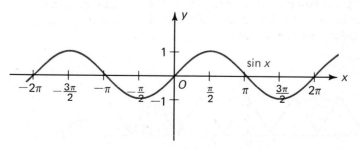

Figure 2-8

We can draw the graph of $y = \cos x$ using a similar method. The final graph of $y = \cos x$ is shown in Figure 2-9.

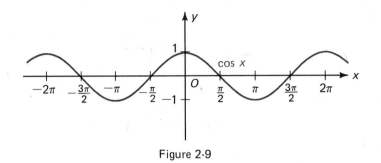

Figure 2-9

The Graphs of $y = \tan x$ and $y = \cot x$

To draw the graph of $y = \tan x$ we use the definition

$$\tan t = \frac{\sin t}{\cos t}.$$

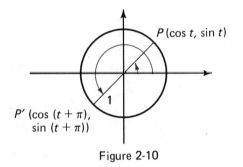

Figure 2-10

In Figure 2-10, angles with radian measure t and $t + \pi$ are illustrated in a unit-circle diagram. Since $t + \pi$ is in Quadrant III, the coordinates of P' are negatives of the coordinates of P. For t in Quadrant I, then,

$$\sin (t + \pi) = -\sin t \quad \text{and}$$

$$\cos (t + \pi) = -\cos t.$$

It can be shown similarly that these equations hold for all real numbers x. Therefore,

$$\tan (x + \pi) = \frac{\sin (x + \pi)}{\cos (x + \pi)} = \frac{-\sin x}{-\cos x} = \frac{\sin x}{\cos x} = \tan x$$

for all real numbers x in the domain of $\tan x$. This means that $\tan x$ is periodic with fundamental period π.

As mentioned on page 50, $\tan x$ has a vertical asymptote at $x = \dfrac{\pi}{2}$. To graph $y = \tan x$, then, we choose values for $0 \leq x \leq \dfrac{\pi}{2}$ (Figure 2-11). Since $\tan x$ is an odd function, we can extend the graph to the domain $-\dfrac{\pi}{2} \leq x \leq \dfrac{\pi}{2}$ (Figure 2-12).

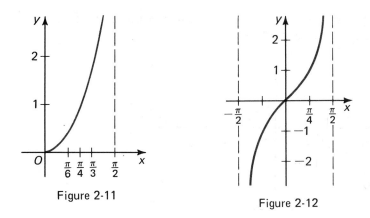

Figure 2-11

Figure 2-12

Because we have drawn the graph for an interval of π, we can use periodicity to complete the graph (Figure 2-13).

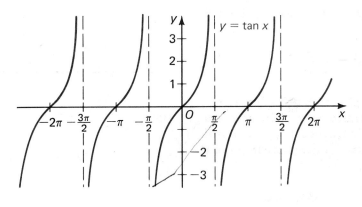

Figure 2-13

We can draw $y = \cot x$ in a similar manner. The graph of $y = \cot x$ is shown in Figure 2-14.

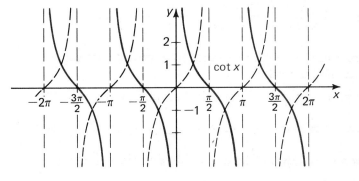

Figure 2-14

The Graphs of y = sec x and y = csc x

We can draw the graphs of $y = \sec x$ and $y = \csc x$ by relating $\sec x$ and $\csc x$ to their reciprocal functions, $\cos x$ and $\sin x$, respectively. For instance,

1. whenever $\cos x = \pm 1$, $\sec x = \pm 1$ also;

2. when $\cos x = 0$, $\sec x$ is undefined;

3. when the values of $\cos x$ are small and positive, the values of $\sec x$ are large and positive;

4. when the values of $\cos x$ are small and negative, the values of $\sec x$ are large and negative.

The function $\csc x$ is likewise related to $\sin x$. The completed graphs of $y = \sec x$ and $y = \csc x$ are shown in Figures 2-15 and 2-16, respectively.

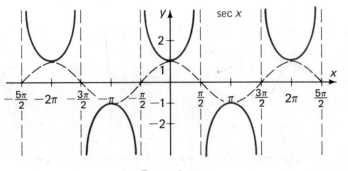

Figure 2-15

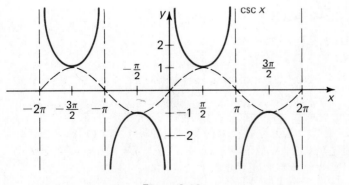

Figure 2-16

Exercises 2-4

By plotting multiples of $\frac{\pi}{4}$ and $\frac{\pi}{6}$, draw a detailed graph of the given function on the given domain.

A 1. $y = \cos x$; $0 \le x \le 2\pi$

2. $y = \sin x$; $-2\pi \le x \le 0$

3. $y = -\sin x$; $0 \le x \le 2\pi$

4. $y = -\cos x$; $0 \le x \le 2\pi$

5. $y = -\tan x$; $0 \le x \le 2\pi$

6. $y = -\cot x$; $0 \le x \le 2\pi$

7. $y = 2 \sec x$; $-\pi \le x \le \pi$

8. $y = \frac{1}{2} \csc x$; $-\pi \le x \le \pi$

9. $y = 1 + \cot x$; $-\pi \le x \le \pi$

10. $y = \sec x - 1$; $-\pi \le x \le \pi$

11. $y = |\sin x|$; $0 \le x \le 2\pi$

12. $y = |\cos x|$; $0 \le x \le 2\pi$

B 13. $y = \sin |x|$; $-\pi \le x \le \pi$

14. $y = \cos \left(x - \frac{\pi}{2} \right)$; $0 \le x \le 2\pi$

In Exercises 15–18, show that each of the given statements is true.

15. The function $\cot t$ is a periodic function with period π.

16. The function $\sec t$ is a periodic function with period 2π.

17. $\csc (t + \pi) = -\csc t$

18. $\sec (t + \pi) = -\sec t$

19. Show that for all real numbers the function $f(x) = (\sin x)(\cos x)$ is periodic with period π.

20. Show that for all real numbers the function $f(x) = \cos^2 x - \sin^2 x$ is periodic with period π.

C 21. Let x be a number very close to $\frac{\pi}{2}$, but less than $\frac{\pi}{2}$.

(a) What number is $\sin x$ close to?
(b) What number is $\cos x$ close to?
(c) Is $\tan x = \frac{\sin x}{\cos x}$ very large or very small?
(d) What do (a), (b), and (c) imply about an asymptote of the graph of $y = \tan x$?

22. Repeat Exercise 21 for the case where x is very close to $-\frac{\pi}{2}$, but greater than $-\frac{\pi}{2}$.

Inverse Circular Functions

2-5 *The Inverse Circular Functions*

Let f and g be functions such that $g(y) = x$ if and only if $f(x) = y$. Then g is called the **inverse** of f and is denoted by f^{-1} (f-inverse). That is, when we "solve" $y = f(x)$ for x, we obtain $x = f^{-1}(y)$. Since we usually denote members of the domain of a function by x and not y, we interchange x and y to get $y = f^{-1}(x)$.

For example, let $f(x) = x^3$. Then $y = f(x)$ becomes $y = x^3$. Interchange the variables and solve for x as follows.

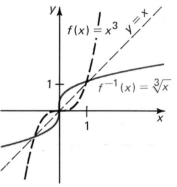

$$x = y^3$$

$$y = \sqrt[3]{x}$$

Therefore $\qquad f^{-1}(x) = \sqrt[3]{x}.$

The definition of an inverse function implies that a point (a, b) is on the graph of a function f if and only if the point (b, a) is on the graph of f^{-1}. The graphs of f and f^{-1} are reflections of each other in the line $y = x$ (Figure 2-17).

Figure 2-17

Not every function has an inverse function. For example, consider the function $f(x) = x^2$. For this function

$$f(3) = 9 \text{ and } f(-3) = 9.$$

If $f^{-1}(x)$ existed, it would be true that

$$f^{-1}(9) = 3 \text{ and } f^{-1}(9) = -3.$$

But this contradicts the definition of a function. In general, a function f has an inverse if and only if the function is **one-to-one**; that is, if and only if $f(x_1) = f(x_2)$ implies that $x_1 = x_2$.

None of the circular functions are one-to-one. Indeed, because of periodicity, a circular function associates the same y-value with infinitely many different x-values. Therefore, before we can define an inverse for a circular function, we must modify the function by restricting its domain.

We define a new function **Sin x** by:

$$\text{Sin } x = \sin x, \quad -\frac{\pi}{2} \le x \le \frac{\pi}{2}$$

The graph of Sin x is illustrated in Figure 2-18. Since Sin x is one-to-one, we can define the inverse of Sin x, which we denote as $\text{Sin}^{-1} x$ (or Arcsin x), by:

$\text{Sin}^{-1} x = $ the unique real number y such that

$$\sin y = x \quad \text{and} \quad -\frac{\pi}{2} \le y \le \frac{\pi}{2}$$

We can think of the expression $\text{Sin}^{-1} x$ as the measure of the angle between $-\frac{\pi}{2}$ and $\frac{\pi}{2}$ whose sine is x. In fact, the name Arcsin x suggests "the arc whose sine is x."

Both Sin x and $\text{Sin}^{-1} x$ are graphed in Figure 2-19.

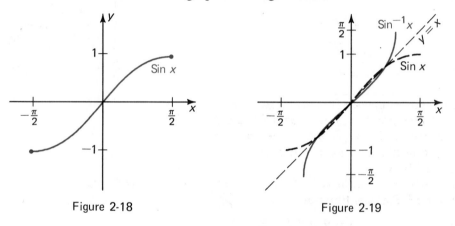

Figure 2-18   Figure 2-19

It is important to note that $\text{Sin}^{-1} x$ *never* means $\dfrac{1}{\sin x}$.

Example 1   Find (a) $\text{Sin}^{-1}\left(-\dfrac{1}{2}\right)$ and (b) $\cos\left(\text{Sin}^{-1}\dfrac{3}{5}\right)$.

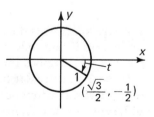

Solution   (a) Let $t = \text{Sin}^{-1}\left(-\dfrac{1}{2}\right)$. Then we
              have

$$\sin t = -\frac{1}{2} \quad \text{and} \quad -\frac{\pi}{2} \le t \le \frac{\pi}{2}.$$

But since $\sin t < 0$, we must have $-\dfrac{\pi}{2} \le t < 0$. Draw a unit-circle diagram. The reference angle for t is $\dfrac{\pi}{6}$ so $t = -\dfrac{\pi}{6}$, that is,

$$\text{Sin}^{-1}\left(-\frac{1}{2}\right) = -\frac{\pi}{6}.$$

(Solution continued on page 58.)

(b) Let $t = \operatorname{Sin}^{-1} \dfrac{3}{5}$. Then we have

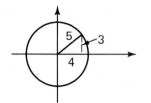

$$\sin t = \frac{3}{5} \quad \text{and} \quad -\frac{\pi}{2} \le t \le \frac{\pi}{2}.$$

But since $\sin t > 0$ we must have $0 < t \le \dfrac{\pi}{2}$. Draw a diagram. From page 25 in Chapter 1, then, $\cos t = \dfrac{4}{5}$. Thus,

$$\cos\left(\operatorname{Sin}^{-1}\frac{3}{5}\right) = \frac{4}{5}.$$

We define the function **Cos** x by:

$$\boxed{\operatorname{Cos} x = \cos x,\ 0 \le x \le \pi}$$

Then:

$$\boxed{\begin{array}{c} \operatorname{Cos}^{-1} x = \text{the unique real number } y \text{ such that} \\ \cos y = x \quad \text{and} \quad 0 \le y \le \pi \end{array}}$$

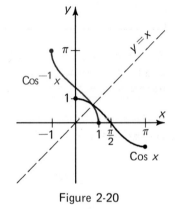

Figure 2-20

The graphs of $\operatorname{Cos} x$ and $\operatorname{Cos}^{-1} x$ are shown in Figure 2-20.

Example 2   Evaluate (a) $\operatorname{Cos}^{-1}\left(-\dfrac{1}{2}\right)$ and (b) $\operatorname{Cos}^{-1}\left(\sin \dfrac{5\pi}{6}\right)$.

Solution   (a) Let $t = \operatorname{Cos}^{-1}\left(-\dfrac{1}{2}\right)$. Then $\cos t = -\dfrac{1}{2}$ and $0 \le t \le \pi$. From the unit-circle diagram we see that the reference angle for t is $\dfrac{\pi}{3}$. Therefore,

$$t = \pi - \frac{\pi}{3} = \frac{2\pi}{3}, \text{ and}$$

$$\operatorname{Cos}^{-1}\left(-\frac{1}{2}\right) = \frac{2\pi}{3}.$$

(b) $\operatorname{Cos}^{-1}\left(\sin \dfrac{5\pi}{6}\right) = \operatorname{Cos}^{-1}\left(-\dfrac{1}{2}\right) = \dfrac{2\pi}{3}.$

We define the function **Tan** x by:

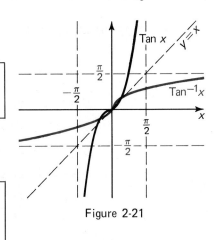

Figure 2-21

$$\text{Tan } x = \tan x, \quad -\frac{\pi}{2} < x < \frac{\pi}{2}$$

Then:

$\text{Tan}^{-1} x$ = the unique real number y such that

$$\tan y = x \text{ and } -\frac{\pi}{2} < y < \frac{\pi}{2}$$

The graphs of Tan x and $\text{Tan}^{-1} x$ are shown in Figure 2-21. Notice that the domain of $\text{Tan}^{-1} x$ is the entire set of real numbers.

The functions $\text{Sec}^{-1} x$, $\text{Csc}^{-1} x$, and $\text{Cot}^{-1} x$ can be defined likewise. The graphs of these functions are shown in Figure 2-22.

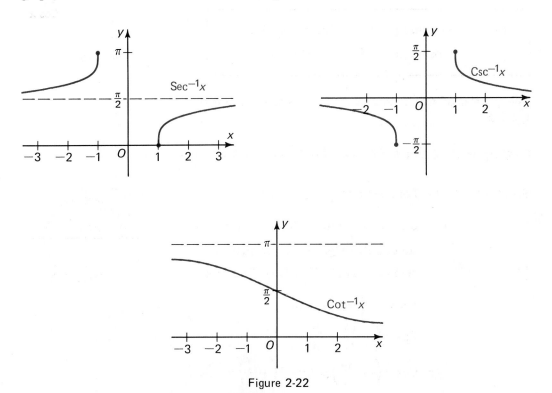

Figure 2-22

Exercises 2-5

Evaluate.

A 1. $\text{Cos}^{-1}\dfrac{\sqrt{3}}{2}$

2. $\text{Sin}^{-1}\dfrac{\sqrt{2}}{2}$

3. $\text{Sin}^{-1}(-1)$

4. $\text{Cos}^{-1}(-1)$

5. $\text{Tan}^{-1}1$

6. $\text{Tan}^{-1}(-1)$

7. $\text{Sin}^{-1}\left(-\dfrac{\sqrt{2}}{2}\right)$

8. $\text{Tan}^{-1}\sqrt{3}$

9. $\text{Cos}^{-1}\left(\dfrac{1}{2}\right)$

10. $\text{Tan}^{-1}\left(-\dfrac{\sqrt{3}}{3}\right)$

11. $\text{Cos}^{-1}\left(-\dfrac{\sqrt{2}}{2}\right)$

12. $\text{Cos}^{-1}0$

13. $\cos\left(\text{Sin}^{-1}\dfrac{\sqrt{3}}{2}\right)$

14. $\sin\left(\text{Cos}^{-1}\left(-\dfrac{1}{2}\right)\right)$

15. $\sin\left(\text{Cos}^{-1}\dfrac{5}{13}\right)$

16. $\cos\left(\text{Sin}^{-1}\dfrac{8}{17}\right)$

17. $\text{Sin}^{-1}\left(\cos\dfrac{7\pi}{6}\right)$

18. $\text{Cos}^{-1}\left(\sin\dfrac{5\pi}{4}\right)$

19. $\text{Cos}^{-1}\left(\sin\dfrac{\pi}{6}\right)$

20. $\text{Sin}^{-1}\left(\cos\dfrac{5\pi}{3}\right)$

21. $\text{Tan}^{-1}\left(\sin\dfrac{\pi}{2}\right)$

22. $\text{Tan}^{-1}\left(\cos\dfrac{\pi}{2}\right)$

23. $\text{Sin}^{-1}\left(\sin\dfrac{3\pi}{4}\right)$

24. $\text{Cos}^{-1}\left(\cos\left(-\dfrac{\pi}{3}\right)\right)$

Write a definition for each of the functions in Exercises 25–30.

B 25. $\text{Sec }x$

26. $\text{Csc }x$

27. $\text{Cot }x$

28. $\text{Sec}^{-1}x$

29. $\text{Csc}^{-1}x$

30. $\text{Cot}^{-1}x$

31. State the domain and range of the function $\text{Sec }x$ and draw its graph.

32. State the domain and range of the function $\text{Csc }x$ and draw its graph.

33. State the domain of the function $\text{Sin}^{-1}(\sin x)$. Is it true that $\text{Sin}^{-1}(\sin x) = x$ for all x in the domain? Why?

34. State the domain of $\sin(\text{Sin}^{-1}x)$. Is it true that $\sin(\text{Sin}^{-1}x) = x$ for all x in the domain? Give a reason for your answer.

35. Is it true that $\text{Sin}^{-1}(-x) = -\text{Sin}^{-1}x$ for all x in the domain of $\text{Sin}^{-1}x$? Give a reason for your answer.

36. Is it true that $\text{Cos}^{-1}(-x) = -\text{Cos}^{-1}x$ for all x in the domain of $\text{Cos}^{-1}x$? Give a reason for your answer.

Justify the following common definitions of Sec^{-1} and Csc^{-1}.

C 37. $\text{Sec}^{-1}x = \text{Cos}^{-1}\dfrac{1}{x}$

38. $\text{Csc}^{-1}x = \text{Sin}^{-1}\dfrac{1}{x}$

Challenge

Draw the graph of $y = \text{Tan}^{-1}(\tan x)$ for $0 \leq x \leq 2\pi$.

CHAPTER SUMMARY

1. Angles may be measured in radians as well as in degrees. Degrees and radians are related by the formula $180° = \pi^R$.

2. When sine, cosine, tangent, secant, cosecant, and cotangent are defined for real numbers they are called circular functions.

3. A point P moving in a circular path at a constant speed is in uniform circular motion. The linear speed v of P along an arc s in a unit of time t is given by $s = vt$. The angular speed ω of P with respect to time t at which the angle θ is generated by the radius of the point's circular path is given by $\theta = \omega t$. Linear speed and angular speed are related by the equation $v = r\omega$.

4. Various properties of functions are used in graphing. Even functions are symmetric with respect to the y-axis. Odd functions are symmetric with respect to the origin. An asymptote is a line that a curve approaches more and more closely. A function f is periodic if there is some fixed positive number p such that whenever x is in the domain of f both $x - p$ and $x + p$ are also in the domain, and $f(x - p) = f(x) = f(x + p)$ for all x in the domain.

5. If f and g are functions such that $g(y) = x$ if and only if $f(x) = y$, then g is the inverse of f. The inverse circular functions can be defined as follows:

$\text{Sin}^{-1} x =$ the unique real number y such that $\sin y = x$ and $-\dfrac{\pi}{2} \leq y \leq \dfrac{\pi}{2}$.

$\text{Cos}^{-1} x =$ the unique real number y such that $\cos y = x$ and $0 \leq y \leq \pi$.

$\text{Tan}^{-1} x =$ the unique real number y such that $\tan y = x$ and $-\dfrac{\pi}{2} < y < \dfrac{\pi}{2}$.

$\text{Sec}^{-1} x =$ the unique real number y such that $\sec y = x$ and $0 \leq y \leq \pi,\ y \neq \dfrac{\pi}{2}$.

$\text{Csc}^{-1} x =$ the unique real number y such that $\csc y = x$ and $-\dfrac{\pi}{2} \leq y \leq \dfrac{\pi}{2},\ y \neq 0$.

$\text{Cot}^{-1} x =$ the unique real number y such that $\cot y = x$ and $0 < y < \pi$.

CHAPTER TEST

2-1 1. Convert the following to radian measure. Leave your answers in terms of π.
    (a) $54°$
    (b) $-112°$

2. Convert the following to degree measure.
    (a) $\dfrac{7\pi}{10}$
    (b) $-\dfrac{5\pi}{9}$

2-2 3. Find the exact values of the six circular functions for the following.
    (a) $\dfrac{5\pi}{6}$
    (b) $-\dfrac{2\pi}{3}$

2-3 4. Find the coordinates of the final position of a point P moving counterclockwise in uniform circular motion at $\omega = \dfrac{\pi}{3}$ radians/s if P starts at the point $(5, 0)$ and moves for 14 seconds.

2-4 5. Graph (a) $y = 2 \sin x$ and (b) $y = 1 + \cos x$.

2-5 6. Evaluate (a) $\text{Tan}^{-1}\, 0$ and (b) $\text{Sin}^{-1}\left(\cos \dfrac{\pi}{3}\right)$.

CUMULATIVE REVIEW   *Chapters 1–2*

Chapter 1

1. Give the degree measure of an angle between $0°$ and $360°$ that is coterminal with an angle of $-112°30'$.

2. Find $\sin \theta$ and $\cos \theta$ to four decimal places.

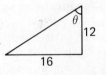

3. Suppose θ is an acute angle in standard position. If the terminal side of θ passes through $(\sqrt{3}, 3)$, find θ without using tables or a calculator.

4. Use tables or a calculator to find each of the following to four significant digits.
    (a) $\sin 13.5°$
    (b) $\cos 62°20'$
    (c) θ if $\cos \theta = 0.8910$
    (d) θ if $\sin \theta = 0.9436$

5. Express $\csc 30° \cdot \cos 45° + \tan 60°$ in simplified radical form.

6. Find the missing sides and angles if $\theta = 32°30'$, $AC = 170$, and $BD = 120$.

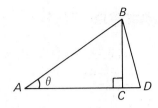

7. The terminal side of an angle in standard position passes through $(-3, \sqrt{7})$. Find all six trigonometric functions of the angle.

8. (a) In which quadrants is the secant positive?
   (b) In which quadrants is the cotangent negative?

9. List all undefined trigonometric functions for specific angles θ, $0° \le \theta \le 180°$. (You should have six answers.)

10. Find $\tan \theta$ if $\cos \theta = -0.4281$ and $360° < \theta < 540°$.

Chapter 2

1. Convert (a) $0°$, (b) $45°$, and (c) $280°$ to radian measure.

2. Convert (a) $\dfrac{3\pi}{2}$, (b) $\dfrac{7\pi}{6}$, and (c) 3π to degree measure.

3. Find the length of an arc of a circle of radius 25 cm that is intercepted by a central angle of $105°$.

4. Find each of the following to two significant digits.
   (a) $\tan 2.8$ (b) $\sin (-1.7)$

5. When a car travels at 88 km/h, a point on the edge of its 28 cm radius wheel has a linear speed of 88 km/h (or 2444 cm/s). Find its angular speed.

6. Given the function $f(x) = |\tan x|$, state whether it is even or odd, and sketch the graph of the function for $-2\pi \le x \le 2\pi$.

7. Graph $y = \sin \left(x + \dfrac{\pi}{2}\right)$, $0 \le x \le 2\pi$, by plotting multiples of $\dfrac{\pi}{6}$ and $\dfrac{\pi}{4}$.

8. Graph $y = |\csc x|$, $-\pi \le x \le \pi$, by plotting multiples of $\dfrac{\pi}{6}$ and $\dfrac{\pi}{4}$.

9. Evaluate (a) $\mathrm{Tan}^{-1}\left(\dfrac{\sqrt{3}}{3}\right)$, (b) $\cot (\mathrm{Sin}^{-1} 0.6)$, (c) $\mathrm{Cos}^{-1}\left(\sec \dfrac{\pi}{3}\right)$, and
   (d) $\mathrm{Sin}^{-1}\left(\sin -\dfrac{7}{6}\pi\right)$.

10. State the domain and range of the function $y = \mathrm{Tan}^{-1} x$ and draw its graph.

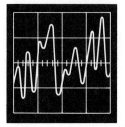

3.14159265358979323846264
3383279 . . .

It has long been known that the ratio of the circumference of a circle to its diameter is always the same, regardless of the size of the circle. This ratio is called π, and its computation has played an important part in the history of mathematics.

Early estimates of π were often related to the problem of constructing a square equal in area to a given circle. About 1700 B.C., Egyptian mathematicians believed that the area of a circle equals that of a square whose side is $\frac{8}{9}$ the diameter of the circle, which gives the value $\pi \approx 3.1605$. (To see this, use the formula Area $= \pi r^2$.)

Around 240 B.C., Archimedes approximated the circumference of a circle by inscribing and circumscribing polygons with successively larger numbers of sides. By finding the perimeters of the polygons, he estimated the value of π to be between $3\frac{10}{71}$ and $3\frac{1}{7}$. Archimedes' idea formed the basis of the integral calculus 1900 years later.

By the seventeenth and eighteenth centuries, mathematicians had found expressions for π that could, theoretically, produce any desired degree of accuracy. Two of the more interesting of these formulas are the following:

$$\pi = \frac{4}{1} - \frac{4}{3} + \frac{4}{5} - \frac{4}{7} + \frac{4}{9} - \frac{4}{11} + \cdots$$

$$\pi = 2\left(\frac{2}{1} \cdot \frac{2}{3} \cdot \frac{4}{3} \cdot \frac{4}{5} \cdot \frac{6}{5} \cdot \frac{6}{7} \cdot \frac{8}{7} \cdot \frac{8}{9} \cdots\right)$$

Neither formula, however, gives a very accurate result until the expressions are carried out to many terms.

The number π can be determined experimentally, as Count Buffon (1707–1788), a French naturalist, discovered. If a needle of length l is dropped at random onto a flat surface on which are drawn parallel lines exactly l units apart, then the probability that the needle will land touching a line is $\dfrac{2}{\pi}$. This means that

$$\frac{\text{number of dropped needles touching a line}}{\text{total number of drops}} \approx \frac{2}{\pi}.$$

Trials of this experiment in which the needle has been dropped thousands of times have produced results that are remarkably close to the known value of π.

A landmark in the direct computation of π was the development of the electronic computer. In 1949, for example, a computer was used to print out the decimal expansion of π to over two thousand places. Other examples include the computation by computer of π to over one hundred thousand decimal places in 1961 and the computation by computer of π to over five hundred thousand decimal places in 1967. Although the given examples demonstrate the power of computers, the computations far exceed the amount of information needed for practical applications. In fact, a thirty-place expansion of π would be sufficient to compute the circumference of the known universe so accurately that the possible error would be too small to be detected.

GRAPHING THE SINE AND COSINE FUNCTIONS

Besides being used to compute values of the functions sine, cosine, tangent, and so on, a computer can also be used to graph them. For example, to graph some functions involving sines and cosines, we can set up a grid such that one line feed represents 0.2 on the *x*-axis and one space represents 0.1 on the *y*-axis.

In the program given below, the *x*-axis is located M = 37 spaces from the left edge, and the graph is plotted between L = 10 and R = 64.

The values of *y* are translated into numbers of spaces from the left edge by line 180.

Note that on a microcomputer without a printer the graph generated by this program will not fit on the display screen. You may wish to let the graph print through and then rerun the program, stopping the graph at various intervals to investigate sections of the graph.

An alternate method that will fit the graph on the display screen, but with less detail, is to change line 140 to the following.

```
140   FOR X1 = −40 TO 40 STEP 8
```

```
10    REM: TO GRAPH FUNCTIONS INVOLVING SIN AND COS:
20    REM: SET UP MARGINS--
30    REM
40    LET L = 10
50    LET R = 64
60    REM: LOCATE X-AXIS M SPACES FROM THE LEFT:
70    LET M = L + (R − L)/2
80    REM: IN LINES 100 AND 160, INSERT
90    REM: FUNCTION TO BE GRAPHED.
100   PRINT "Y = SIN X"
110   PRINT
120   PRINT TAB(2); "X"; TAB(7); "Y"
130   PRINT
140   FOR X1 = −40 TO 40 STEP 2
150   LET X = .1*X1
160   LET Y = SIN(X)
170   PRINT X; TAB(5); INT(Y*100 + .5)/100;
180   LET Y1 = INT(10*Y + .5) + M
190   IF Y1 < L THEN 350
```

(program continued on next page)

```
200   IF Y1 > R THEN 350
210   IF Y1 = M THEN 270
220   IF Y1 < M THEN 250
230   PRINT TAB(M); "!"; TAB(Y1); "*";
240   GOTO 280
250   PRINT TAB(Y1); "*"; TAB(M); "!";
260   GOTO 280
270   PRINT TAB(Y1); "*";
280   IF X <> 0 THEN 310
290   PRINT TAB (M + 5); "Y"
300   GOTO 320
310   PRINT
320   NEXT X1
330   PRINT TAB(M); "X"
340   STOP
350   PRINT "TOO WIDE"
360   END
```

Exercises

1. Copy and RUN the given program. Notice that the *x*-axis is vertical. A line can be drawn through 0 in the X-column and the letter Y on the right side of the print-out to represent the *y*-axis.

In each of the following exercises, make the given changes and RUN the program.

2. 100 PRINT "Y = COS X"
   160 LET Y = COS(X)

3. 100 PRINT "Y = ABS(SIN X)"
   160 LET Y = ABS(SIN(X))

   (See Exercise 11 on page 55.)

4. 100 PRINT "Y = ABS(COS X)"
   160 LET Y = ABS(COS(X))

   (See Exercise 12 on page 55.)

The approximate distance that a hurled football will travel can be found by using a trigonometric formula involving the initial speed and the angle at which the ball was thrown.

PROPERTIES OF TRIGONOMETRIC FUNCTIONS

OBJECTIVES

1. To simplify trigonometric expressions using the trigonometric identities.
2. To solve trigonometric equations.

Trigonometric Identities

3-1 *Simplifying Trigonometric Expressions*

In algebra we often see equations called **identities** that are true for all values of the variable for which both sides of the equation are defined. For instance,

$$\frac{1 - x^2}{1 - x} = 1 + x, \quad x \neq 1.$$

In trigonometry we can make use of the algebraic identities to simplify expressions. For example, we can write

$$\frac{1 - \sin^2 x}{1 - \sin x} = 1 + \sin x \quad \text{for all } x \text{ such that } \sin x \neq 1.$$

But there are other identities unique to trigonometry. Some of these we have already seen.

69

$$\tan x = \frac{\sin x}{\cos x} \qquad (1) \qquad\qquad \cot x = \frac{\cos x}{\sin x} \qquad (2)$$

$$\sec x = \frac{1}{\cos x} \qquad (3) \qquad\qquad \csc x = \frac{1}{\sin x} \qquad (4)$$

$$\tan x = \frac{1}{\cot x} \qquad (5) \qquad\qquad \cot x = \frac{1}{\tan x} \qquad (6)$$

Other identities can be derived from definitions. For example, we know that the point $(\cos t, \sin t)$ lies on the unit circle $x^2 + y^2 = 1$. When we substitute, we obtain $\cos^2 t + \sin^2 t = 1$, or, equivalently,

$$\sin^2 x + \cos^2 x = 1 \qquad (7)$$

Using this identity and identities (1)–(6), we can derive the following:

$$1 + \tan^2 x = \sec^2 x \qquad (8)$$
$$1 + \cot^2 x = \csc^2 x \qquad (9)$$

Identities (7)–(9) are called the **Pythagorean identities** since they are closely related to the Pythagorean theorem. (See Exercises 26 and 27 on page 29.)

The following examples illustrate how identities can be used to simplify trigonometric expressions in terms of a single function.

Example 1   Express $\csc^2 x + \cot^2 x$ in terms of $\sin x$.

Solution   Using identities (2) and (4), we have

$$\csc^2 x + \cot^2 x = \left(\frac{1}{\sin x}\right)^2 + \left(\frac{\cos x}{\sin x}\right)^2$$

$$= \frac{1}{\sin^2 x} + \frac{\cos^2 x}{\sin^2 x}$$

$$= \frac{1 + \cos^2 x}{\sin^2 x}$$

$$= \frac{1 + (1 - \sin^2 x)}{\sin^2 x} \quad \text{by identity (7)}$$

$$= \frac{2 - \sin^2 x}{\sin^2 x}, \text{ or } \frac{2}{\sin^2 x} - 1$$

Example 2   Express $1 - \sin x \cos x \tan x$ in terms of $\cos x$.

Solution   By identity (1), we have

$$1 - \sin x \cos x \tan x = 1 - \sin x \cos x \, \frac{\sin x}{\cos x}$$

$$= 1 - \sin^2 x$$

$$= \cos^2 x \quad \text{by identity (7)}$$

Example 3   Simplify $\dfrac{\tan \theta (1 + \cot^2 \theta)}{\cot \theta}$.

Solution   Using identity (9), we have

$$\frac{\tan \theta (1 + \cot^2 \theta)}{\cot \theta} = \frac{\tan \theta (\csc^2 \theta)}{\cot \theta}$$

$$= \frac{\tan \theta (\csc^2 \theta)}{\dfrac{1}{\tan \theta}} \quad \text{by identity (6)}$$

$$= \tan^2 \theta \csc^2 \theta$$

$$= \frac{\sin^2 \theta}{\cos^2 \theta} \cdot \frac{1}{\sin^2 \theta}$$

$$= \frac{1}{\cos^2 \theta}, \text{ or } \sec^2 \theta$$

Exercises 3-1

Express the following in terms of sin x. Give your answers in simplest form.

A   1. $\tan^2 x$ 2. $\cot^2 x$ 3. $\sec x \tan x$ 4. $\dfrac{\cot x}{\sec x}$

5. $\dfrac{\csc x}{1 + \cot^2 x}$ 6. $\dfrac{\sec^2 x - \tan^2 x}{\csc x}$

7. $\cos x (\tan x + \cot x)$ 8. $\dfrac{1 + \cot x}{\sin x + \cos x}$

Express the following in terms of cos x. Give your answers in simplest form.

9. $1 + \tan^2 x$ 10. $\sec^2 x - 1$

11. $\sec x (\cos x + \sin^2 x \sec x)$ 12. $(1 + \cot x)(1 - \cot x)$

13. $\dfrac{\sin^2 x}{1 + \cos x}$ 14. $\csc^2 x (1 - \cos x)$

15. $\csc x (\csc x + \cot x)$ 16. $\dfrac{\sec x + 1}{\sin^2 x \sec x}$

Express each of the following in terms of a single trigonometric function.

17. $\dfrac{\csc x - \sin x}{\cos x}$

18. $\sec x \csc x - \tan x$

19. $\dfrac{\sin t \cos t}{1 - \cos^2 t}$

20. $\dfrac{1 + \tan^2 t}{\csc t \sec t}$

21. $\sin x(\cos x + \sin x \tan x)$

22. $\csc x(\sec x - \cos x)$

23. $\csc x(1 - \cos x)(1 + \cos x)$

24. $\dfrac{(\sec x - \tan x)(\sec x + \tan x)}{\cos x}$

B 25. $\dfrac{\cos \theta}{1 + \sin \theta} + \tan \theta$

26. $\dfrac{\sec \theta + \csc \theta}{1 + \tan \theta}$

27. $\dfrac{\sec y}{\sin y} - \dfrac{\sec y}{\csc y}$

28. $\dfrac{\sin x + \tan x}{\tan x(\csc x + \cot x)}$

29. $\dfrac{\sec \theta - \cos \theta}{\sin^2 \theta \sec^2 \theta}$

30. $\dfrac{\tan \theta + \cot \theta}{\sec^2 \theta}$

31. $\dfrac{\cot^2 x}{\csc x + 1} + 1$

32. $\dfrac{1}{2}\left(\dfrac{\sin x}{1 - \cos x} + \dfrac{1 - \cos x}{\sin x}\right)$

Use identity (7) and any of identities (1)–(6) on page 70 to derive the following.

33. $1 + \tan^2 x = \sec^2 x$

34. $1 + \cot^2 x = \csc^2 x$

Show that each of the following expressions equals either 0 or 1.

C 35. $\dfrac{1 + \sec x}{\sec x - 1} + \dfrac{1 + \cos x}{\cos x - 1}$

36. $\dfrac{\sec^2 x(1 + \csc x) - \tan x(\sec x + \tan x)}{\csc x(1 + \sin x)}$

37. $\dfrac{\sec x}{1 - \cos x} - \dfrac{\sec x + 1}{\sin^2 x}$

38. $\dfrac{\tan x}{\tan x + \sin x} - \dfrac{1 - \cos x}{\sin^2 x}$

39. $\dfrac{\csc \theta}{1 + \sec \theta} - \dfrac{\cot \theta}{1 + \cos \theta}$

40. $\dfrac{\sin \theta + \cos \theta - 1}{\sin \theta - \cos \theta + 1} - \dfrac{\cos \theta}{\sin \theta + 1}$

Challenge

Show that $\text{Sin}^{-1} x = \text{Tan}^{-1}\left[\dfrac{x}{(1 - x^2)^{\frac{1}{2}}}\right]$, $-1 < x < 1$.

(Hint: Use the fact that $x = \sin t$ for some t, $-\dfrac{\pi}{2} < t < \dfrac{\pi}{2}$, and $\sin^2 t + \cos^2 t = 1$.)

3-2 *Proving Identities*

In this section we will use the identities presented in Section 3-1 to prove other identities. The procedures for doing this can be best illustrated by examples.

Example 1 Prove the identity $\sec x - \cos x = \sin x \tan x$

Solution
$$\sec x - \cos x = \frac{1}{\cos x} - \cos x \quad \text{by identity (3)}$$

$$= \frac{1 - \cos^2 x}{\cos x} \quad \text{using algebra}$$

$$= \frac{\sin^2 x}{\cos x} \quad \text{by identity (7)}$$

$$= \sin x \frac{\sin x}{\cos x} \quad \text{using algebra}$$

$$= \sin x \tan x \quad \text{by identity (1)}$$

Notice that in Example 1 we first wrote down just one side of the given identity and then used known identities to transform it until it was shown to be equivalent to the other side of the identity. When proving an identity we must be careful to consider each side of the identity as an independent expression, that is, we must not assume that the two sides are equal until we prove them to be.

Example 2 Prove that $\dfrac{\sin x}{\csc x} + \dfrac{\cos x}{\sec x} = \sec^2 x - \tan^2 x$.

Solution
Starting with the left side of the given equality, we have
$$\frac{\sin x}{\csc x} + \frac{\cos x}{\sec x} = \frac{\sin x}{\frac{1}{\sin x}} + \frac{\cos x}{\frac{1}{\cos x}}$$

$$= \sin^2 x + \cos^2 x = 1$$

Since this last expression cannot be further simplified, we work separately with the expression on the right.

$$\sec^2 x - \tan^2 x = (1 + \tan^2 x) - \tan^2 x \quad \text{by identity (8)}$$

$$= 1$$

Because both sides have independently been shown to equal 1, they are equal to each other.

If no other approach to proving an identity seems promising, it may be helpful to rewrite one or both sides of the identity in terms of sine or cosine.

Example 3 Prove that $\tan \theta + \cot \theta = \sec \theta \csc \theta$

Solution
$$\tan \theta + \cot \theta = \frac{\sin \theta}{\cos \theta} + \frac{\cos \theta}{\sin \theta}$$

$$= \frac{\sin^2 \theta + \cos^2 \theta}{\cos \theta \sin \theta}$$

$$= \frac{1}{\cos \theta \sin \theta}$$

$$= \frac{1}{\cos \theta} \cdot \frac{1}{\sin \theta}$$

$$= \sec \theta \csc \theta$$

Exercises 3-2

Prove each identity.

A 1. $\dfrac{x^2 - 1}{x} - \dfrac{2x - 1}{2} = \dfrac{1}{2} - \dfrac{1}{x}$

2. $\dfrac{1}{x - 1} - \dfrac{1}{x + 1} = \dfrac{2}{x^2 - 1}$

3. $\dfrac{1}{1 + y} - \dfrac{y^2}{1 + y} = 1 - y$

4. $\dfrac{1 + t^2}{1 - t^2} - \dfrac{t}{1 - t} = \dfrac{1}{1 + t}$

5. $\dfrac{1}{x}(x - 1)^2 \doteq x - 2 + \dfrac{1}{x}$

6. $\dfrac{x^2 - 2x + 1}{x^2 - 1} = \dfrac{x - 1}{x + 1}$

7. $\sec^2 x(1 - \cos^2 x) = \tan^2 x$

8. $\dfrac{\sin^2 x + \cos^2 x}{\tan x} = \cot x$

9. $\cos x(\sec x - \cos x) = \sin^2 x$

10. $\sin x(\csc x + \sin x \sec^2 x) = \sec^2 x$

11. $\cos \theta + \sin \theta \tan \theta = \sec \theta$

12. $\sec \theta(\csc \theta - \cot \theta \cos \theta) = \tan \theta$

13. $\dfrac{1}{1 + \tan^2 x} + \dfrac{1}{1 + \cot^2 x} = 1$

14. $\dfrac{\sec^2 x - 1}{\csc^2 x - 1} = \tan^4 x$

15. $\sec^2 x + \csc^2 x = \sec^2 x \csc^2 x$

16. $\csc t - \sin t = \cos t \cot t$

17. $(\tan x + \sin x)(1 - \cos x) = \sin^2 x \tan x$

18. $(\cot x - \cos x)(\csc x + 1) = \cos x \cot^2 x$

19. $(1 - \cos \theta)(\csc \theta + \cot \theta) = \sin \theta$

20. $(\sec \theta + 1)(\csc \theta - \cot \theta) = \tan \theta$

21. $\dfrac{\cos x}{1 - \sin x} - \dfrac{\cos x}{1 + \sin x} = 2 \tan x$

22. $\dfrac{1}{\sec t - 1} + \dfrac{1}{\sec t + 1} = 2 \cot t \csc t$

23. $\cos x(1 + \tan x)^2 = \sec x + 2 \sin x$

24. $\dfrac{(\sin x + \cos x)^2}{\sin x} = \csc x + 2 \cos x$

B 25. $\dfrac{1 + \sec x}{\tan x} + \dfrac{\tan x}{1 + \sec x} = 2 \csc x$

26. $\dfrac{\cos x}{1 + \sin x} + \dfrac{1 + \sin x}{\cos x} = 2 \sec x$

27. $\dfrac{\sec \theta}{\sec \theta - 1} - \dfrac{\sec \theta + 1}{\tan^2 \theta} = 1$

28. $\dfrac{\sec \theta - 1}{1 - \cos \theta} = \sec \theta$

29. $\dfrac{\csc x - 1}{\cot x} + \dfrac{\cot x}{\csc x + 1} = \dfrac{2 \cos x}{1 + \sin x}$

30. $\dfrac{\sec x - 1}{\tan x} + \dfrac{\tan x}{\sec x + 1} = \dfrac{2 \sin x}{1 + \cos x}$

31. $\dfrac{1 - \cos x}{\sin x} = \dfrac{\sin x}{1 + \cos x}$ $\left(\text{Hint: It suffices to show that}\right.$
$$\dfrac{1 - \cos x}{\sin x} - \dfrac{\sin x}{1 + \cos x} = 0. \bigg)$$

32. $\dfrac{\tan t}{1 + \sec t} = \dfrac{\sec t - 1}{\tan t}$ (Hint: See the hint for Exercise 31.)

33. $(\sec \theta - \cos \theta)^2 = \tan^2 \theta - \sin^2 \theta$

34. $\dfrac{\tan^2 x}{1 - \cos x} = \sec x + \sec^2 x$

35. $\dfrac{1 + \sin x}{1 - \sin x} = (\sec x + \tan x)^2$

C 36. $\dfrac{\sec x + \tan x}{\sec x - \tan x} = \dfrac{1 + 2 \sin x + \sin^2 x}{\cos^2 x}$

37. $(\sin x + \csc x)^2 + (\cos x - \sec x)^2 = \sec^2 x \csc^2 x + 1$

38. $(1 + \tan t)^2 + (1 + \cot t)^2 = (\sec t + \csc t)^2$

39. Use the given unit-circle diagram to give a geometric proof of the identity in Exercise 38 for an acute angle t. (Hint: Show that $AN = \tan t$, $AO = \sec t$, $CM = \cot t$, and $CO = \csc t$.)

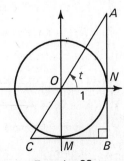

Exercise 39

3-3   *Trigonometric Sum and Difference Formulas*

If x_1 and x_2 are any two real numbers, we can find the sine, cosine, and tangent of $x_1 + x_2$ or $x_1 - x_2$ if we know the values of $\sin x_1$, $\sin x_2$, $\cos x_1$, $\cos x_2$, $\tan x_1$, and $\tan x_2$.

Cosine Formulas

To prove the formula for $\cos (x_1 - x_2)$, let s and t be any two real numbers such that $0 \leq s - t < 2\pi$. (The general result follows from this case. See Exercise 37 on page 80.) Let arcs of lengths s and t be drawn on the unit circle beginning with the point $(1, 0)$, and ending with the points P and Q, respectively (Figure 3-1). The length of arc PQ is therefore $s - t$.

On another unit circle, let an arc RA of length $s - t$ be drawn beginning with $(1, 0)$ (Figure 3-2).

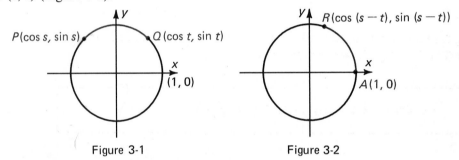

Figure 3-1   Figure 3-2

Since arcs PQ and RA are equal, the chords PQ and RA are equal. Hence,

$$(RA)^2 = (PQ)^2.$$

Recall that the distance d between any two points (x_1, y_1) and (x_2, y_2) in a co-ordinate plane can be given by the formula $d = \sqrt{(x_1 - x_2)^2 + (y_1 - y_2)^2}$. Expressing $(RA)^2$ and $(PQ)^2$ in terms of the distance formula, we have

$$[\cos (s - t) - 1]^2 + [\sin (s - t) - 0]^2 = (\cos s - \cos t)^2 + (\sin s - \sin t)^2.$$

Simplifying this equation we get

$$\cos^2 (s - t) + \sin^2(s - t) - 2 \cos (s - t) + 1 =$$
$$\cos^2 s + \sin^2 s + \cos^2 t + \sin^2 t - 2 \cos s \cos t - 2 \sin s \sin t.$$

By identity (7) on page 70 we obtain

$$2 - 2 \cos (s - t) = 2 - 2 \cos s \cos t - 2 \sin s \sin t,$$

which simplifies to $\cos (s - t) = \cos s \cos t + \sin s \sin t$, or equivalently,

$$\cos (x_1 - x_2) = \cos x_1 \cos x_2 + \sin x_1 \sin x_2 \qquad (10)$$

Example 1 Find cos 15° in simplest radical form.

Solution Express 15° as a difference of two angles whose sine and cosine we know. For example $15° = 45° - 30°$.

By identity (10), we have

$$\cos 15° = \cos 45° \cos 30° + \sin 45° \sin 30°$$

$$= \frac{\sqrt{2}}{2} \cdot \frac{\sqrt{3}}{2} + \frac{\sqrt{2}}{2} \cdot \frac{1}{2}$$

$$= \frac{\sqrt{6} + \sqrt{2}}{4}$$

We can derive a formula for $\cos(x_1 + x_2)$ by using the fact that $x_1 + x_2 = x_1 - (-x_2)$, and applying identity (10).

$$\cos(x_1 + x_2) = \cos(x_1 - (-x_2))$$

$$= \cos x_1 \cos(-x_2) + \sin x_1 \sin(-x_2)$$

Using the fact that sine is an odd function and cosine is an even function, we obtain the formula for $\cos(x_1 + x_2)$.

$$\cos(x_1 + x_2) = \cos x_1 \cos x_2 - \sin x_1 \sin x_2 \qquad (11)$$

Sine Formulas

To derive the formulas for $\sin(x_1 + x_2)$ and $\sin(x_1 - x_2)$, we first need the following identities, which are proved in Exercises 27 and 28 on page 80.

$$\cos\left(\frac{\pi}{2} - x\right) = \sin x \qquad (12)$$

$$\sin\left(\frac{\pi}{2} - x\right) = \cos x \qquad (13)$$

By identity (12),

$$\sin(x_1 + x_2) = \cos\left[\frac{\pi}{2} - (x_1 + x_2)\right]$$

$$= \cos\left[\left(\frac{\pi}{2} - x_1\right) - x_2\right]$$

$$= \cos\left(\frac{\pi}{2} - x_1\right)\cos x_2 + \sin\left(\frac{\pi}{2} - x_1\right)\sin x_2 \quad \text{by identity (10)}.$$

Applying identities (12) and (13), we obtain a formula for $\sin(x_1 + x_2)$.

$$\sin(x_1 + x_2) = \sin x_1 \cos x_2 + \cos x_1 \sin x_2 \qquad (14)$$

The following formula for $\sin(x_1 - x_2)$ can be derived using identity (14). (See Exercise 29 on page 80.)

$$\sin(x_1 - x_2) = \sin x_1 \cos x_2 - \cos x_1 \sin x_2 \qquad (15)$$

Example 2 Find $\sin(x_1 + x_2)$ if $\sin x_1 = \dfrac{3}{5}$, $\sin x_2 = \dfrac{5}{13}$, $0 < x_1 < \dfrac{\pi}{2}$, and

$$\frac{\pi}{2} < x_2 < \pi.$$

Solution   To use identity (14), we first use identity (7) to find $\cos x_1$ and $\cos x_2$.

$$\left(\frac{3}{5}\right)^2 + \cos^2 x_1 = 1 \text{ so } \cos x_1 = \frac{4}{5} \text{ since } 0 < x_1 < \frac{\pi}{2}.$$

Similarly, $\cos x_2 = -\dfrac{12}{13}$.

Applying identity (14), we have

$$\sin(x_1 + x_2) = \frac{3}{5}\left(-\frac{12}{13}\right) + \frac{4}{5}\left(\frac{5}{13}\right)$$

$$= -\frac{16}{65}$$

Tangent Formulas

We can derive a formula for $\tan(x_1 + x_2)$ directly from the fact that

$$\tan x = \frac{\sin x}{\cos x}.$$

$$\tan(x_1 + x_2) = \frac{\sin(x_1 + x_2)}{\cos(x_1 + x_2)}$$

$$= \frac{\sin x_1 \cos x_2 + \cos x_1 \sin x_2}{\cos x_1 \cos x_2 - \sin x_1 \sin x_2}$$

To get an expression in terms of the tangent function only, we divide the numerator and denominator by $\cos x_1 \cos x_2$.

$$\tan (x_1 + x_2) = \dfrac{\dfrac{\sin x_1 \cos x_2}{\cos x_1 \cos x_2} + \dfrac{\cos x_1 \sin x_2}{\cos x_1 \cos x_2}}{\dfrac{\cos x_1 \cos x_2}{\cos x_1 \cos x_2} - \dfrac{\sin x_1 \sin x_2}{\cos x_1 \cos x_2}}$$

$$= \dfrac{\dfrac{\sin x_1}{\cos x_1} + \dfrac{\sin x_2}{\cos x_2}}{1 - \dfrac{\sin x_1}{\cos x_1} \cdot \dfrac{\sin x_2}{\cos x_2}}$$

Therefore,

$$\tan (x_1 + x_2) = \dfrac{\tan x_1 + \tan x_2}{1 - \tan x_1 \tan x_2} \qquad (16)$$

for all x_1 and x_2 such that $\cos x_1 \neq 0$, $\cos x_2 \neq 0$, $\cos (x_1 + x_2) \neq 0$ since tangent is undefined for these values.

The following formula for $\tan (x_1 - x_2)$ can be deduced from (16) using the fact that tangent is an odd function. (See Exercise 30 on page 80.)

$$\tan (x_1 - x_2) = \dfrac{\tan x_1 - \tan x_2}{1 + \tan x_1 \tan x_2} \qquad (17)$$

Exercises 3-3

Give the exact value of each expression in simplest radical form.

A   1. $\sin 105°$ 2. $\cos 165°$ 3. $\sin 15°$ 4. $\cos 345°$

     5. $\tan 75°$ 6. $\tan 165°$ 7. $\cos (-75°)$ 8. $\sin (-75°)$

     9. $\cos \left(-\dfrac{\pi}{12} \right)$ 10. $\sin \dfrac{13\pi}{12}$ 11. $\tan \left(-\dfrac{5\pi}{12} \right)$ 12. $\tan \dfrac{11\pi}{12}$

Find each of the following if $\sin x_1 = \dfrac{4}{5}$, $\sin x_2 = -\dfrac{12}{13}$, $0 < x_1 < \dfrac{\pi}{2}$, and $\dfrac{3\pi}{2} < x_2 < 2\pi$.

     13. $\cos (x_1 + x_2)$ 14. $\sin (x_1 + x_2)$ 15. $\sin (x_1 - x_2)$ 16. $\cos (x_1 - x_2)$

Find each of the following if $\cos s = -\dfrac{3}{5}$, $\cos t = -\dfrac{15}{17}$, $\dfrac{\pi}{2} < s < \pi$, and $\pi < t < \dfrac{3\pi}{2}$.

17. $\sin (s - t)$ 18. $\cos (s - t)$ 19. $\cos (s + t)$

20. $\sin (s + t)$ 21. $\tan (s + t)$ 22. $\tan (s - t)$

Prove each identity.

23. $\cos \left(\dfrac{3\pi}{2} - x\right) + \cos \left(\dfrac{3\pi}{2} + x\right) = 0$ 24. $\sin (\pi + x) + \sin (\pi - x) = 0$

25. $\tan \left(x + \dfrac{\pi}{4}\right) = \dfrac{\cos x + \sin x}{\cos x - \sin x}$ 26. $\tan \left(\dfrac{\pi}{4} - x\right) = \dfrac{\cos x - \sin x}{\cos x + \sin x}$

B 27. Use identity (10) on page 76 to prove that $\cos \left(\dfrac{\pi}{2} - x\right) = \sin x$.

28. Use identity (12) on page 77 to prove that $\sin \left(\dfrac{\pi}{2} - x\right) = \cos x$.

29. Use identity (14) on page 78 to prove that
$$\sin (x_1 - x_2) = \sin x_1 \cos x_2 - \cos x_1 \sin x_2.$$

30. Use the fact that tangent is an odd function to prove that
$$\tan (x_1 - x_2) = \dfrac{\tan x_1 - \tan x_2}{1 + \tan x_1 \tan x_2}.$$

31. Prove that $\tan \left(\dfrac{\pi}{2} - x\right) = \cot x$.   32. Prove that $\sec \left(\dfrac{\pi}{2} - x\right) = \csc x$.

The identities given in Exercises 33–36 are useful in transforming a product of sines and cosines into a sum or vice versa.

33. Prove that $\cos (x_1 - x_2) + \cos (x_1 + x_2) = 2 \cos x_1 \cos x_2$.

34. Prove that $\sin (x_1 + x_2) - \sin (x_1 - x_2) = 2 \cos x_1 \sin x_2$.

35. Using the substitution $u = x_1 + x_2$ and $v = x_1 - x_2$, show that
$$\cos u + \cos v = 2 \cos \left(\dfrac{u + v}{2}\right) \cos \left(\dfrac{u - v}{2}\right) \text{ follows from Exercise 33.}$$

36. Using the same substitution as in Exercise 35, show that
$$\sin u - \sin v = 2 \cos \left(\dfrac{u + v}{2}\right) \sin \left(\dfrac{u - v}{2}\right) \text{ follows from Exercise 34.}$$

C 37. Prove identity (10) for all real numbers s and t. [Hint: There is an integer k such that $0 \le s - (t + 2k\pi) < 2\pi$. Note that $\cos (s - t) = \cos (s - t - 2k\pi)$.]

Prove each of the following identities.

38. $\dfrac{\sin(\alpha - \beta)}{\sin \beta} + \dfrac{\cos(\alpha - \beta)}{\cos \beta} = \dfrac{\sin \alpha}{\sin \beta \cos \beta}$

39. $\dfrac{\cos(\alpha + \beta)}{\sin \beta} + \dfrac{\sin(\alpha + \beta)}{\cos \beta} = \dfrac{\cos \alpha}{\sin \beta \cos \beta}$

Evaluate each expression.

40. $\sin\left(\text{Cos}^{-1}\dfrac{3}{5} + \text{Cos}^{-1}\dfrac{5}{13}\right)$

41. $\cos\left(\text{Sin}^{-1}\dfrac{15}{17} - \text{Sin}^{-1}\dfrac{4}{5}\right)$

42. $\sin\left(\text{Sin}^{-1}\left(-\dfrac{3}{5}\right) - \text{Cos}^{-1}\dfrac{12}{13}\right)$

43. $\cos\left(\text{Cos}^{-1}\left(-\dfrac{5}{13}\right) + \text{Cos}^{-1}\dfrac{24}{25}\right)$

44. $\cos\left(\text{Tan}^{-1}\dfrac{3}{4} + \text{Tan}^{-1}\dfrac{7}{24}\right)$

45. $\sin\left(\text{Tan}^{-1}(-1) + \text{Tan}^{-1}\sqrt{3}\right)$

Exercise 46 will prove identity (14) on page 78 directly for α and β such that $\alpha + \beta < 90°$.

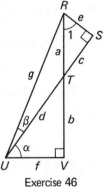

46. (a) Show that $\angle 1 = \alpha$.
    (b) Show that $ab = cd$ by using similar triangles.
    (c) Use triangle RST to find expressions for $\sin \alpha$ and $\cos \alpha$.
    (d) Find expressions for $\sin \beta$ and $\cos \beta$.
    (e) Find an expression for $\sin(\alpha + \beta)$.
    (f) Use steps (c) and (d) to give an expression for $\sin \alpha \cos \beta + \cos \alpha \sin \beta$.
    (g) Show that the expression in step (f) is equivalent to the one in step (e). (Hint: Use step (b).)

Exercise 46

3-4 *Double-Angle and Half-Angle Formulas*

Double-Angle Formulas

The trigonometric addition formulas presented in Section 3-3 can be used to derive formulas for $\sin 2x$, $\cos 2x$, and $\tan 2x$.

$$\sin 2x = \sin(x + x) = \sin x \cos x + \cos x \sin x$$

Therefore,

$$\sin 2x = 2 \sin x \cos x \qquad (18)$$

Similarly, it can be shown that

$$\cos 2x = \cos^2 x - \sin^2 x \qquad (19)$$

$$\tan 2x = \frac{2 \tan x}{1 - \tan^2 x} \qquad (20)$$

Using the identity $\sin^2 x + \cos^2 x = 1$, we can write identity (19) in two other forms.

$$\cos 2x = 1 - 2 \sin^2 x \qquad (19a)$$

$$\cos 2x = 2 \cos^2 x - 1 \qquad (19b)$$

Example 1   Prove the identity $\sin 3x = 3 \sin x - 4 \sin^3 x$.

Solution   Note that $\sin 3x = \sin (2x + x)$. Applying identity (14) on page 78,

$$\sin 3x = \sin 2x \cos x + \cos 2x \sin x$$

By identity (18) and identity (19a),

$$\sin 3x = (2 \sin x \cos x) \cos x + (1 - 2 \sin^2 x) \sin x$$
$$= 2 \sin x \cos^2 x + \sin x - 2 \sin^3 x$$

Finally, by identity (7) on page 70,

$$\sin 3x = 2 \sin x(1 - \sin^2 x) + \sin x - 2 \sin^3 x$$
$$= 3 \sin x - 4 \sin^3 x.$$

Half-Angle Formulas

To derive a formula for $\cos \frac{x}{2}$, we use identity (19b) as follows.

$$\cos 2 \left(\frac{x}{2}\right) = 2 \cos^2 \left(\frac{x}{2}\right) - 1 \quad \text{so}$$

$$\cos^2 \left(\frac{x}{2}\right) = \frac{1 + \cos x}{2}$$

Therefore,

$$\cos \frac{x}{2} = \pm \sqrt{\frac{1 + \cos x}{2}} \qquad (21)$$

$\left(\text{The expression } \pm\sqrt{\dfrac{1 + \cos x}{2}} \text{ means } \sqrt{\dfrac{1 + \cos x}{2}} \text{ or } -\sqrt{\dfrac{1 + \cos x}{2}}.\right)$

We can choose the proper sign once we know the quadrant of $\dfrac{x}{2}$.

Similarly, we can use identity (19a) to obtain a formula for $\sin \dfrac{x}{2}$.

$$\sin \frac{x}{2} = \pm\sqrt{\frac{1 - \cos x}{2}} \qquad (22)$$

Example 2 Find $\cos \dfrac{\theta}{2}$ if $\cos \theta = -\dfrac{31}{81}$ and $180° < \theta < 360°$.

Solution

$$\cos \frac{\theta}{2} = \pm\sqrt{\frac{1 + \cos \theta}{2}}$$

$$= \pm\sqrt{\frac{1 + \left(-\dfrac{31}{81}\right)}{2}}$$

$$= \pm\sqrt{\frac{\dfrac{50}{81}}{2}} = \pm\sqrt{\frac{25}{81}} = \pm\frac{5}{9}$$

Since $180° < \theta < 360°$, $90° < \dfrac{\theta}{2} < 180°$.

Thus $\cos \dfrac{\theta}{2}$ is negative and $\cos \dfrac{\theta}{2} = -\dfrac{5}{9}$.

To derive a formula for $\tan \dfrac{x}{2}$, we use the identity $\tan x = \dfrac{\sin x}{\cos x}$.

$$\tan \frac{x}{2} = \frac{\sin \dfrac{x}{2}}{\cos \dfrac{x}{2}}$$

$$= \frac{\sin \dfrac{x}{2}}{\cos \dfrac{x}{2}} \cdot \frac{2 \cos \dfrac{x}{2}}{2 \cos \dfrac{x}{2}}$$

$$= \frac{2 \sin \dfrac{x}{2} \cos \dfrac{x}{2}}{1 + \left(2 \cos^2 \dfrac{x}{2} - 1\right)}$$

By identity (18), the numerator of this expression equals $\sin x$ and by identity (19b), the denominator equals $1 + \cos x$.
Therefore,

$$\tan \frac{x}{2} = \frac{\sin x}{1 + \cos x} \qquad (23a)$$

We can obtain an alternative formula for $\tan \frac{x}{2}$ from identity (23a).

$$\tan \frac{x}{2} = \frac{\sin x}{1 + \cos x} \cdot \frac{1 - \cos x}{1 - \cos x}$$

$$= \frac{\sin x(1 - \cos x)}{1 - \cos^2 x}$$

$$= \frac{\sin x(1 - \cos x)}{\sin^2 x} \qquad \text{by identity (7)}$$

Therefore,

$$\tan \frac{x}{2} = \frac{1 - \cos x}{\sin x} \qquad (23b)$$

An additional formula for $\tan \frac{x}{2}$ is as follows. (See Exercise 40 on page 85.)

$$\tan \frac{x}{2} = \pm \sqrt{\frac{1 - \cos x}{1 + \cos x}} \qquad (24)$$

Exercises 3-4

For each angle θ satisfying the given condition in the given quadrant, find (a) $\sin 2\theta$ and (b) $\cos 2\theta$.

A 1. $\sin \theta = \frac{4}{5}$; I

2. $\cos \theta = \frac{12}{13}$; I

3. $\cos \theta = -\frac{5}{13}$; II

4. $\sin \theta = -\frac{3}{5}$; III

5. $\sin \theta = -\frac{3}{4}$; III

6. $\cos \theta = \frac{2}{3}$; IV

7–12. For the angles in Exercises 1–6 find tan 2θ.

13–18. For the angles in Exercises 1–6 find tan 4θ.

Use a half-angle formula to evaluate each expression.

19. $\sin 15°$ 20. $\cos 75°$ 21. $\cos 67.5°$

22. $\sin 157.5°$ 23. $\tan 75°$ 24. $\tan 67.5°$

25. $\cos \dfrac{x}{2}$, if $\cos x = -\dfrac{8}{25}$ and $0 < x < \pi$.

26. $\sin \dfrac{x}{2}$, if $\cos x = -\dfrac{31}{49}$ and $\pi < x < 2\pi$.

27. $\tan \dfrac{x}{2}$, if $\sin x = \dfrac{24}{25}$ and $0 < x < \dfrac{\pi}{2}$.

28. $\cos \dfrac{x}{2}$, if $\sin x = -\dfrac{4\sqrt{5}}{9}$ and $\dfrac{3\pi}{2} < x < 2\pi$.

29. Use identity (11) on page 77 to derive the identity

$$\cos 2x = \cos^2 x - \sin^2 x.$$

30. Use identity (16) on page 79 to derive the identity $\tan 2x = \dfrac{2\tan x}{1 - \tan^2 x}$.

Prove each of the following identities.

B 31. $\dfrac{\cos 2x + 1}{\sin 2x} = \cot x$ 32. $\dfrac{\sin 2x}{1 - \cos 2x} = \cot x$

33. $(\sin \theta + \cos \theta)^2 = 1 + \sin 2\theta$

34. $\cos^4 \theta - \sin^4 \theta = \cos 2\theta$

35. $\cot \theta - \tan \theta = 2 \cot 2\theta$

36. $\cot \theta + \tan \theta = 2 \csc 2\theta$

37. $\sin 4x = 4(\sin x \cos^3 x - \sin^3 x \cos x)$

38. $\cos 4x = 8 \cos^4 x - 8 \cos^2 x + 1$

39. $\cot 2\theta = \dfrac{1}{2}\left(\cot \theta - \dfrac{1}{\cot \theta}\right)$ $\left(\text{Hint: } \cot 2\theta = \dfrac{\cos 2\theta}{\sin 2\theta}\right)$

40. $\tan \dfrac{x}{2} = \pm\sqrt{\dfrac{1 - \cos x}{1 + \cos x}}$

C 41. $\dfrac{\cos 2x + \sin^4 x}{\sin 2x} = \dfrac{1}{2} \cot x \cos^2 x$

42. $\sec 2x = \dfrac{\sec^2 x}{2 - \sec^2 x}$ 43. $\csc^2 2x = \dfrac{\csc^4 x}{4(\csc^2 x - 1)}$

Exercise 44 outlines a way to derive the formula for $\cos \dfrac{\theta}{2}$ for an acute angle θ. It is thought to be similar to the method used by Hipparchus of Nicaea (page 1) to construct the first known table of trigonometric functions in intervals of 7.5°.

44. (a) Explain why $\angle PAB = \dfrac{\theta}{2}$.

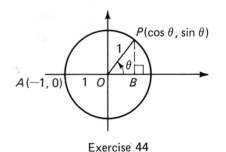

(b) Explain why $\cos \angle PAB = \dfrac{1 + \cos \theta}{PA}$.

(c) Use the distance formula to find PA in simplest radical form.

(d) Substitute the value for PA found in (c) in the equation in (b) to get the formula

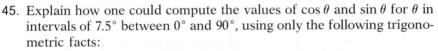

$$\cos \frac{\theta}{2} = \sqrt{\frac{1 + \cos \theta}{2}}.$$

Exercise 44

45. Explain how one could compute the values of $\cos \theta$ and $\sin \theta$ for θ in intervals of 7.5° between 0° and 90°, using only the following trigonometric facts:
   1. the half-angle formulas
   2. the values of sine and cosine of 30° and 45°
   3. the identities $\sin (90° - \theta) = \cos \theta$ and $\cos (90° - \theta) = \sin \theta$

46. Prove that if γ in the given diagram is the angle between two nonvertical lines L_1 and L_2, then

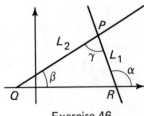

$$\tan \gamma = \frac{m_1 - m_2}{1 + m_1 m_2}$$

where m_1 is the slope of L_1 and m_2 is the slope of L_2. (Hint: If θ is the angle made by the graph of $y = mx + b$ and the positive x-axis, then $\tan \theta = m$.)

Exercise 46

Trigonometric Equations

3-5 *Solving Trigonometric Equations*

A **trigonometric equation** is simply an equation involving one or more circular or trigonometric functions. We solve a trigonometric equation just as we would any algebraic equation.

Example 1   Solve $2 \sin^2 t - \sin t = 1$ for $0 \le t < 2\pi$.

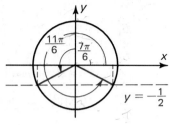

Solution   First solve for $\sin t$.

$$2 \sin^2 t - \sin t = 1$$
$$2 \sin^2 t - \sin t - 1 = 0$$
$$(2 \sin t + 1)(\sin t - 1) = 0$$

Therefore, $\sin t = -\dfrac{1}{2}$ or $\sin t = 1$.

On a unit-circle diagram, draw the line $y = -\dfrac{1}{2}$ to help locate values of t such that $\sin t = -\dfrac{1}{2}$. For $0 \le t < 2\pi$, $t = \dfrac{7\pi}{6}, \dfrac{11\pi}{6}$. For $\sin t = 1$, $t = \dfrac{\pi}{2}$. The solution is

$$t = \frac{\pi}{2}, \frac{7\pi}{6}, \frac{11\pi}{6}.$$

Example 2   Find the general solution of $2 \sin^2 t - \sin t = 1$.

Solution   Since the domain of t is not specified, we must consider all values for t. From Example 1, $\sin t = -\dfrac{1}{2}$ or $\sin t = 1$. Since sine has a period of 2π, the general solution of $\sin t = -\dfrac{1}{2}$ is $t = \dfrac{7\pi}{6} + 2k\pi$, $\dfrac{11\pi}{6} + 2k\pi$ for any integer k. The general solution of $\sin t = 1$ is $t = \dfrac{\pi}{2} + 2k\pi$. The general solution of $2 \sin^2 t - \sin t = 1$ is therefore

$$t = \frac{\pi}{2} + 2k\pi, \frac{7\pi}{6} + 2k\pi, \frac{11\pi}{6} + 2k\pi.$$

When dividing an equation by an expression containing a variable, we must, of course, exclude values for which the expression equals zero. Therefore, before dividing, we must check that the values for which the expression equals zero are not solutions of the original equation.

Example 3   Solve $\sin x = \sqrt{3} \cos x$ for $0 \le x < 2\pi$.

Solution   Before dividing by $\cos x$, check whether or not $\cos x = 0$ yields a solution. If $\cos x = 0$, then $x = \dfrac{\pi}{2}, \dfrac{3\pi}{2}$. Now $\sin \dfrac{\pi}{2} = 1$ and $1 \ne \sqrt{3} \cdot 0$, so $\dfrac{\pi}{2}$ is not a solution. Similarly, $\dfrac{3\pi}{2}$ is not a solution. Therefore we can divide by $\cos x$ as follows.

$$\frac{\sin x}{\cos x} = \frac{\sqrt{3} \cos x}{\cos x}, \text{ so}$$

$$\tan x = \sqrt{3} \text{ and } x = \frac{\pi}{3}, \frac{4\pi}{3}.$$

If an equation is solved by squaring both sides, we must check to see whether all solutions satisfy the original equation.

Example 4 Solve $\tan x + \sec x = 1$ for $0 \le x \le 2\pi$.

Solution We proceed as follows.

$$\sec x + \tan x = 1$$
$$\sec x = 1 - \tan x$$
$$\sec^2 x = (1 - \tan x)^2$$
$$= 1 - 2\tan x + \tan^2 x$$
$$1 + \tan^2 x = 1 - 2\tan x + \tan^2 x$$
$$2\tan x = 0$$
$$\tan x = 0$$
$$x = 0,\ \pi,\ 2\pi$$

Checking in the original equation we have:

$$\tan(0) + \sec(0) = 0 + 1 = 1$$
$$\tan \pi + \sec \pi = 0 - 1 \ne 1$$
$$\tan 2\pi + \sec 2\pi = 0 + 1 = 1$$

Therefore we must reject the value $x = \pi$, which was actually introduced in the squaring process. The solution is $x = 0,\ 2\pi$.

The expression or variable upon which a function operates is called the **argument** of the function. (For example, in the function $\sin 2x$, the argument of the function is $2x$.) When a trigonometric equation contains more than one function or argument, a good procedure for solving the equation is to transform it, if possible, into an equation containing only one function of one argument.

Example 5 Solve $\sin \theta + \cos 2\theta = 0$ for $0° \le \theta < 360°$.

Solution Begin by transforming the equation so that it contains only one function of one argument.

$$\sin \theta + \cos 2\theta = \sin \theta + (1 - 2\sin^2 \theta) \quad \text{by identity (19a)}$$

We have

$$\sin \theta + (1 - 2\sin^2 \theta) = 0, \text{ or } 2\sin^2 \theta - \sin \theta - 1 = 0.$$

From Example 1 the solution is $\theta = 90°,\ 210°,\ 330°$.

If only one argument appears in the equation, we do not need to transform the equation.

Example 6 Solve $6\sin 2\theta = 3$ for $0° \le \theta < 360°$.

Solution Dividing both sides by 6, we have $\sin 2\theta = \dfrac{1}{2}$.

Since we want all values of θ such that $0° \leq \theta < 360°$, we must consider values of 2θ such that $0° \leq 2\theta < 720°$. Therefore

$$2\theta = 30°,\ 150°,\ 390°,\ 510° \text{ and } \theta = 15°,\ 75°,\ 195°,\ 255°.$$

Example 7 illustrates a method for solving an equation that cannot easily be put in terms of a single function.

Example 7 Solve $\sin 2x = \sin x$ for $0 \leq x < 2\pi$.

Solution Using identity (18), we have

$$2 \sin x \cos x = \sin x$$
$$2 \sin x \cos x - \sin x = 0$$

Factoring, we have

$$\sin x(2 \cos x - 1) = 0$$

Therefore $\sin x = 0$ or $2 \cos x - 1 = 0$, so

$$x = 0,\ \pi,\ \frac{\pi}{3},\ \frac{5\pi}{3}.$$

Note that if we had divided both sides of the equation $2 \sin x \cos x = \sin x$ by $\sin x$, we would have lost the solutions $x = 0$, π (unless we first checked whether values for which $\sin x = 0$ satisfy the original equation).

Exercises 3-5

Solve each equation for $0° \leq \theta < 360°$.

A 1. $\sin \theta = -\cos \theta$

2. $2\sqrt{3} \cos \theta - 6 \sin \theta = 0$

3. $\sin \theta + 2 \cos \theta = 0$

4. $4 \sec \theta - \csc \theta = 0$

5. $4 \sin^2 \theta - 3 = 0$

6. $2 \sin \theta = \csc \theta$

7. $1 - 3 \cos \theta = \sin^2 \theta$

8. $\tan^2 \theta = 2 \sec \theta - 1$

9. $\cot^2 \theta = 3(\csc \theta - 1)$

10. $2 \cos^2 \theta + \sin \theta = 1$

11. $\tan \theta = 2 \sin \theta$

12. $\sqrt{2} \sin \theta = \cot \theta$

Solve each equation for $0 \leq x < 2\pi$.

13. $\cos 2x = \sin x$

14. $\cos 2x = -\cos x$

15. $\sin 2x = -\sin x$

16. $\sin 2x = \cos x$

17. $\sin 2x = -\cos 2x$

18. $2 \sin^2 2x = 1$

Give the general solution for each equation.

19. $\sin 2x = \cos 4x$

20. $\tan\left(x - \dfrac{\pi}{4}\right) = 2\sin\left(x - \dfrac{\pi}{4}\right)$

21. $4(\sin x + 1) = 3\csc x$

22. $\tan x + \cot x = -2$

23. $1 + \cos x = 4\sin^2 x$

24. $1 + 2\cot^2 x + \csc x = 0$

Solve for $0 \le x < 2\pi$. (Hint: Use the trigonometric addition formulas or the double-angle formulas.)

B 25. $4\sin x \cos x = \sqrt{3}$

26. $4\sin x \cos x = -\sqrt{2}$

27. $\cos 2x \cos x + \sin 2x \sin x = -\dfrac{1}{2}$

28. $2\cos 3x \cos x - 2\sin 3x \sin x = \sqrt{3}$

29. $\sqrt{2}(\sin x + \cos x) = \sqrt{3}$ (Hint: Square both sides.)

30. $\sqrt{2}(\sin x - \cos x) = \sqrt{2}$

Exercises 31 and 32 use the following information. The approximate distance s in meters that an object will travel if given an initial linear speed v_0 at an angle of elevation θ is given by the formula

$$s = \frac{v_0^2 \sin\theta \cos\theta}{5},$$

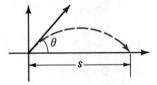

Exercises 31 and 32

where v_0 is in meters per second.

31. At what angle must a football be thrown at 20 m/s in order to travel 20 m? (Disregard the height of the person throwing the ball.)

32. For what value of θ will the football in Exercise 31 travel the farthest? How far can the football travel?

33. The area of a right triangle is $\dfrac{1}{2}$ and the hypotenuse has length 2. Find the angles of the triangle.

34. Solve the equation $\sin\theta + \cos\theta = \sqrt{\dfrac{2 + \sqrt{3}}{2}}$ by squaring both sides.

Solve each equation for $0° \le \theta < 360°$.

C 35. $2(\cos^4\theta - \sin^4\theta) = 1$

36. $4\cos^4\theta - 4\cos^2\theta = -\dfrac{1}{2}$ (Hint: Add 1 to both sides.)

37. $\sqrt{1 - \cos 2\theta} = 2\sin^2\theta$

38. $\sqrt{\cos 2\theta + 1} = 2\cos^2\theta$

3-6 *The Expression A cos θ + B sin θ*

An expression that occurs frequently in trigonometry and its applications is

$$A \cos \theta + B \sin \theta,$$

where A and B are nonzero real numbers. For example, the force necessary to move an object at a constant speed up an inclined plane is given by an expression of this type. (See Exercise 27 on page 93.)

It often is convenient to write the expression in a different form, namely,

$$A \cos \theta + B \sin \theta = C \cos (\theta - \phi), \text{ where } C = \sqrt{A^2 + B^2}$$

and ϕ is such that $\cos \phi = \dfrac{A}{C}$ and $\sin \phi = \dfrac{B}{C}$ (25)

We can derive identity (25) as follows. If $C = \sqrt{A^2 + B^2}$, then

$$\left(\frac{A}{C}\right)^2 + \left(\frac{B}{C}\right)^2 = \frac{A^2 + B^2}{C^2} = \frac{C^2}{C^2} = 1.$$

Hence the point $\left(\dfrac{A}{C}, \dfrac{B}{C}\right)$ is on the unit circle. Therefore, there exists an angle ϕ such that

$$\cos \phi = \frac{A}{C} \quad \text{and} \quad \sin \phi = \frac{B}{C}.$$

We have

$$A \cos \theta + B \sin \theta = C \left(\frac{A}{C} \cos \theta + \frac{B}{C} \sin \theta\right)$$
$$= C(\cos \phi \cos \theta + \sin \phi \sin \theta)$$
$$= C \cos (\theta - \phi) \quad \text{by identity (10)}.$$

Example 1   Express $\cos \theta - \sqrt{3} \sin \theta$ in the form $C \cos (\theta - \phi)$ for some ϕ, $-180° < \phi < 180°$.

Solution $\cos \theta - \sqrt{3} \sin \theta = \cos \theta + (-\sqrt{3}) \sin \theta$
Therefore, $A = 1$, $B = -\sqrt{3}$, and $C = \sqrt{(1)^2 + (-\sqrt{3})^2} = 2$, so that
$\cos \phi = \dfrac{1}{2}$ and $\sin \phi = -\dfrac{\sqrt{3}}{2}$.
Hence $\phi = -60°$ and we have

$$\cos \theta - \sqrt{3} \sin \theta = C \cos (\theta - \phi)$$
$$= 2 \cos (\theta - (-60°))$$
$$= 2 \cos (\theta + 60°)$$

Example 2   Solve $-6\cos\theta + 8\sin\theta = 5\sqrt{2}$ to the nearest degree, for $0° \le \theta < 360°$.

Solution   Express $-6\cos\theta + 8\sin\theta$ in the form $C\cos(\theta - \phi)$.
$A = -6$, $B = 8$, $C = \sqrt{(-6)^2 + 8^2} = 10$, so $\cos\phi = -0.6$ and $\sin\phi = 0.8$.
Since $\cos\phi < 0$ and $\sin\phi > 0$, ϕ is in Quadrant II. Use of a table or calculator then gives $\phi = 127°$ to the nearest degree. Therefore, $-6\cos\theta + 8\sin\theta = 10\cos(\theta - 127°)$, and the equation becomes

$$10\cos(\theta - 127°) = 5\sqrt{2}.$$

We have
$$\cos(\theta - 127°) = \frac{5\sqrt{2}}{10} = \frac{\sqrt{2}}{2},$$

so that the reference angle for $(\theta - 127°)$ is $45°$. Since $\cos(\theta - 127°) > 0$, the angle $(\theta - 127°)$ is either in Quadrant I or Quadrant IV.
Thus $\theta - 127° = 45°$ or $\theta - 127° = -45°$.
Solving for θ, $\theta = 172°$ or $82°$.

Exercises 3-6

Rewrite each expression in the form $C\cos\theta$ for $0° < \theta < 360°$.

A   1. $\cos 20° + \sin 20°$ 2. $\cos 125° - \sin 125°$

3. $-\sqrt{3}\cos 100° + \sin 100°$ 4. $-\sqrt{2}\cos 40° - \sqrt{6}\sin 40°$

Rewrite each expression in the form $C\cos(\theta - \phi)$ for $-180° \le \phi < 180°$.
Find ϕ to the nearest degree.

5. $2\cos\theta - 2\sin\theta$ 6. $-3\cos\theta + 3\sin\theta$

7. $2\cos\theta + 2\sqrt{3}\sin\theta$ 8. $\dfrac{\sqrt{3}}{4}\cos\theta - \dfrac{1}{4}\sin\theta$

9. $4\cos\theta - 3\sin\theta$ 10. $\dfrac{3}{2}\cos\theta + 2\sin\theta$

11. $-5\cos\theta + 12\sin\theta$ 12. $\sqrt{5}\cos\theta - 2\sin\theta$

13. $-\cos\theta - 2\sqrt{2}\sin\theta$ 14. $3\cos\theta + \sqrt{7}\sin\theta$

Solve each equation for $0° \le \theta < 360°$ to the nearest degree.

B   15. $\sqrt{2}\sin\theta + \sqrt{2}\cos\theta = 1$ 16. $\sin\theta - \cos\theta = \dfrac{\sqrt{2}}{2}$

17. $\sqrt{3}\cos\theta - \sin\theta = \dfrac{1}{2}$ 18. $3\cos\theta + 3\sqrt{3}\sin\theta = \dfrac{3}{5}$

19. $4 \cos \theta + 3 \sin \theta = \dfrac{5}{2}$

20. $\cos \theta - 2\sqrt{6} \sin \theta = 3$

21. If the expression $A \cos \theta + B \sin \theta$ is rewritten in the form $C \sin (\theta + \phi)$ using the identity $\sin (\alpha + \beta) = \sin \alpha \cos \beta + \cos \alpha \sin \beta$, give equations that define C and ϕ.

22. If the expression $A \cos \theta + B \sin \theta$ is rewritten in the form $C \sin (\theta - \phi)$, give equations that define C and ϕ.

23–24. Use the formula developed in Exercise 21 to solve the equations in Exercises 15 and 16.

25–26. Use the formula developed in Exercise 22 to solve the equations in Exercises 17 and 18.

C 27. The force necessary to accelerate a 1 kg mass 1 m/s² is called a newton (N). The force F in newtons necessary to keep an object moving at a constant speed up an incline at an angle θ with the horizontal is given (approximately) by

$$F = 10m(\sin \theta + k \cos \theta),$$

where m is the mass of the object in kilograms and k is the coefficient of friction. Suppose a carton on a factory ramp has a mass of 12 kg and a coefficient of friction of $\dfrac{5}{12}$. If a force of 65 N is applied to the carton, and the carton moves at a constant speed, what is the angle of the ramp?

Exercise 27

28. The force F in newtons necessary to keep an object with mass m (in kilograms) moving at constant speed on a horizontal surface is given (approximately) by

$$F = \frac{10mk}{\cos \theta + k \sin \theta},$$

Exercise 28

where θ is the angle at which the force is exerted and k is the coefficient of friction. At what angle must a force of 80 N be exerted if a sled of mass 8.5 kg is to be kept moving at constant speed on a horizontal surface, given that the coefficient of friction is $\dfrac{8}{17}$?

29. Rectangle $ABCD$ is inscribed in circle O of radius 1. For what value of θ will the rectangle have a perimeter of $2\sqrt{6}$?

30. Show that the rectangle in Exercise 29 will have maximum perimeter if $\theta = 45°$.

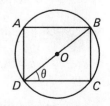

Exercise 29

Basic Trigonometric Identities

$$\tan x = \frac{\sin x}{\cos x} \qquad (1)$$

$$\cot x = \frac{\cos x}{\sin x} \qquad (2)$$

$$\sec x = \frac{1}{\cos x} \qquad (3)$$

$$\csc x = \frac{1}{\sin x} \qquad (4)$$

$$\tan x = \frac{1}{\cot x} \qquad (5)$$

$$\cot x = \frac{1}{\tan x} \qquad (6)$$

The Pythagorean Identities

$$\sin^2 x + \cos^2 x = 1 \qquad (7)$$

$$1 + \tan^2 x = \sec^2 x \qquad (8)$$

$$1 + \cot^2 x = \csc^2 x \qquad (9)$$

Sum and Difference Formulas

$$\cos(x_1 - x_2) = \cos x_1 \cos x_2 + \sin x_1 \sin x_2 \qquad (10)$$

$$\cos(x_1 + x_2) = \cos x_1 \cos x_2 - \sin x_1 \sin x_2 \qquad (11)$$

$$\cos\left(\frac{\pi}{2} - x\right) = \sin x \qquad (12)$$

$$\sin\left(\frac{\pi}{2} - x\right) = \cos x \qquad (13)$$

$$\sin(x_1 + x_2) = \sin x_1 \cos x_2 + \cos x_1 \sin x_2 \qquad (14)$$

$$\sin(x_1 - x_2) = \sin x_1 \cos x_2 - \cos x_1 \sin x_2 \qquad (15)$$

$$\tan(x_1 + x_2) = \frac{\tan x_1 + \tan x_2}{1 - \tan x_1 \tan x_2} \qquad (16)$$

$$\tan(x_1 - x_2) = \frac{\tan x_1 - \tan x_2}{1 + \tan x_1 \tan x_2} \qquad (17)$$

Double-Angle Formulas

$$\sin 2x = 2 \sin x \cos x \qquad (18)$$

$$\cos 2x = \cos^2 x - \sin^2 x \qquad (19)$$

$$\cos 2x = 1 - 2\sin^2 x \qquad (19a)$$

$$\cos 2x = 2\cos^2 x - 1 \qquad (19b)$$

$$\tan 2x = \frac{2 \tan x}{1 - \tan^2 x} \qquad (20)$$

Half-Angle Formulas

$$\cos \frac{x}{2} = \pm \sqrt{\frac{1 + \cos x}{2}} \qquad (21)$$

$$\sin \frac{x}{2} = \pm \sqrt{\frac{1 - \cos x}{2}} \qquad (22)$$

$$\tan \frac{x}{2} = \frac{\sin x}{1 + \cos x} \qquad (23a)$$

$$\tan \frac{x}{2} = \frac{1 - \cos x}{\sin x} \qquad (23b)$$

$$\tan \frac{x}{2} = \pm \sqrt{\frac{1 - \cos x}{1 + \cos x}} \qquad (24)$$

$$A \cos \theta + B \sin \theta = C \cos(\theta - \phi) \text{ where } C = \sqrt{A^2 + B^2}, \qquad (25)$$

$$\cos \phi = \frac{A}{C}, \text{ and } \sin \phi = \frac{B}{C}.$$

CHAPTER SUMMARY

1. Equations that are true for all values of the variable are referred to as identities.

2. Identities can be proved by transforming one side into the other or by transforming both sides separately into the same expression.

3. Sum and difference formulas and double-angle and half-angle formulas for the sine, cosine, and tangent functions are listed on page 94.

4. Trigonometric equations are equations involving one or more circular functions.

5. The expression $A \cos \theta + B \sin \theta$ can be written in the form $C \cos (\theta - \phi)$, where $C = \sqrt{A^2 + B^2}$ and ϕ is such that $\cos \phi = \dfrac{A}{C}$ and $\sin \phi = \dfrac{B}{C}$.

CHAPTER TEST

3-1   1. Express $\sin x \cos^2 x + \sin^2 x$ in simplest form.

3-2   2. Prove $\dfrac{1 + 2 \sin x \cos x}{\sin x + \cos x} = \sin x + \cos x.$

3-3   3. Give the exact value of each expression in simplest radical form.

　　　　(a) $\cos 255°$　　(b) $\tan \left(-\dfrac{7\pi}{12} \right)$

3-4   4. Given that $\cos \theta = \dfrac{4}{5}$ and $\dfrac{3\pi}{2} < \theta < 2\pi$, find $\sin 2\theta$ and $\cos 2\theta$.

　　　　5. Use a half-angle formula to evaluate each expression.

　　　　(a) $\sin 22.5°$　　(b) $\tan \dfrac{\pi}{12}$

3-5   6. Solve $2 \sin x \sec x = \sec x$ for $0 \le x < 2\pi$.

3-6   7. Solve $\sin \theta - \sqrt{3} \cos \theta = 1$ to the nearest degree for $0° \le \theta < 360°$.

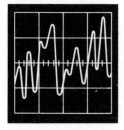

Hyperbolic Functions

There is a pair of functions that are related to the "unit hyperbola" $x^2 - y^2 = 1$ in much the same way that the sine and cosine are related to the unit circle $x^2 + y^2 = 1$. These functions are called the **hyperbolic sine,** written sinh, and the **hyperbolic cosine,** written cosh. They are defined by

$$\sinh t = \frac{e^t - e^{-t}}{2} \text{ and } \cosh t = \frac{e^t + e^{-t}}{2},$$

where e is a constant approximately equal to 2.71828. (The value of e need not concern us here.)

It is not difficult to check that if

$$x = \cosh t \text{ and } y = \sinh t,$$

then $x^2 - y^2 = 1$. Thus, $(\cosh t, \sinh t)$ is a point on the hyperbola $x^2 - y^2 = 1$ just as $(\cos t, \sin t)$ is a point on the circle $x^2 + y^2 = 1$. (See the accompanying diagram.)

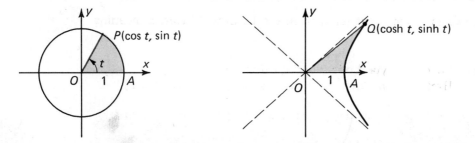

Other similarities are listed below; the hyperbolic identities can be proved by using the definition of the hyperbolic functions given above.

Circular Functions

$$\cos^2 t + \sin^2 t = 1$$
$$\sin (t_1 + t_2) = \sin t_1 \cos t_2 + \cos t_1 \sin t_2$$
$$\cos (t_1 + t_2) = \cos t_1 \cos t_2 - \sin t_1 \sin t_2$$

$$\cos 2t = \cos^2 t - \sin^2 t$$
$$= 2 \cos^2 t - 1$$
$$= 1 - 2 \sin^2 t$$

$$\sin 2t = 2 \sin t \cos t$$

Hyperbolic Functions

$\cosh^2 t - \sinh^2 t = 1$ (from $x^2 - y^2 = 1$)
$\sinh (t_1 + t_2) = \sinh t_1 \cosh t_2 + \cosh t_1 \sinh t_2$
$\cosh (t_1 + t_2) = \cosh t_1 \cosh t_2 + \sinh t_1 \sinh t_2$

$$\cosh 2t = \cosh^2 t + \sinh^2 t$$
$$= 2\cosh^2 t - 1$$
$$= 1 + 2\sinh^2 t$$

$\sinh 2t = 2\sinh t \cosh t$

It is *not* the case in the hyperbola diagram that t is a measure of the angle AOQ. It can be shown with the help of integral calculus that

$$t = 2 \text{ (area of the hyperbolic sector } AOQ)$$

This is another similarity because if we refer to the circle diagram, we see that the area of the sector $AOP = \dfrac{t}{2\pi} \cdot \pi r^2 = \dfrac{t}{2\pi} \cdot \pi \cdot 1^2 = \dfrac{t}{2}$. Thus in the circular-function case,

$$t = 2 \text{ (area of the circular sector } AOP).$$

Hyperbolic functions can be used to describe certain physical phenomena. For example, if a perfectly flexible chain is suspended from two points (for instance, a power line hanging between two poles), it will assume a shape that can be described by an equation of the form $y = a\cosh\dfrac{x}{a}$. Such a curve is called a **catenary,** from the Latin word *catena* meaning chain.

Challenge: Prove the hyperbolic identities listed above.
(Hint: Use the laws of exponents: e.g., $e^t \cdot e^{-t} = e^0 = 1$.)

Graphing Inverse Circular Functions

The graphing program on pages 66–67 can be modified to graph other trigonometric functions.

As an example, let us change the program so that it will graph a portion of tan x. First, delete line 10. Then substitute the following steps in the program.

```
100   PRINT "Y = TAN X"

140   FOR X1 = −40 TO 40 STEP 2

160   LET Y = TAN(X)
```

The following is a RUN of the program. As in Chapter 2, notice that the x-axis is vertical. The portion of the graph of tan x shown in the print-out is from $x = -4$ to $x = -1.8$. At this point, the graph exceeds the bounds of the program. Since all the x-values are negative, the y-axis is not shown.

As noted in Chapter 2, the graphs generated may not fit on a micro-computer display screen. Again, you may wish to run the graph through and then rerun the program, stopping at various intervals.

```
Y = TAN X

     X     Y

-4     -1.16              *          *           !
-3.8  -.77                     *           !
-3.6  -.49                        *        !
-3.4  -.26                          *  !
-3.2  -.06                            *!
-3     .14                            !*
-2.8   .36                            !   *
-2.6   .6                             !     *
-2.4   .92                            !         *
-2.2  1.37                            !            *
-2    2.19                            !                 *
-1.8  4.29  TOO WIDE

END
```

Exercises

First, delete line 10.

1. Try these modifications to graph a portion of $\text{Tan}^{-1} x$.

    ```
    100   PRINT "Y = ARCTAN X"
    140   FOR X1 = 0 TO 40 STEP 4
    160   LET Y = ATN(X)
    ```

2. The graph of $\text{Cot}^{-1} x$ can be represented by the negative of the graph of $\text{Tan}^{-1} x$ after it has been moved vertically upward $\frac{\pi}{2}$ units. Try the following modifications to graph a portion of $\text{Cot}^{-1} x$.

    ```
    100   PRINT "Y = ARCCOT X"
    140   FOR X1 = 0 TO 40 STEP 4
    160   LET Y = −ATN(X) + 3.14159/2
    ```

3. Since the only inverse trigonometric function in BASIC is the inverse tangent function, we can use the formula developed in the Challenge on page 72 in Section 3-1 to investigate the inverse sine function.

    ```
    100   PRINT "Y = ARCSIN X"
    140   FOR X1 = −9 TO 9 STEP 2
    160   LET Y = ATN(X/SQR(1 − X*X))
    ```

4. The graph of $\text{Cos}^{-1} x$ can be represented by the negative of the graph of $\text{Sin}^{-1} x$ after it has been moved vertically upward $\frac{\pi}{2}$ units. Try the following modifications to graph a portion of $\text{Cos}^{-1} x$.

    ```
    100   PRINT "Y = ARCCOS X"
    140   FOR X1 = −9 TO 9 STEP 2
    160   LET Y = −ATN(X/SQR(1 − X*X)) + 3.14159/2
    ```

The law of cosines can be used in navigation to find the distances between ships.

OBLIQUE TRIANGLES

OBJECTIVES

1. To solve oblique triangles using the law of sines and the law of cosines.
2. To find the areas of oblique triangles.

Solving Oblique Triangles

4-1 *The Law of Cosines*

Consider these two problems.

Problem 1 Ships A and B leave port C at the same time and sail on straight paths making an angle of 60° with each other. How far apart are the ships at the end of one hour if the speed of ship A is 25 km/h and that of ship B is 15 km/h?

Problem 2 The sides of a triangle have lengths 10 cm, 9 cm, and 3 cm. Find its largest angle to the nearest 10′.

Both of these problems can be solved by using the **law of cosines,** which states that in any triangle ABC (Figure 4-1):

$$c^2 = a^2 + b^2 - 2ab \cos C$$
$$b^2 = a^2 + c^2 - 2ac \cos B$$
$$a^2 = b^2 + c^2 - 2bc \cos A$$

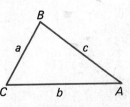

Figure 4-1

It is sufficient to prove the first of these formulas. Introduce a coordinate system with the origin at the vertex C, and the x-axis along the side CA (Figure 4-2). The coordinates (x, y) of B are then $(a \cos C, a \sin C)$. Using the distance formula, we have

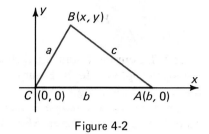

$$c^2 = (AB)^2 = (x - b)^2 + y^2$$
$$= (a \cos C - b)^2 + (a \sin C)^2$$
$$= a^2(\cos^2 C + \sin^2 C) - 2ab \cos C + b^2.$$

Since $\cos^2 C + \sin^2 C = 1$, we have

$$c^2 = a^2 + b^2 - 2ab \cos C.$$

Figure 4-2

Note that the law of cosines is a generalization of the Pythagorean theorem. (See Exercise 11, page 103.)

Example 1   Solve Problem 1 concerning ships A and B leaving port C at the same time.

Solution   At the end of one hour the paths have lengths 25 km and 15 km and they form a 60° angle as shown in the diagram. We wish to find the length c of the third side of the triangle that the paths determine. Using the law of cosines we have

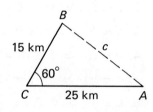

$$c^2 = 25^2 + 15^2 - 2 \cdot 25 \cdot 15 \cos 60°$$
$$= 625 + 225 - 2 \cdot 25 \cdot 15 \cdot \frac{1}{2} = 475$$

Therefore $c = \sqrt{475} = 21.8$ km.

Example 2   Solve Problem 2 concerning the triangle with sides of lengths 10 cm, 9 cm, and 3 cm.

Solution   Let $a = 10$ cm, $b = 9$ cm, and $c = 3$ cm as shown in the diagram. Since the largest angle of a triangle is opposite the longest side, we wish to find $\angle A$. We can use the formula

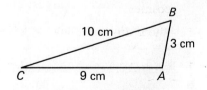

$$a^2 = b^2 + c^2 - 2bc \cos A$$
$$10^2 = 9^2 + 3^2 - 2 \cdot 9 \cdot 3 \cos A$$
$$100 = 81 + 9 - 54 \cos A$$

(Solution continued on next page.)

Therefore,

$$54 \cos A = -10 \text{ and } \cos A = -\frac{10}{54} = -0.1852.$$

Hence, $\angle A = 100.67°$ or $100°40'$.

Both Examples 1 and 2 involve an **oblique triangle,** that is, a triangle having no right angle. In Example 1 the triangle is determined by two sides and the included angle. We abbreviate this case by SAS (side, angle, side). In Example 2 the triangle is determined by three sides, that is, the SSS case. The remaining cases, SSA and ASA (or SAA or AAS) are considered in the next two sections.

Exercises 4-1

Complete Exercises 1–8 for $\triangle ABC$ with the given parts.

A 1. $a = 8$, $b = 3$, $\angle C = 60°$, $c = ?$

2. $a = 5$, $c = 21$, $\angle B = 60°$, $b = ?$

3. $a = 16$, $c = 5$, $\angle B = 120°$, $b = ?$

4. $b = 3$, $c = 5$, $\angle A = 120°$, $a = ?$

5. $a = 6$, $b = 14$, $c = 16$, $\angle B = ?$

6. $a = 38$, $b = 32$, $c = 10$, $\angle A = ?$

7. $a = 11$, $b = 5$, $c = 7$, $\angle A = ?$

8. $a = 3$, $b = 6$, $c = 8$, $\angle C = ?$

9. Find the lengths of the diagonals of a parallelogram with sides of lengths 6 and 8, and one angle of measure 60°.

10. Find the lengths of the diagonals of a parallelogram with sides 10 and 50, and one angle of measure $104°50'$.

11. Show that the Pythagorean theorem is a special case of the law of cosines.

12. Show that in $\triangle ABC$, if $a^2 > b^2 + c^2$, then $\angle A$ is obtuse.

13. Show that in $\triangle ABC$, $\angle A = \text{Cos}^{-1}\left(\dfrac{b^2 + c^2 - a^2}{2bc}\right).$

14. Use Exercise 13 to show that if $a > b + c$, then no triangle can be formed with sides of lengths a, b, and c. $\left(\text{Hint: If } a > b + c, \text{ then}\right.$

$$\dfrac{b^2 + c^2 - a^2}{2bc} < \dfrac{b^2 + c^2 - (b + c)^2}{2bc}.\Big)$$

In Exercises 15–18 leave your answers in simplest radical form. Exercises 15 and 16 refer to the given diagram.

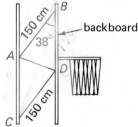

15. The supports for a basketball backboard are parallel and meet the backboard at an angle of 38° as shown. If the distance from B to D is 140 cm, and the supports are each 150 cm long, find the length AD of the diagonal brace.

16. What should be the length of the diagonal brace in Exercise 15 if it is placed so that it connects points B and C?

Exercises 15 and 16

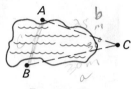

Exercise 17

17. Points A and B on opposite banks of a pond are sighted from point C as shown in the given diagram. If $\angle C = 30°$, $AC = 25$ m, and $BC = 30$ m, how far apart are points A and B?

18. The pneumatic door-closer for a storm door extends from the outer end of a bracket 14 cm long to a point on the door 80 cm from the hinge, as shown in the diagram. If the door is to make a maximum angle of 60° with the bracket when the door is open, find the maximum length of the door-closer.

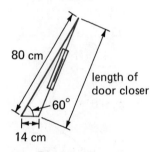

Exercise 18

In Exercises 19–22 find the length of $\overline{BD}$ in the given diagram. Leave your answer in simplest radical form.

Example $AB = 15$, $BC = 13$, $AD = 3$, $DC = 11$

Solution First use the law of cosines in $\triangle ABC$ to find $\cos A$.
$$13^2 = 15^2 + (3 + 11)^2 - 2(15)(3 + 11) \cos A$$
so $\cos A = \dfrac{3}{5}$.

Use the value of $\cos A$ and apply the law of cosines to $\triangle ABD$.
$$(BD)^2 = 15^2 + 3^2 - 2(15)(3)\left(\frac{3}{5}\right)$$
$$BD = 6\sqrt{5}$$

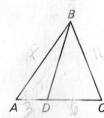

Exercises 19–22

B 19. $AB = 6$, $BC = 4$, $AD = 3$, $DC = 2$

20. $AB = 5$, $BC = 9$, $AD = 2$, $DC = 6$

21. $AB = 15$, $BC = 20$, $AD = 4$, $DC = 21$

22. $AB = 7$, $BC = 11$, $AD = 3$, $DC = 6$

23. Find the length of the median to one of the legs of an isosceles triangle with sides of lengths 8, 8, and 4.

24. Find the length of the median to one of the legs of an isosceles triangle with sides of lengths 18, 18, and 6.

25. Use the law of cosines to find two possible values for a in $\triangle ABC$ in which $b = 2$, $c = 4$, and $\cos B = 0.875$.

26. Repeat Exercise 25 if $b = 7$, $c = 8$, and $\angle B = 60°$.

27. The outer girders of a triangular module of a TV transmitter tower have lengths 6 m and 10 m, respectively, and make an obtuse angle of 120° as shown in the given diagram. The cross-brace of the module (dashed lines in diagram) is a median of the triangle. Find the length of the cross-brace.

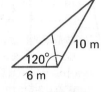

Exercise 27

28. The legs of a folding chair have the lengths shown in the diagram below. If $CD = 30$ cm, how deep (AB) is the seat?

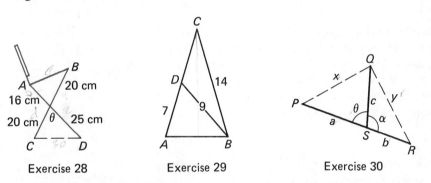

Exercise 28   Exercise 29   Exercise 30

C 29. In the diagram above, $\triangle ABC$ is isosceles with legs $\overline{AC}$ and $\overline{BC}$, each of length 14 and median $\overline{DB}$ of length 9. Find the length of the base $\overline{AB}$. (Hint: Use $\triangle BCD$ to find $\cos C$.)

30. In the diagram above, imagine $\overline{PR}$ and $\overline{QS}$ as rods that can pivot at point S. Show that no matter what angle θ is, the quantity $bx^2 + ay^2$ is a constant. (Hint: $\alpha = 180° - \theta$.)

31. Use the law of cosines and the given diagram to prove that if the median to the longest side of a triangle has length half that of the side, then the triangle is a right triangle. (Hint: Use the converse of the Pythagorean theorem.)

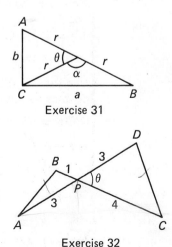

Exercise 31

Exercise 32

32. In the given diagram, $AB = DC$, $AP = 3$, $BP = 1$, $CP = 4$, and $DP = 3$. Find AB.

33. A sloping wall of a building is 5 m long. It is set back 3 m from, and rises above, a vertical wall of height 4 m as shown in the given diagram. What length will a ladder need to be in order to reach point A at the top of the sloping wall? (Hint: Find AB and BC separately.)

34. What angle does the seat of the chair in Exercise 28 make with the horizontal?

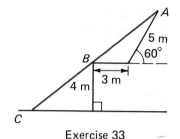

Exercise 33

Challenge

In a bicycle wheel, there are 18 equally spaced holes around the center of each side of the hub, and 36 equally spaced holes around the rim. In the diagram, r is the radius from the center O on one side of the hub to a hole in the same side of the hub, and r' is the radius from O to a hole in the wheel rim. If spokes $\overline{AB}$ and $\overline{CD}$ are of equal length with 4 holes between A and C and one hole between B and D, find an expression for the spoke length AB in terms of r and r'. (Hint: Consider triangle AOB.)

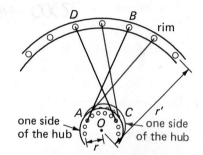

4-2 The Law of Sines

Problem A weather balloon B is directly over a 2000 m airstrip extending from A to C in Figure 4-3. The angles of elevation from A to B and from C to B are 62° and 31°, respectively. Find the distances from A to B and from C to B.

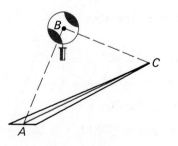

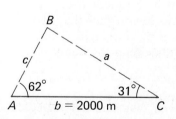

Figure 4-3

In the given problem we have a triangle ABC in which we know the length of one side, AC (or b), and two angles; that is, the triangle is determined by the ASA case. To solve such a triangle we use the **law of sines,** which states that in any triangle ABC:

$$\frac{\sin A}{a} = \frac{\sin B}{b} = \frac{\sin C}{c}$$

To prove the first equality, draw the altitude $\overline{CD}$ to $\overline{AB}$ (Figure 4–4). From right triangle ADC we have $CD = b \sin A$, and from right triangle BDC, $CD = a \sin B$. Thus, $b \sin A = a \sin B$. Dividing by ab gives

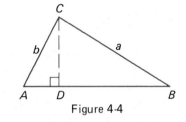

Figure 4-4

$$\frac{\sin A}{a} = \frac{\sin B}{b}.$$

The other equalities follow similarly.

Example Solve the problem concerning the weather balloon B directly over a 2000 m airstrip.

Solution First note that $\angle B = 180° - (62° + 31°) = 87°$ (Figure 4-3).
To find the distance a, we use $\dfrac{\sin A}{a} = \dfrac{\sin B}{b}$, or

$$a = \frac{b \sin A}{\sin B} = \frac{2000 \sin 62°}{\sin 87°} = 1768 \text{ m.}$$

To find c, use $\dfrac{\sin C}{c} = \dfrac{\sin B}{b}$, or

$$c = \frac{b \sin C}{\sin B} = \frac{2000 \sin 31°}{\sin 87°} = 1031 \text{ m.}$$

Exercises 4-2

Solve each $\triangle ABC$ having the given angles and the given side. Give lengths to the nearest tenth.

A 1. $\angle A = 30°$, $\angle B = 70.5°$, $a = 9$

2. $\angle A = 41°50'$, $\angle C = 60°$, $a = 40$

3. $\angle B = 24°50'$, $\angle C = 15°40'$, $b = 84$

4. $\angle B = 14°50'$, $\angle C = 58°$, $b = 8$

5. $\angle A = 30°40'$, $\angle B = 117°$, $a = 17$

6. $\angle A = 115.5°$, $\angle C = 30°$, $a = 36$

7. $\angle B = 125°50'$, $\angle C = 34°10'$, $a = 27$

8. $\angle A = 22°20'$, $\angle B = 127°40'$, $c = 10$

9. $\angle A = 30°$, $\angle C = 52°$, $b = 20$

10. $\angle B = 34°20'$, $\angle C = 114°20'$, $a = 13$

11. In $\triangle ABC$, if $\angle A = 60°$, $\angle B = 45°$, and $b = 2\sqrt{6}$, find a in simplest radical form without using tables.

12. In $\triangle PQR$, if $\angle P = 45°$, $\angle Q = 15°$, and $r = 6\sqrt{3}$, find p in simplest radical form without using tables.

B 13. Show that in $\triangle ABC$, $a = \dfrac{c \sin A}{\sin (A + B)}$.

14. Show that the altitude h to $\overline{AB}$ in $\triangle ABC$ is given by $h = \dfrac{c \sin A \sin B}{\sin (A + B)}$.

15. Solve Exercise 19 on page 23 using the law of sines. (Hint: Use the Pythagorean theorem to find the third side of the triangle.)

16. Solve Exercise 20 on page 23 using the law of sines. (Hint: Use the law of sines to find the side opposite β.)

17. If points A and B at the top of two support towers for a bridge are sighted from points C and D on shore, the angles of elevation are as shown in the given diagram. If C and D are 1.25 km apart and $\overline{CD}$ is parallel to $\overline{AB}$, how far apart are A and B?

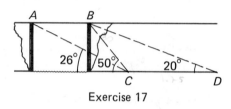

Exercise 17

18. Points P and Q on the far bank of a river are sighted from points A and B on the near bank. The lines of sight form the angles shown in the given diagram. If the banks are parallel and $AB = 25$ m, find PQ.

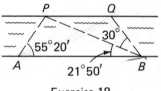

Exercise 18

In Exercises 19 and 20 we derive equations called Mollweide's equations, which are useful in checking the solution of a triangle.

C 19. In the given diagram $\triangle ABC$ is any triangle in which $a > b$. Side b has been marked off along side a, and $\overline{AD}$ has been drawn.

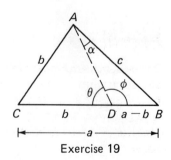

Exercise 19

(a) Given that θ is an exterior angle of $\triangle ADB$ and $\angle A = \alpha + \angle CAD$, show that
$$\alpha = \frac{\angle A - \angle B}{2}.$$

(b) Given that ϕ is an exterior angle of $\triangle ACD$ and that ϕ and θ are supplementary, show that
$$\phi = \frac{1}{2}\angle C + 90°.$$

(c) Show that $\dfrac{a-b}{c} = \dfrac{\sin \alpha}{\sin \phi}$.

(d) Show that $\dfrac{a-b}{c} = \dfrac{\sin \frac{1}{2}(A-B)}{\cos \frac{1}{2}C}$.

20. In the given diagram, $\triangle ABC$ is any triangle and $\overline{BC}$ has been extended by length b.

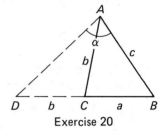

Exercise 20

(a) Show that $\alpha = \dfrac{\angle A - \angle B}{2} + 90°$.

(b) Show that $\angle D = \dfrac{1}{2} \angle C$.

(c) Show that $\dfrac{a+b}{c} = \dfrac{\sin \alpha}{\sin D}$.

(d) Show that $\dfrac{a+b}{c} = \dfrac{\cos \frac{1}{2}(A-B)}{\sin \frac{1}{2}C}$.

21. Use the results of Exercises 19 and 20 to derive the formula

$$\frac{a-b}{a+b} = \frac{\tan \frac{1}{2}(A-B)}{\tan \frac{1}{2}(A+B)}, \quad \text{called the } \textbf{law of tangents.}$$

(Hints: 1. $\dfrac{\angle C}{2} = \dfrac{180° - (\angle A + \angle B)}{2}$, since $\angle A + \angle B + \angle C = 180°$.

2. $\cot(90° - \theta) = \tan \theta$.)

Challenge

In the given diagram points A and B are on the same meridian on Earth, and their latitudes differ by θ. Point O is the center of Earth (assumed to be spherical), and point P is a celestial object. Show that the distance AP can be found by means of the formula

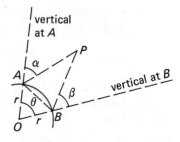

$$AP = \frac{2r \sin \dfrac{\theta}{2} \sin \left(90° - \beta + \dfrac{\theta}{2}\right)}{\sin(\alpha + \beta - \theta)}.$$

4-3   *The Ambiguous Case*

The remaining oblique-triangle case is the SSA case, that is, when two sides and a nonincluded angle are known. It is called the *ambiguous case* because there may be two sets of solutions, or there may be one or none. Some figures will help explain why this is so.

Suppose that in a triangle ABC we are given sides a and b and angle A. We can have any one of the four subcases illustrated in Figure 4-5.

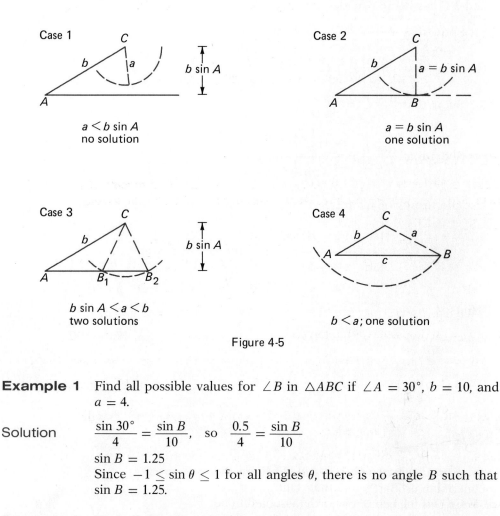

Figure 4-5

Example 1   Find all possible values for $\angle B$ in $\triangle ABC$ if $\angle A = 30°$, $b = 10$, and $a = 4$.

Solution $\dfrac{\sin 30°}{4} = \dfrac{\sin B}{10}$,   so $\dfrac{0.5}{4} = \dfrac{\sin B}{10}$

$\sin B = 1.25$

Since $-1 \le \sin \theta \le 1$ for all angles θ, there is no angle B such that $\sin B = 1.25$.

Example 2   Find all possible values for $\angle B$ in $\triangle ABC$ if $\angle A = 30°$, $b = 10$, and $a = 5$.

Solution $\dfrac{\sin 30°}{5} = \dfrac{\sin B}{10}$ so $\dfrac{0.5}{5} = \dfrac{\sin B}{10}$

From this last equation we get $\sin B = 1$. Therefore, $\angle B = 90°$ is the only solution.

Example 3 Find all possible values for $\angle B$ in $\triangle ABC$ if $\angle A = 30°$, $b = 10$, and $a = 6$. Estimate your answer to the nearest 10'.

Solution $\dfrac{\sin 30°}{6} = \dfrac{\sin B}{10}$, so $\dfrac{0.5}{6} = \dfrac{\sin B}{10}$

Thus, $\angle B = 56°30'$. But since $\sin(180° - \theta) = \sin \theta$, $\angle B = 56°30'$ or $180° - 56°30' = 123°30'$.

Example 4 Find all possible values for $\angle B$ in $\triangle ABC$ if $\angle A = 30°$, $b = 10$, and $a = 15$. Estimate your answer to the nearest 10'.

Solution $\dfrac{\sin 30°}{15} = \dfrac{\sin B}{10}$, so $\dfrac{0.5}{15} = \dfrac{\sin B}{10}$

and $\sin B = 0.3333$. Thus $\angle B = 19°30'$.

[Note that $(180° - 19°30') + 30° > 180°$.]

Exercises 4-3

For each $\triangle ABC$, state all possible values for $\angle B$ to the nearest 10' or tenth of a degree. Sketch a diagram to illustrate each solution. If no triangle can be formed, so state.

A 1. $\angle A = 52°$, $a = 40$, $b = 25$

2. $\angle A = 30°$, $a = 20$, $b = 37$

3. $\angle A = 14°50'$, $a = 64$, $b = 125$

4. $\angle A = 58°$, $a = 106$, $b = 62.5$

5. $\angle A = 20.5°$, $a = 9$, $b = 10$

6. $\angle A = 30°$, $a = 20$, $b = 9$

7. $\angle A = 15°40'$, $a = 54$, $b = 200$

8. $\angle A = 9°$, $a = 8$, $b = 60$

9. $\angle A = 35°10'$, $a = 144$, $b = 238$

10. $\angle A = 77°10'$, $a = 39$, $b = 40$

11. $\angle A = 60.5°$, $a = 10$, $b = 15$

12. $\angle A = 83.5°$, $a = 25$, $b = 24$

13. $\angle A = 24°20'$, $a = 103$, $b = 212$

14. $\angle A = 30°40'$, $a = 13$, $b = 22.5$

B 15. Show that if sides a and b and an acute angle A are given, no triangle can be formed having these parts if $a < b \sin A$.

16. Show that if sides a and b and an acute angle A are given, at least one triangle can be formed having these parts if $a \geq b \sin A$.

Exercises 17 and 18 refer to the given diagram.

17. Find $\angle DBC$ to the nearest 10' or tenth of a degree if $AB = 13$, $BD = BC = 8$, and $\angle A = 30°$.

18. Given the measurements in Exercise 17, find $\angle ABD$ to the nearest 10' or tenth of a degree.

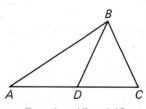

Exercises 17 and 18

Exercises 19 and 20 refer to the given diagram.

19. Given lengths a and b and the measure of θ, suppose two triangles can be formed as illustrated. Show that the altitudes to $\overline{AB}$ and $\overline{A'B'}$ both equal $a \sin B$.

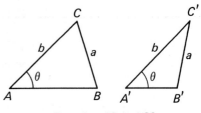

20. Use the result of Exercise 19 to show that the areas of $\triangle ABC$ and $\triangle A'B'C'$ are in the ratio $\dfrac{\sin C}{\sin C'}$.

Exercises 19 and 20.

C 21. Show that the ratio of the altitudes to $\overline{AC}$ and $\overline{A'C'}$ in Exercise 19 is $\dfrac{\sin (B + A)}{\sin (B - A)}$.

Area Formulas for Oblique Triangles

4-4 *The Area of an Oblique Triangle*

Recall that we can find the area of a triangle by means of the formula Area $= \dfrac{1}{2} bh$ if we know the length of a side b of the triangle and the length of the altitude h to side b. We can also find the area of a triangle uniquely determined by SAS, ASA, or SSS.

Suppose, for example, the lengths of a and b and the measure of angle C are given (SAS) in triangle ABC in Figure 4-6. Since $\sin C = \dfrac{h}{a}$, $h = a \sin C$. Substituting this value in the formula Area $= \dfrac{1}{2} bh$, we get

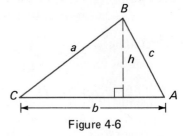

Figure 4-6

$$\text{Area} = \frac{1}{2} b(a \sin C), \text{ or}$$

$$\text{Area} = \frac{1}{2} ab \sin C$$

Similarly,

$$\text{Area} = \frac{1}{2}bc \sin A$$

$$\text{Area} = \frac{1}{2}ac \sin B$$

Example 1 Find the area of $\triangle ABC$ in which $a = 25$, $c = 12$, and $\angle B = 50°$. Estimate your answer to the nearest tenth.

Solution   Use the formula Area $= \frac{1}{2}ac \sin B$.

Then Area $= \frac{1}{2}(25)(12)(\sin 50°)$

$= 114.9$ square units.

Example 2 Find the area of a regular pentagon inscribed in a circle of radius 7. Estimate your answer to the nearest tenth.

Solution   Draw a diagram. Pentagon $ABCDE$ can be partitioned into five congruent triangles. Therefore,

$$\text{Area} = 5\left[\frac{1}{2}(OA)(OB)(\sin \angle AOB)\right]$$

$$= 5\left[\frac{1}{2}(7)(7)(\sin 72°)\right]$$

$$= 5(23.30) = 116.5 \text{ square units.}$$

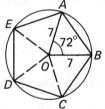

Example 3 Find the area of $\triangle ABC$ with $a = 13$, $b = 14$, and $c = 15$. Estimate your answer to the nearest tenth.

Solution   Note that the triangle is uniquely determined by SSS. Use the law of cosines to find $\cos A$.

$a^2 = b^2 + c^2 - 2bc \cos A$

$13^2 = 14^2 + 15^2 - 2(14)(15) \cos A$, so $\cos A = \frac{3}{5}$.

Since $\sin^2 A + \cos^2 A = 1$,

$\sin^2 A + \left(\frac{3}{5}\right)^2 = 1$ and $\sin A = \pm\frac{4}{5}$.

But $0° < \angle A < 180°$, so $\sin A = \frac{4}{5}$.

Therefore,

Area $= \frac{1}{2}(14)(15)\left(\frac{4}{5}\right) = 84.0$ square units.

To find the area of a triangle uniquely determined by ASA or AAS, we can use the law of sines to find a second side of the triangle and then apply one of the area formulas.

Other Triangle Area Formulas (Optional)

To find the area of a triangle uniquely determined by SSS, AAS, or ASA directly, without resorting to the law of sines or the law of cosines, one of the following formulas can be used.

$$\text{Area} = \sqrt{s(s - a)(s - b)(s - c)},$$
$$\text{where } s = \frac{a + b + c}{2} \qquad \text{(Hero's formula)}$$

$$\text{Area} = \frac{1}{2} a^2 \frac{\sin B \sin C}{\sin A}$$

$$\text{Area} = \frac{1}{2} b^2 \frac{\sin A \sin C}{\sin B}$$

$$\text{Area} = \frac{1}{2} c^2 \frac{\sin A \sin B}{\sin C}$$

Compare the following example with Example 3.

Example 4   Use Hero's formula to find the area, to the nearest tenth, of $\triangle ABC$ with $a = 13$, $b = 14$, and $c = 15$.

Solution $s = \dfrac{13 + 14 + 15}{2} = 21$

Therefore,
$$\text{Area} = \sqrt{21(21 - 13)(21 - 14)(21 - 15)}$$
$$= \sqrt{7056} = 84.0 \text{ square units.}$$

Example 5   Find the area, to the nearest tenth, of $\triangle ABC$ with $\angle A = 50°$, $\angle B = 30°$, and $c = 20$.

Solution   Use the formula
$$\text{Area} = \frac{1}{2} c^2 \frac{\sin A \sin B}{\sin C}.$$

We have $\angle C = 180° - 50° - 30° = 100°$. Therefore,
$$\text{Area} = \frac{1}{2} (20)^2 \frac{\sin 50° \sin 30°}{\sin 100°}$$
$$= 77.8 \text{ square units.}$$

Exercises 4-4

Find the area of $\triangle ABC$ with the given sides and angles. Estimate your answer to the nearest tenth.

A 1. $a = 15$, $b = 8$, $\angle C = 30°$ 2. $b = 6$, $c = 14$, $\angle A = 150°$

   3. $a = 18$, $c = 25$, $\angle B = 148°40'$ 4. $a = 8$, $b = 15$, $\angle C = 67°40'$

   5. $b = 30$, $c = 75$, $\angle A = 165°10'$ 6. $a = 125$, $c = 40$, $\angle B = 22°20'$

   7. $a = 13$, $b = 15$, $c = 4$ 8. $a = 11$, $b = 5$, $c = 6\sqrt{2}$

   9. $a = 6$, $\angle A = 18°$, $\angle B = 38°10'$ 10. $b = 8$, $\angle B = 16°40'$, $\angle C = 145°$

   11. $a = 12$, $b = 4$, $\angle A = 45°50'$ 12. $b = 16$, $c = 8$, $\angle B = 100°$

Find the area of a regular polygon with the given number of sides n inscribed in a circle of radius r. Estimate your answer to the nearest tenth.

13. $n = 12$; $r = 4$ 14. $n = 9$; $r = 9$ 15. $n = 15$; $r = 8$ 16. $n = 8$; $r = 9$

Exercises 17–19 give a sequential proof of Hero's formula.

B 17. Prove that the area of $\triangle ABC$ is given by the formula

$$\text{Area} = \frac{1}{2}ab\sqrt{1 - \left(\frac{a^2 + b^2 - c^2}{2ab}\right)^2}$$

$$= \frac{1}{4}\sqrt{4a^2b^2 - (a^2 + b^2 - c^2)^2}.$$

(Hint: Use the law of cosines to solve for cos C.)

18. Show that in $\triangle ABC$ with $s = \dfrac{a + b + c}{2}$,

$$s(s - c) = \frac{(a + b)^2 - c^2}{4} \quad \text{and}$$

$$(s - a)(s - b) = \frac{c^2 - (a - b)^2}{4}.$$

19. Use the results of Exercise 18 to show that

$$s(s - a)(s - b)(s - c) = \frac{4a^2b^2 - (a^2 + b^2 - c^2)^2}{16},$$

and use this and the result of Exercise 17 to prove Hero's formula.

20. Prove that the area of $\triangle ABC$ is given by the formula

$$\text{Area} = \frac{1}{2}c^2\,\frac{\sin A \sin B}{\sin C}.$$

21. Find the length of the third side of a triangle whose area is 6 square units and two of whose sides have lengths 3 and 5, respectively.

22. Find the length of side b of $\triangle ABC$ whose area is $24\sqrt{3}$ and which has $a = 6$ and $\angle B = 120°$.

C 23. A pyramid has a square base 314 cm on each side as shown in the given diagram. If each lateral edge makes an angle of 22°20′ with a diagonal of the base, find the lateral area of the pyramid to the nearest ten square centimeters.

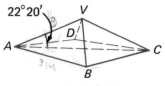

Exercise 23

24. (a) Give a formula for the area of a regular n-sided polygon inscribed in a circle of radius r, in terms of r and n.
    (b) Change the formula in part (a) so that any angles are given in radian measure. Then use the approximation $\sin x \approx x$ for small angles in radian measure to show that as n gets large, the area of the polygon approaches πr^2. Explain this fact geometrically.

CHAPTER SUMMARY

1. The law of cosines relates the measure of one angle of a triangle and all three sides. The law states that
$$a^2 = b^2 + c^2 - 2bc \cos A$$
$$b^2 = a^2 + c^2 - 2ac \cos B$$
$$c^2 = a^2 + b^2 - 2ab \cos C.$$

2. The law of sines relates the measures of the angles of a triangle to the sides opposite them. The law states that
$$\frac{\sin A}{a} = \frac{\sin B}{b} = \frac{\sin C}{c}.$$

3. Given two sides, a and b, and an angle A not included between the sides, we must consider the possibility that there may be two different triangles fitting the given information. There will be
   (a) no solution if $a < b \sin A$
   (b) one solution if $a = b \sin A$
   (c) two solutions if $b \sin A < a < b$
   (d) one solution if $a \geq b$.

4. The area of any triangle may be found using the following formulas.

$$\text{Area} = \frac{1}{2}ab \sin C$$

$$\text{Area} = \frac{1}{2}bc \sin A$$

$$\text{Area} = \frac{1}{2}ac \sin B$$

If two sides and an included angle are not given, the law of sines or the law of cosines should be used to obtain a pair of sides and their included angle. Other area formulas are:

$$\text{Area} = \sqrt{s(s-a)(s-b)(s-c)}, \text{ where } s = \frac{a+b+c}{2}$$

$$\text{Area} = \frac{1}{2}a^2\frac{\sin B \sin C}{\sin A}$$

$$\text{Area} = \frac{1}{2}b^2\frac{\sin A \sin C}{\sin B}$$

$$\text{Area} = \frac{1}{2}c^2\frac{\sin A \sin B}{\sin C}$$

CHAPTER TEST

In Exercises 1–2, find the missing side or angle.

4-1 1. $a = 25$, $c = 40$, $\angle B = 60°$, $b = ?$

2. $a = 12$, $b = 18$, $c = 15$, $\angle A = ?$

In Exercises 3–6, solve each triangle.

4-2 3. $\angle A = 30°$, $b = 28$, $a = 20$

4. $\angle B = 50°$, $b = 25$, $c = 35$

4-3 5. $\angle A = 55°20'$, $b = 18$, $c = 23$

6. $a = 15$, $b = 18$, $c = 22$

4-4 7. Find the area of a regular polygon of 18 sides inscribed in a circle of radius 9 cm.

8. Find the area of quadrilateral *ABCD* to the nearest square centimeter. (Hint: first solve △*ACD*.)

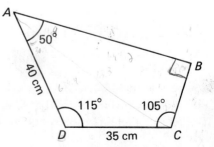

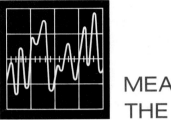

MEASURING THE EARTH

Surveying, the accurate measurement of land and water regions, has been important to civilization since ancient times. For hundreds of years before the twentieth century, land surveying methods remained virtually unchanged. The basic procedure of *plane surveying* (that is, taking measurements small enough so that the curvature of the earth can be ignored) is called *triangulation*. The region to be measured is divided into triangles, and all the angles are measured, using either a *theodolite* or a *transit*. Each of these instruments is basically a telescope mounted so that its direction can be accurately read from a scale marked in degrees. Since measurement of distances was, until recently, more difficult than angle measurement, a mathematical method is employed to calculate all distances from a single measurement of the length of one line segment, called the *base line*. (A second measurement, however, is often used as a check.) This method is based on the law of sines. For example, in the diagram if $\overline{BC}$ is chosen as the base line, DB can be calculated using the equation

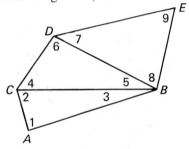

$$\frac{DB}{\sin \angle 4} = \frac{BC}{\sin \angle 6}.$$

Once DB is known, BE can be calculated using the equation

$$\frac{BE}{\sin \angle 7} = \frac{DB}{\sin \angle 9}.$$

Continuing in this manner, we can calculate the length of a side of any triangle, so long as that triangle can be connected to triangle BCD by a sequence of triangles, each of which has one side in common with the next triangle in the sequence. Also, all the angle measurements must be known.

Until recently, the measurement of distances in plane surveying was accomplished by means of a simple chain or steel tape. These instruments, however, are subject to expansion and contraction with changes in temperature. They must therefore be used only when temperature can be accurately determined, and their data must be corrected accordingly. In 1946, however, an instrument called a *geodimeter* was invented. This device measures distances by sending out pulses of light and timing their journey to and from a remote point.

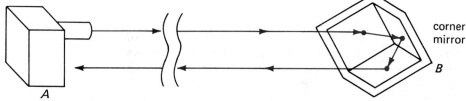

To see how the geodimeter works, let us assume that the light transmitter is located at point *A*. Light pulses are sent out at regular intervals and are reflected in a corner mirror at point *B*. These reflected light pulses travel in the direction exactly opposite to their initial direction. The phase shift between the original pulse and the reflected pulse is then measured. By measuring several such phase shifts, for pulses of different frequencies, the distance between *A* and *B* can be computed very accurately.

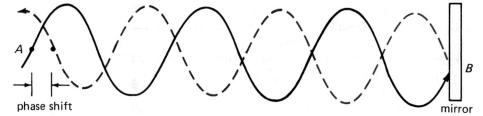

The major drawback of the geodimeter is that it requires the two points whose distance apart is to be measured to be visible to each other. A device that overcomes this problem is called a *tellurometer*. Instead of using visible light pulses, this device uses modulated microwaves to detect the phase shift. These waves can penetrate fog, haze, and even certain types of foliage. The tellurometer produces a carrier wave of about 3,000,000,000 Hz or 3 gigaherz and modulates the wave to produce a pulse. The carrier wave thus forms another wave of much lower frequency in much the same way as an AM radio signal carries a lower audible frequency (see pages 142–143). This pulse is then returned to its starting point by means of a retransmitter located at the second point.

Within the last few years the same principle of phase shift has been used in a new inertial reference system for airplanes. This system informs the pilot of very minute changes in the altitude and heading of the plane. The system uses laser beams traveling in opposite directions around a triangular path less than a meter long and is capable of measuring variations as small as 0.0005 degrees.

OBLIQUE TRIANGLES AND THE LAW OF COSINES

The program given below provides a solution of a triangle when two sides and the included angle are given or when three sides are given.

When two sides and the included angle are given, the law of cosines is used to find the third side. Then the three sides are used to find another angle. When the three sides are given, two angles are found.

Since the only inverse trigonometric function that **BASIC** has is the inverse tangent function, we use the following formulas. (See Exercises 3–5 on page 121.)

$$\tan \frac{C}{2} = \sqrt{\frac{(c + a - b)(c - a + b)}{(a + b + c)(a + b - c)}} \qquad \text{(lines 550–570)}$$

$$\tan \frac{A}{2} = \sqrt{\frac{(a + b - c)(a - b + c)}{(b + c + a)(b + c - a)}} \qquad \text{(lines 600–620)}$$

```
230   REM***DEFINE THE RATIO OF RADIANS TO DEGREES
240   LET R=3.14159/180
250   REM
260   REM***HERE THE CHOICES ARE PRESENTED:
270   REM
280   PRINT "TO SOLVE A TRIANGLE GIVEN:"
290   PRINT "(1) A1, B1, C (2) A1, B1, C1"
300   PRINT "(ANGLES IN DEGREES)"
310   PRINT
320   PRINT "TYPE NUMBER 1 OR 2";
330   INPUT N
340   REM
350   REM***INPUT DATA
360   REM
370   PRINT "INPUT A1";
380   INPUT A1
390   PRINT "INPUT B1";
400   INPUT B1
410   IF N=2 THEN 470
420   PRINT "INPUT C";
430   INPUT C
440   LET C1=SQR(A1*A1+B1*B1-2*A1*B1*COS(C*R))
```

(program continued on next page)

```
450    PRINT "C1="; INT(100*C1+.5)/100
460    GOTO 570
470    PRINT "INPUT C1";
480    INPUT C1
490    IF A1+B1 <=C1 THEN 650
500    IF B1+C1 <= A1 THEN 650
510    IF C1+A1 <= B1 THEN 650
520    LET N3=(C1+A1−B1)*(C1−A1+B1)
530    LET D3=(A1+B1+C1)*(A1+B1−C1)
540    LET T3=SQR(N3/D3)
550    LET C=2*ATN(T3)/R
560    PRINT "C="; INT(100*C+.5)/100; "DEGREES"
570    LET N1=(A1+B1−C1)*(A1−B1+C1)
580    LET D1=(B1+C1+A1)*(B1+C1−A1)
590    LET T1=SQR(N1/D1)
600    LET A=2*ATN(T1)/R
610    PRINT "A="; INT(100*A+.5)/100; "DEGREES"
620    LET B=180−A−C
630    PRINT "B="; INT(100*B+.5)/100; "DEGREES"
640    GOTO 660
650    PRINT "NO TRIANGLE"
660    END
```

Exercises

1. Use the program to solve (a) Example 1 and (b) Example 2 on page 102.

2. If A is an angle of a triangle, show that $\tan \dfrac{A}{2} = \sqrt{\dfrac{1 - \cos A}{1 + \cos A}}$.

    (Since $0° < \angle A < 180°$, $\dfrac{A}{2}$ is an acute angle, and $\tan \dfrac{A}{2}$ is positive.)

3. From the law of cosines, show that $1 - \cos A = \dfrac{(a + b - c)(a - b + c)}{2bc}$ and

    $1 + \cos A = \dfrac{(b + c + a)(b + c - a)}{2bc}$, and therefore,

    $$\tan \frac{A}{2} = \sqrt{\frac{(a + b - c)(a - b + c)}{(b + c + a)(b + c - a)}}.$$

4. Show that $\tan \dfrac{B}{2} = \sqrt{\dfrac{(b + a - c)(b - a + c)}{(a + c + b)(a + c - b)}}$.

5. Show that $\tan \dfrac{C}{2} = \sqrt{\dfrac{(c + a - b)(c - a + b)}{(a + b + c)(a + b - c)}}$.

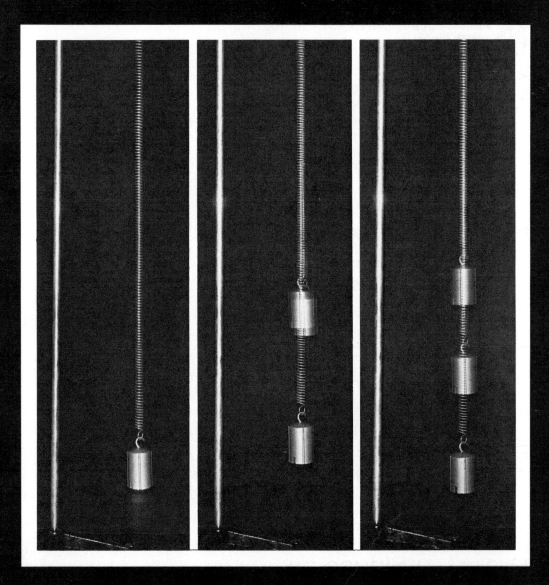

When an object attached to a spring is pulled down-ward and released, its path can be described by a trig-onometric function of time. The graph of such a func-tion resembles a sine curve and is called a sinusoid.

SINUSOIDAL VARIATION

OBJECTIVES

1. To graph $y = a \sin b(x - c) + d$ using the concepts of period, amplitude, phase shift, and vertical shift.
2. To apply sinusoidal variation to problems involving simple harmonic motion.
3. To graph functions by addition of ordinates.

Graphing Sinusoids

5-1 *Period and Amplitude*

Many scientific applications of trigonometry have no obvious connection with angles. For example, consider an object attached to a spring that is suspended from something stationary (such as a ceiling). If the object is pulled downward and then released, it will *oscillate* up and down. If we neglect the effects of friction and air resistance, the object will repeat the same oscillations indefinitely.

Figure 5-1 illustrates several positions (*y* in centimeters) of an oscillating object, like the one described, at various times (*t* in seconds).

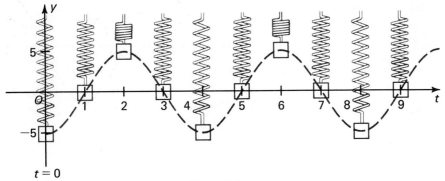

Figure 5-1

Notice that the object makes one complete oscillation in 2 seconds, covering a vertical range of 10 centimeters. The position *y* of the object is clearly a function of time *t*. A graph of this function is shown in Figure 5-2. We have replaced *t* by the more customary *x*.

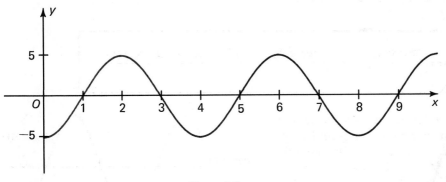

Figure 5-2

Laws of physics imply that this function is given by an equation of the form

$$y = a \sin b(x - c) + d \quad \text{or} \quad y = a \cos b(x - c) + d,$$

where *a*, *b*, *c*, and *d* are constants. The graph of such a function is called a **sinusoid,** and we say that *y* **varies sinusoidally** with *x*.

Example 1   Graph $y = 2 \sin x$.

Solution   Set up a table that compares values of $2 \sin x$ with values of $\sin x$. Draw a dashed sine curve as a guide.

(The first line of the table is read: "As *x* varies from 0 to $\dfrac{\pi}{2}$, sin *x* varies from 0 to 1 and $2 \sin x$ varies from 0 to 2.")

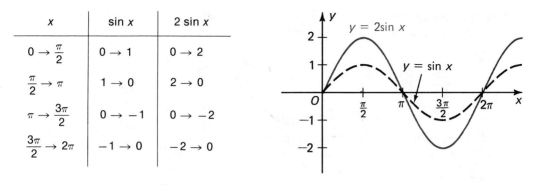

x	$\sin x$	$2 \sin x$
$0 \to \dfrac{\pi}{2}$	$0 \to 1$	$0 \to 2$
$\dfrac{\pi}{2} \to \pi$	$1 \to 0$	$2 \to 0$
$\pi \to \dfrac{3\pi}{2}$	$0 \to -1$	$0 \to -2$
$\dfrac{3\pi}{2} \to 2\pi$	$-1 \to 0$	$-2 \to 0$

The function in Example 1 has the same fundamental period of 2π as the function $y = \sin x$, but its greatest and least values are 2 and -2 respectively, rather than 1 and -1. In order to be able to discuss variation of function values more easily, we introduce the concept of *amplitude* of a periodic function $f(x)$, defined as follows:

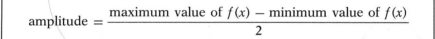

$$\text{amplitude} = \frac{\text{maximum value of } f(x) - \text{minimum value of } f(x)}{2}$$

We can consider the amplitude as the *maximum deviation of the graph from its central axis* (Figure 5-3). In Example 1, the amplitude $= \dfrac{2-(-2)}{2} = \dfrac{4}{2} = 2.$

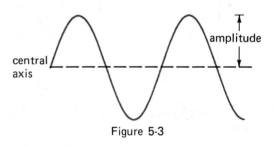

Figure 5-3

Since the maximum and minimum values of both the sine and cosine functions are 1 and -1, respectively, we can conclude:

The amplitude of any function of the form   (1)

$$y = a \sin (f(x)) \quad \text{or} \quad y = a \cos (f(x))$$

is $|a|$.

Example 2 Graph $y = \sin 2x$.

Solution Set up a table and draw a dashed sine curve as a guide.

x	$2x$	$\sin 2x$
$0 \to \dfrac{\pi}{4}$	$0 \to \dfrac{\pi}{2}$	$0 \to 1$
$\dfrac{\pi}{4} \to \dfrac{\pi}{2}$	$\dfrac{\pi}{2} \to \pi$	$1 \to 0$
$\dfrac{\pi}{2} \to \dfrac{3\pi}{4}$	$\pi \to \dfrac{3\pi}{2}$	$0 \to -1$
$\dfrac{3\pi}{4} \to \pi$	$\dfrac{3\pi}{2} \to 2\pi$	$-1 \to 0$
$\pi \to \dfrac{5\pi}{4}$	$2\pi \to \dfrac{5\pi}{2}$	$0 \to 1$
$\dfrac{5\pi}{4} \to \dfrac{3\pi}{2}$	$\dfrac{5\pi}{2} \to 3\pi$	$1 \to 0$
$\dfrac{3\pi}{2} \to \dfrac{7\pi}{4}$	$3\pi \to \dfrac{7\pi}{2}$	$0 \to -1$
$\dfrac{7\pi}{4} \to 2\pi$	$\dfrac{7\pi}{2} \to 4\pi$	$-1 \to 0$

The function in Example 2 has the same amplitude, 1, as the function $y = \sin x$, but its period is π rather than 2π. If we consider the functions $y = \sin bx$ and $y = \cos bx$, for constant $b > 0$, both will complete one full cycle when $bx = 2\pi$, or when $x = \dfrac{2\pi}{b}$. We can conclude:

The period of any function of the form (2)

$$y = a \sin bx \quad \text{or} \quad y = a \cos bx, \quad b > 0$$

is $\dfrac{2\pi}{b}$.

Combining statements (1) and (2), we have the following theorem.

Any function of the form $y = a \sin bx$ or $y = a \cos bx$, $b > 0$, has amplitude $|a|$ and period $\dfrac{2\pi}{b}$.

Example 3 Graph $y = -2 \cos 3x$.

Solution   The function has amplitude 2 and period $\frac{2\pi}{3}$. Draw the graph of $y = 2 \cos 3x$ and reflect it in the x-axis. The graph of $y = 2 \cos 3x$ is actually the cosine graph "compressed" horizontally and "stretched" vertically.

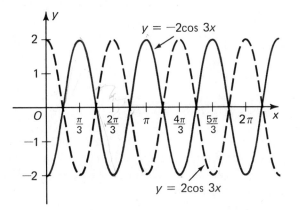

Exercises 5-1

Graph each function for $0 \le x \le 2\pi$.

A   1. $y = 5 \cos x$ 2. $y = 3 \sin x$ 3. $y = -2 \sin x$

     4. $y = -\frac{1}{2} \cos x$ 5. $y = 3 \sin 4x$ 6. $y = 2 \cos 3x$

     7. $y = 4 \cos \frac{1}{2} x$ 8. $y = 3 \sin \frac{x}{3}$ 9. $y = \frac{5}{2} \cos \frac{3}{2} x$

   10. $y = 4 \sin \frac{3}{4} x$ 11. $y = -3 \sin 2x$ 12. $y = -\sin 3x$

   13. $y = \frac{1}{2} \sin \frac{1}{2} x$ 14. $y = \frac{2}{3} \cos \frac{3}{2} x$ 15. $y = \cos \pi x$

   16. $y = \pi \sin 2\pi x$ 17. $y = |2 \sin x|$ 18. $y = |2 \cos x|$

Graph each function for $-2\pi \le x \le 2\pi$.

B 19. $y = \sin |x|$ 20. $y = \sin 2|x|$

Give the domain and range of each of the following functions, and graph each function.

21. $y = 2 \, \text{Sin}^{-1} x$ 22. $y = 2 \, \text{Cos}^{-1} x$ 23. $y = \text{Sin}^{-1} \dfrac{1}{2} x$ 24. $y = \text{Cos}^{-1} \dfrac{1}{2} x$

Graph each of the following.

C 25. $y = 2 \, \text{Sin}^{-1} \dfrac{1}{2} x$ 26. $y = 4 \, \text{Sin}^{-1} 2x$ 27. $y = 2 \, \text{Cos}^{-1} 2x$

28. $y = 3 \, \text{Cos}^{-1} \dfrac{x}{2}$ 29. $y = x \sin x$ 30. $y = x \cos x$

5-2   *Phase Shift and Vertical Shift*

Recall from algebra that the graph of $y = (x - 2)^2$ can be obtained by shifting the graph of $y = x^2$ two units horizontally to the right (Figure 5-4). Similarly, the graph of $y = x^2 + 3$ can be obtained by raising the graph of $y = x^2$ three units vertically upward (Figure 5-5).

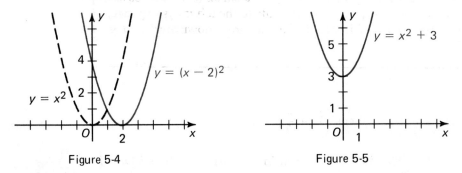

Figure 5-4   Figure 5-5

A shift of the graph of a function in which the final position of the graph is parallel to or coincident with its original position is called a **translation**. The combined effect of both a vertical and a horizontal translation is illustrated by the graph of $y = (x - 2)^2 + 3$ in Figure 5-6.

In fact, for all functions in the form $y = f(x - c) + d$ for constants a, b, c, and d, we have the following rule.

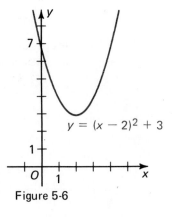

Figure 5-6

The graph of $y = f(x - c) + d$ can be ob-   (1)
tained by translating the graph of $y = f(x)$ c
units to the right if c is positive or $|c|$ units to
the left if c is negative, and d units upward if
d is positive or $|d|$ units downward if d is
negative.

In the case of sinusoidal variation, the number c is called the **phase shift** (or phase angle) and the number d is called the **vertical shift.**

Example 1   Graph $y = \sin\left(x - \dfrac{\pi}{2}\right) + 2.$

Solution   First draw a dashed sine curve as a guide. Next, translate this graph 2 units upward and $\dfrac{\pi}{2}$ units to the right.

Since we can write the equation of any sinusoidal variation in the form $y = a \sin b(x - c) + d$ or $y = a \cos b(x - c) + d$ for constants a, b, c, and d, we can restate rule (1) for sinusoidal functions.

> The graph of $y = a \sin b(x - c) + d$ (or $y = a \cos b(x - c) + d$) can be obtained by translating the graph of $y = a \sin bx$ (or $y = a \cos bx$) c units to the right if c is positive or $|c|$ units to the left if c is negative, and d units upward if d is positive or $|d|$ units downward if d is negative.

Example 2   Graph $y = 2 \cos\left(3x + \dfrac{\pi}{2}\right) + 1.$

Solution   First rewrite the function in the form $y = 2 \cos 3\left(x - \left(-\dfrac{\pi}{6}\right)\right) + 1.$

Therefore, $a = 2$, $b = 3$, $c = -\dfrac{\pi}{6}$, and $d = 1$. Next, sketch the graph of the corresponding function $y = 2 \cos 3x$ with phase shift and vertical shift both equal to zero. The amplitude is 2 and the period is $\dfrac{2\pi}{3}$.

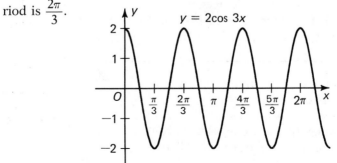

(Solution continued on page 130.)

Next, since $c = -\dfrac{\pi}{6}$, translate the graph $\dfrac{\pi}{6}$ units to the left. Since $d = 1$, translate the graph 1 unit upward.

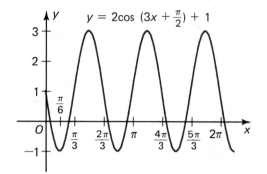

$$y = 2\cos\left(3x + \dfrac{\pi}{2}\right) + 1$$

Exercises 5-2

Graph all three functions in each exercise on one set of axes. Label the graph of each function.

A   1. (a) $y = x^2$ (b) $y = (x + 1)^2$ (c) $y = x^2 - 4$

2. (a) $y = \dfrac{x^3}{2}$ (b) $y = \dfrac{(x - 1)^3}{2}$ (c) $y = \dfrac{-x^3}{2} - 4$

3. (a) $y = \sin x$ (b) $y = \sin x - 2$ (c) $y = \sin (x + \pi)$

4. (a) $y = \cos x$ (b) $y = \cos\left(x - \dfrac{\pi}{2}\right)$ (c) $y = \cos x + 3$

5. (a) $y = 2\cos x$ (b) $y = 2\cos\left(x - \dfrac{\pi}{2}\right)$ (c) $y = 2\cos\left(x + \dfrac{\pi}{2}\right)$

6. (a) $y = 3\sin x$ (b) $y = 3\sin x + 1$ (c) $y = 3\sin x - 1$

7. (a) $y = \sin x$ (b) $y = \cos\left(x - \dfrac{\pi}{2}\right)$ (c) $y = -\sin (x - \pi)$

8. (a) $y = 2\cos x$ (b) $y = 2\sin\left(x - \dfrac{\pi}{2}\right)$ (c) $y = -2\cos (x - \pi)$

Graph each of the following.

9. $y = 4\sin\left(x + \dfrac{\pi}{2}\right) - 1$ 10. $y = 5\cos (2x - \pi) - 2$

11. $y = 2\cos\left(2x - \dfrac{3\pi}{2}\right) + 2$ 12. $y = 3\sin (x + \pi) + \dfrac{1}{2}$

13. $y = \dfrac{1}{2}\cos(4x - \pi) - \dfrac{1}{2}$

14. $y = 4\sin\left(\dfrac{1}{2}x + \dfrac{3\pi}{2}\right) + 3$

B 15. $y = \sin\left(3x + \dfrac{\pi}{2}\right) + 2$

16. $y = 2\cos(3x - \pi) - 1$

17. $y = -\sin\left(2x - \dfrac{\pi}{3}\right) - \dfrac{1}{2}$

18. $y = -2\cos(4x - 2\pi) + 3$

Rewrite each function in the form $y = c\cos(x - \phi)$, and graph the function.

Example $y = \sqrt{2}(\cos x + \sin x)$

Solution   Recall from Section 3-6, page 91, that $a\cos x + b\sin x = c\cos(x - \phi)$, where $c = \sqrt{a^2 + b^2}$ and ϕ is the measure of an angle such that $\cos\phi = \dfrac{a}{c}$ and $\sin\phi = \dfrac{b}{c}$. If the given function is written in the form $a\cos x + b\sin x$, then $a = b = \sqrt{2}$. Therefore, $c = \sqrt{a^2 + b^2} = \sqrt{4} = 2$, and $\cos\phi = \dfrac{\sqrt{2}}{2}$; $\sin\phi = \dfrac{\sqrt{2}}{2}$. Hence, $\phi = \dfrac{\pi}{4}$ and the function can be expressed as $y = 2\cos\left(x - \dfrac{\pi}{4}\right)$.

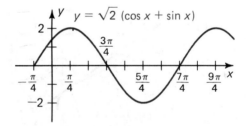

19. $y = \sqrt{3}\cos x + \sin x$

20. $y = 2\cos x - 2\sqrt{3}\sin x$

21. $y = 2\sqrt{2}(\cos x - \sin x)$

22. $y = -\sqrt{2}(\cos x + \sin x)$

23. $y = -3\cos x + 3\sqrt{3}\sin x$

24. $y = \dfrac{\sqrt{3}\cos x - \sin x}{4}$

25. Use the fact that $\sin\theta = \cos\left(\theta - \dfrac{\pi}{2}\right)$ to express the function $y = \sin\left(2x - \dfrac{\pi}{3}\right)$ in terms of the cosine function.

26. Repeat Exercise 25 for the function $y = 2\sin\left(\dfrac{x}{2} + \dfrac{\pi}{4}\right)$.

27. Express the function $y = \cos(3x - \pi)$ in terms of the sine function.

28. Repeat Exercise 27 for the function $y = 3 \cos \left(2x + \dfrac{\pi}{2} \right)$.

C 29. Give the domain and range of the function $y = \operatorname{Sin}^{-1}(x - 1) + \dfrac{\pi}{2}$ and draw its graph.

30. Give the domain and range of the function $y = \operatorname{Cos}^{-1}(x + 1) - \dfrac{\pi}{2}$ and draw its graph.

Challenge

Graph the following.

(a) $y = \sin \dfrac{1}{x}, x \neq 0$ $\left(\text{Hint: Consider values of } x \text{ such as } \dfrac{2}{\pi}, \dfrac{2}{2\pi}, \dfrac{2}{3\pi}, \dfrac{2}{4\pi}, \right.$ $\dfrac{2}{5\pi}, \cdots \Big)$

(b) $y = x \sin \dfrac{1}{x}, x \neq 0$ (Use the hint for (a).)

5-3   *Simple Harmonic Motion*

The motion of an object suspended from a spring as described at the beginning of this chapter is an example of simple harmonic motion. **Simple harmonic motion** is the motion that occurs when an object is displaced from an equilibrium (at rest) position, and the force tending to restore it to this position is directly proportional to the displacement. The equilibrium position is usually given the coordinate 0 in a coordinate system. If the displacement is denoted by x, the restoring force is $-kx$ for some positive constant k. The negative sign indicates that the displacement and the restoring force are in opposite directions.

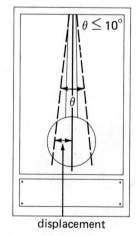

displacement

Figure 5-7

Some other approximations of simple harmonic motion are the motion of a tine of a tuning fork, the vertical motion of a floating object that has been pulled a short distance below its equilibrium position and released, and the motion of a clock pendulum if the angle through which the pendulum swings is relatively small (Figure 5-7).

It can be shown (using calculus) that all types of simple harmonic motion are sinusoidal variations as functions of time. In fact, if we assign the *y*-coordinate 0 to the equilibrium position of an object in simple harmonic motion, then the position *y* of the object at time *t* can be given by an equation of the form

$$y = a \sin (\omega t + \beta) \text{ or } y = a \cos (\omega t + \beta),$$

where $\omega > 0$ and a and β are constants. The fundamental period of the function is $\frac{2\pi}{\omega}$. Furthermore, $\frac{2\pi}{\omega}$ represents the time necessary for the object to complete one "round trip" in its path. In this context the number $\frac{2\pi}{\omega}$ is often denoted by the capital letter *T* and is called the **period** of the motion.

To denote oscillations per unit of time, we define the **frequency** *f* of a simple harmonic motion by the equation:

$$f = \frac{1}{T} = \frac{\omega}{2\pi}$$

Example An object suspended from a spring is pulled downward 5 cm from its equilibrium position and released. It oscillates sinusoidally, making one complete (up-and-down) oscillation every 2 seconds.
(a) Give an equation that defines the position *y* of the object as a function of time *t*.
(b) Sketch the graph of the function.
(c) Give the frequency of the motion.

Solution (a) We can write the equation in the form $y = a \sin (\omega t + \beta)$ or $y = a \cos (\omega t + \beta)$. Then $a = 5$ since the object has a maximum displacement of 5 cm, and $T = 2$ since the period of the motion is given to be 2 seconds. But $T = \frac{2\pi}{\omega}$; therefore, $\omega = \pi$.

To solve for β, we note that when $t = 0$, $y = -5$. Therefore we can substitute in the equation $y = 5 \sin (\pi t + \beta)$ to find β.

$$-5 = 5 \sin (\pi \cdot 0 + \beta)$$
$$-1 = \sin \beta$$
$$\beta = -\frac{\pi}{2}$$

The equation of the motion is $y = 5 \sin \left(\pi t - \frac{\pi}{2} \right)$.

(Solution continued on page 134.)

(b) Rewriting the equation in (a) in the form $y = a \sin b(t - c) + d$, we have

$$y = 5 \sin \pi \left(t - \frac{1}{2}\right).$$

From Section 2-4, we can now graph the equation. The graph is actually the sine graph "compressed" horizontally and "stretched" vertically. (Note how the graph shows that a "one-way trip" is completed in one second.)

(c) Since $T = 2$, we have $f = \dfrac{1}{T} = \dfrac{1}{2}$ cycle per second.

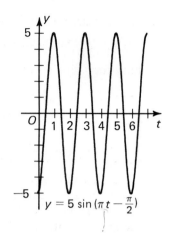

$$y = 5 \sin\left(\pi t - \frac{\pi}{2}\right)$$

Exercises 5-3

For each of the following cases of simple harmonic motion (Exercises 1–10), (a) write an equation for the position y of the object as a function of time t in the form $y = a \sin(\omega t + \beta)$, (b) graph the function, and (c) find the frequency of the oscillation. Neglect the effects of friction and air resistance.

A 1. An object attached to a spring is pulled downward 4 cm from its equilibrium position and released. It makes one complete oscillation every 4 seconds.

2. The object in Exercise 1 is pulled downward 3 cm from its equilibrium position and released. It makes one complete oscillation every $\dfrac{1}{2}$ second.

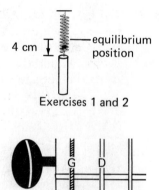

Exercises 1 and 2

3. A point on the G-string of a violin oscillates through a vertical range of 0.02 cm. Starting at the lowest point of the range, it makes one complete oscillation every 0.005 second.

4. A point on the D-string of a violin oscillates through a vertical range of 0.01 cm. Starting at the highest point of the range, it makes one complete oscillation every $\dfrac{1}{300}$ s.

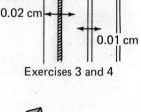

Exercises 3 and 4

5. The water line of a buoy starts from a position 30 cm below the surface of the water and bobs up to a maximum position 30 cm above the surface in 1 second.

6. The buoy in Exercise 5 covers the same vertical range in the same amount of time, but starts from a position 15 cm below the surface of the water and bobs up.

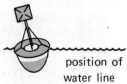

position of water line

7. A point on a pendulum swings one way from an angular position $-\frac{\pi}{4}$ to an angular position $-\frac{3\pi}{4}$ in 20 seconds. At $t = 0$ the pendulum's position is $-\frac{\pi}{4}$. $\left(\text{Hint: } y \text{ varies from } \frac{\pi}{4} \text{ to } -\frac{\pi}{4}.\right)$

8. Suppose the pendulum in Exercise 7 makes a complete back-and-forth swing in 20 seconds and is at a position of $-\frac{3\pi}{4}$ when $t = 0$.

9. Imagine that a frictionless hole is drilled through Earth as shown in the given diagram. If a ball were dropped into one end of the hole, it would roll past point P at the center of the hole, to the other end of the hole, and, by virtue of its momentum, back to its starting point. Starting at the left end of the hole with a position coordinate of $-a$, the ball oscillates in simple harmonic motion, making a one-way trip in 42 minutes.

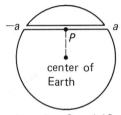

Exercises 9 and 10

10. Suppose that when $t = 0$ the ball in Exercise 9 is at the position $\frac{a}{2}$ and is moving to the right in the given diagram.

11–20. Express the functions in Exercises 1–10 in the form
$$y = a \cos(\omega t + \beta).$$

B 21. The line voltage in a house circuit is a sinusoidal function of time, varying from a maximum of 120 V (volts) to a minimum of -120 V (the current changes direction) with a frequency of 60 hz ("hertz" or cycles per second).
(a) Give an equation in the form $V = a \sin(\omega t + \beta)$ that defines the voltage V as a function of time t.
(b) Find β if $V = 120$ when $t = \frac{1}{144}$ second.

22. (a) Express the function in Exercise 21 using the cosine function.
(b) Find β if $V = -120$ when $t = \frac{1}{144}$ second.

Exercises 23 and 24 refer to the following information.

The given diagram shows a point P on a circle of radius a. Consider P as moving counterclockwise in uniform circular motion (Section 2-3) with angular velocity ω. Let M be the projection of P on the x-axis so that as P travels a distance of $2\pi a$ beginning at $(a, 0)$, M travels along the x-axis from $(a, 0)$ to $(-a, 0)$ and back again to $(a, 0)$.

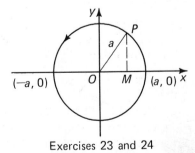

Exercises 23 and 24

23. Write an equation that describes the motion of M. (Hint: The coordinate of M on the x-axis equals the abscissa of P.)

24. Use Exercise 23 to explain why M is in simple harmonic motion.

25. Show that the graph of $y = a \cos \omega t$ is $\dfrac{\pi}{2\omega}$ units to the left of the graph of $y = a \sin \omega t$.

26. The graph of $y = a \cos (\omega t + \beta)$ is how many units to the left of the graph of $y = a \sin (\omega t + \beta)$?

C 27. An object in the simple harmonic motion $y = a \sin (\omega t + \beta)$ has a velocity given by the equation $v = a\omega \cos (\omega t + \beta)$. For what values of t is the velocity a maximum?

28. Suppose that an object is moving according to the equation $y = \sqrt{3} \sin 2t$. Use Exercise 27 to find the maximum velocity of the object.

Graphing Combinations of Sinusoids

5-4 *Graphing by Addition of Ordinates*

Periodic motion cannot always be described by simple functions of the form $y = a \sin (\omega t + \beta)$ or $y = a \cos (\omega t + \beta)$. It can, however, sometimes be described by the sums of such functions.

Consider, for example, the sound of a violin string. The quality of the sound is determined by the different vibrations of the string. The note of lowest frequency is called the **fundamental.** The string simultaneously produces other weaker notes, called **overtones.** The first and usually the strongest overtone is the note an octave higher than the fundamental. This overtone has a frequency twice that of the fundamental. Since $F = \dfrac{\omega}{2\pi}$, an equation describing the combined sound of the fundamental and the first overtone would be of the form

$$y = a \sin \omega t + b \sin 2\omega t.$$

The method of graphing such a function, as illustrated by the following example, is called addition of ordinates. In order to locate the y-value for a particular x-value, we add the y-coordinates that correspond to that x-value.

Example   Graph the function $y = \sin x + \sin 2x$ for $0 \le x \le 2\pi$.

Solution   Graph the functions $y = \sin x$ and $y = \sin 2x$ on one set of axes.

(Solution continued on next page.)

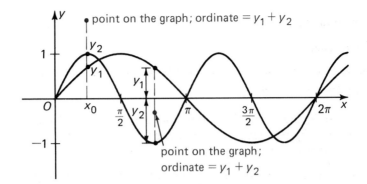

Starting from either graph, say, at y_1, measure off the y-coordinate y_2 of the other graph, up if y_2 is positive, down if y_2 is negative. Repeat this procedure for as many x-values as you need to draw the final graph.

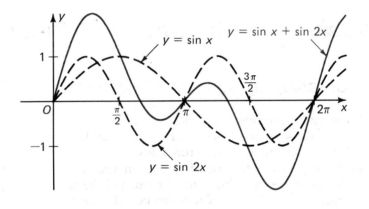

A graph showing all the combined sounds of a violin string can be produced on an electronic device called an oscilloscope (Figure 5-8, taken from Example 3 on page 2). The frequency of the function graphed corresponds to the pitch of the sound, and the amplitude corresponds to the loudness. The shape of the graph corresponds to the quality of the sound. A difference in the quality of two sounds, for example, enables us to distinguish between two instruments, even when they are playing notes of the same pitch and loudness.

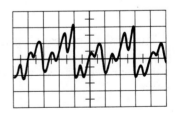

Figure 5-8

Exercises 5-4

Graph each function for $0 \le x \le 2\pi$.

A 1. $y = \sin x + \cos x$ 2. $y = -\sin x + \cos x$ 3. $y = \sin x - \cos x$

4. $y = -2(\sin x + \cos x)$ 5. $y = 2\sin x + \cos x$ 6. $y = \sin x - 2\cos x$

7. $y = -\sin x + 2\cos x$ 8. $y = -2\sin x - \cos x$ 9. $y = \cos x + \cos 2x$

10. $y = \cos x - \cos 2x$ 11. $y = \sin x - \sin 2x$ 12. $y = \sin 2x - \sin x$

B 13. $y = x + \sin x$ 14. $y = x - \cos x$

15. $y = \sin x + |\sin x|$ 16. $y = \cos x + |\sin x|$

In Exercises 17–22, use one of the following identities to change each function so that it is expressed as a sum of sine and cosine functions instead of a product, and graph the function.

$$-\cos(\alpha + \beta) + \cos(\alpha - \beta) = 2\sin\alpha\sin\beta$$
$$\sin(\alpha + \beta) + \cos(\alpha - \beta) = 2\sin\alpha\cos\beta$$
$$\sin(\alpha + \beta) - \cos(\alpha - \beta) = 2\cos\alpha\sin\beta$$
$$\cos(\alpha + \beta) + \cos(\alpha - \beta) = 2\cos\alpha\cos\beta$$

17. $y = 2\sin 2x \cos x$ 18. $y = 2\cos 2x \cos x$ 19. $y = \sin 3x \cos x$

20. $y = \sin 3x \sin x$ 21. $y = \cos\dfrac{3x}{2}\cos\dfrac{x}{2}$ 22. $y = \cos\dfrac{3x}{2}\sin\dfrac{x}{2}$

C 23. Prove that if $f(x)$ is a periodic function with fundamental period $\dfrac{2\pi}{p}$ and $g(x)$ is a periodic function with fundamental period $\dfrac{2\pi}{q}$, where p and q are positive integers, then the function $f(x) + g(x)$ is periodic with period $\dfrac{2\pi}{\text{the greatest common divisor of } p \text{ and } q}$.

CHAPTER SUMMARY

1. The sine or cosine functions may be used to represent sinusoidal variation. The amplitude of such a variation is defined by

$$\text{amplitude} = \frac{|\text{maximum value} - \text{minimum value}|}{2}.$$

We can adjust the period and amplitude of a function by the appropriate choice of a and b in the equations $y = a \sin bx$ or $y = a \cos bx$. Each equation will have an amplitude of $|a|$ and a period of $\dfrac{2\pi}{|b|}$.

2. A sinusoidal function can be shifted or translated by the use of constants c and d in the equations

$$y = a \sin b(x - c) + d \quad \text{or} \quad y = a \cos b(x - c) + d.$$

The value of c produces a horizontal or phase shift under which the graph is shifted $|c|$ units to the right if c is positive or $|c|$ units to the left if c is negative. The value of d produces a vertical shift under which the graph is shifted $|d|$ units upward if d is positive or $|d|$ units downward if d is negative.

3. Simple harmonic motion may be represented by the equations

$$y = a \sin (\omega t + b) \quad \text{or} \quad y = a \cos (\omega t + b)$$

where ω is the angular speed. The period of such a motion is given by $T = \dfrac{2\pi}{\omega}$

and the frequency is defined to be $f = \dfrac{1}{T} = \dfrac{\omega}{2\pi}$.

4. More complicated periodic motion may require the addition of several functions, each representing a simple harmonic motion. Such functions may be graphed by adding the ordinates of each simpler function.

CHAPTER TEST

5-1   1. Graph $y = -2 \cos \dfrac{x}{2}$ for $-2\pi \le x \le 2\pi$.

5-2   2. Graph $y = \sin \left(x - \dfrac{\pi}{4} \right) + 1$ for $0 \le x \le 2\pi$.

3. Graph $y = -\cos (2x + \pi) + 2$ for $0 \le x \le 2\pi$.

5-3   4. A point P moves counterclockwise around a circle of radius 5 with an angular velocity of $\dfrac{\pi}{4}$ radians/s. If P is at $(5, 0)$ at time $t = 0$ and the circle has center $(0, 0)$, (a) write an equation for the height of P above the x-axis as a function of time, (b) find the period, amplitude, and frequency of the motion of P, and (c) graph the equation.

5-4   5. Graph $y = |\sin x| + \cos \left(x - \dfrac{\pi}{2} \right)$ for $0 \le x \le 4\pi$.

CUMULATIVE REVIEW *Chapters 3–5*

Chapter 3

1. Express $\tan x (\sin x - \sec x \cot x)$ in terms of $\cos x$.

2. Express $\dfrac{\sin \theta}{1 + \cos \theta} + \cot \theta$ in terms of a single trigonometric function.

Prove each identity.

3. $\dfrac{(\sin x + \cos x)^2}{\cos x} = \sec x + 2 \sin x$

4. $\dfrac{\cot x}{\csc x - 1} = \dfrac{1 + \sin x}{\cos x}$

5. If $\sin x = -\dfrac{3}{5}$ and $\pi < x < \dfrac{3\pi}{2}$, find each of the following.

  (a) $\sin \left(x + \dfrac{\pi}{6} \right)$ (b) $\cos \left(x - \dfrac{\pi}{6} \right)$ (c) $\tan \left(x + \dfrac{\pi}{6} \right)$

6. Prove $\cos 3x = \cos x \,(\cos^2 x - 3 \sin^2 x)$.

7. Evaluate $\tan 22.5°$ using identities.

Solve each of the following for $0 \le x < 2\pi$.

8. $1 + 2 \tan^2 x + \sec x = 0$

9. $1 - \sin 2x = \cos x - 2 \sin x$

10. Rewrite $8 \sin \theta - 6 \cos \theta$ in $C \cos (\theta - \phi)$ form for $-180° \le \phi \le 180°$.

Chapter 4

1. In $\triangle ART$, $\angle R = 135°$ and $\angle T = 15°$. If $AT = 6\sqrt{3}$, find RT in simplest radical form.

2. Find all possible values for $\angle B$ in $\triangle ABC$ if $\angle A = 27°$, $BC = 15$, and $AB = 24$. Round answers to the nearest 10' or tenth of a degree.

3. In order to determine the angle formed by two walls, a meter stick is placed as shown with one end touching one wall 20 cm from the intersection of the walls. If the stick touches the other wall at a point 105 cm from the intersection, find the angle formed by the walls to the nearest degree.

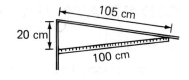

4. Find the area of $\triangle ABC$ if $\angle B = 30°$, $a = 10$, and $c = 15$.

5. Find the area of a triangle with sides of 25 cm, 39 cm, and 40 cm.

6. A floodlight mounted on a wall illuminates a triangular area as shown. If $AB = 14$ m, $\angle B = 41°$, and $\angle A = 100°$, find the area illuminated by the light to the nearest square meter.

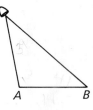

Chapter 5

In Exercises 1–4, (a) give the period and amplitude of each function and (b) graph the function over the interval $-2\pi \le x \le 2\pi$.

1. $y = 3 \sin 2x$

2. $y = 4 \cos \dfrac{x}{2}$

3. $y = 5 - 2 \sin 3x$

4. $y = \cos\left(2x - \dfrac{\pi}{3}\right)$

5. A child on a swing reaches a height of 2.5 m at the peak of the swing. At rest (time $t = 0$) the swing is 0.5 m above the ground. If one complete trip (back and forth) takes 2 seconds,

   (a) write an equation for the height above the ground as a function of time,

   (b) graph the function,

   (c) find the frequency of the motion, and

   (d) rewrite your equation using a different trigonometric function. You may assume that the swing reaches its maximum height on its first swing.

6. Graph $y = \sin\left(x + \dfrac{\pi}{2}\right) + \sin 2x$ for $0 \le x \le 2\pi$.

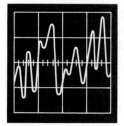

RADIO WAVES

The radio waves broadcast by any given station must be capable of transmitting *sounds* of *varying* frequencies. The basic problem confronting designers of radio transmitters is that the radio waves must also have a certain *fixed* frequency that is assigned to the station. The problem can be solved by regulating waves called **carrier** waves, whose graphs are sinusoids.

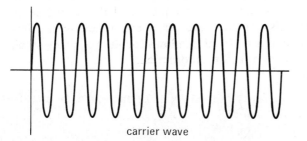

carrier wave

A carrier wave can be regulated by modulating (varying) either its amplitude or its frequency. In amplitude modulation (AM), the amplitude of the carrier wave is changed periodically. An auxiliary curve, called the envelope (represented by the dashed line in the diagram),

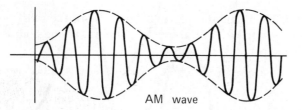

AM wave

is formed that has the same frequency as the sound wave (music or words) being transmitted. An equation of such a curve as a function of time t for a musical tone of pure pitch, for example, might be of the form

$$y = (a \sin pt)(\sin qt)$$

where q, the frequency of the carrier wave, would be a large number in relation to p, the frequency of the sound being transmitted.

In frequency modulation (FM), the amplitude of the carrier wave is held constant and the frequency itself is modulated by a relatively small amount in accordance with the sound wave being transmitted. An equa-

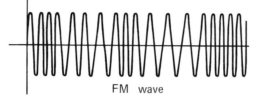

FM wave

tion of this type of curve for a pure musical tone, for example, might be of the form

$$y = a \sin (qt + b \sin pt)$$

in which q would again represent the frequency of the carrier wave (a large number in relation to p), and p the frequency of the sound being transmitted. In FM transmission, the receiver is still tuned to the frequency of the carrier wave, but the tuning allows for the relatively small variations in this frequency due to the modulation.

GRAPHING SINUSOIDS

The program given on pages 66 and 67 can be used to graph various sinusoids.

First, change line 140 to:

    140   FOR X1 = 0 TO 64 STEP 2

Next, change lines 100 and 160 to graph the given function.

The printout on the next page shows the graph of the example on page 136. Lines 100 and 160 were changed as follows.

    100   PRINT "Y = SIN X + SIN 2X"
    160   LET Y = SIN (X) + SIN (2 * X)

Exercises

Change lines 100 and 160 to graph the following functions for $0 \leq x \leq 2\pi$.

1. $y = \sin 2x$

2. $y = 2 \sin x$

3. $y = .5 \sin x$

4. $y = \sin 3x$

5. $y = \sin x + 1$

6. $y = \sin (x + 3.14159 \div 4)$

7. Example 1, page 129

8. Example 2, page 129

9–14. Graph the functions of Exercises 9–12 and 15, 16 on page 138.

Change line 140 to

    140   FOR X1 = 0 TO 126 STEP 2

and graph the following functions.

15. $y = \sin (.5x)$

16. $y = 2 \sin (.5x)$

Y=SIN X + SIN 2X

```
     X       Y

     0       0
    .2      .59
    .4     1.11
    .6     1.5
    .8     1.72
    1      1.75
    1.2    1.61
    1.4    1.32
    1.6     .94
    1.8     .53
    2       .15
    2.2   -.14
    2.4   -.32
    2.6   -.37
    2.8   -.3
    3     -.14
    3.2    .06
    3.4    .24
    3.6    .35
    3.8    .36
    4      .23
    4.2   -.02
    4.4   -.37
    4.6   -.77
    4.8  -1.17
    5    -1.5
    5.2  -1.71
    5.4  -1.75
    5.6  -1.61
    5.8  -1.29
    6     -.82
    6.2   -.25
    6.4    .35
```

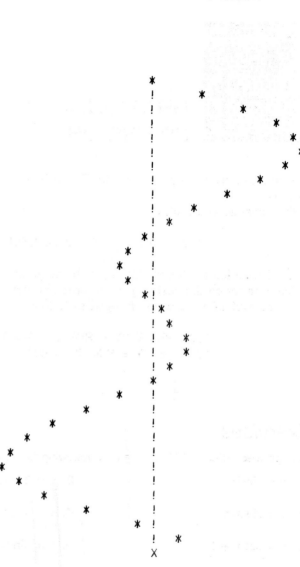

END

The forces acting on a steel beam being hoisted by
cables can be represented by means of vectors.

VECTORS IN THE PLANE

OBJECTIVES

1. To define the basic vector operations.
2. To apply vectors to navigation, force, work, and energy.
3. To define the dot product.

Vector Operations and Applications

6-1 *Basic Vector Operations*

Many quantities, such as speed, mass, and length, are called **scalar quantities** because they can be described by a single real number, or **scalar.** Quantities that have both *magnitude* (size) and *direction,* such as velocity, force, and displacement, are called **vector quantities.**

A natural way to represent a vector quantity geometrically is by an arrow whose length is the magnitude of the quantity and whose direction is the direction of the quantity. We call these arrows vectors and use boldface letters to represent them.* The tail of the arrow is called the **initial point** of the vector and the tip is called the **terminal point.** We shall regard two vectors as equal if they have the same length and the same direction. Thus, in Figure 6-1, **a** = **b** and **F** = **G**.

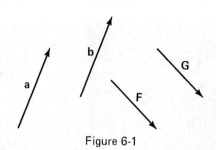

Figure 6-1

*In handwritten work, it is often convenient to represent a vector by a letter with a small arrow over it, like $\vec{a}$.

As the following examples show, vectors may be added, and they may also be multiplied by scalars.

Example 1   A directional rocket attached to a space vehicle is used to correct the course of the vehicle. This rocket produces a thrust **R** that combines with the thrust **E** of the vehicle's engines to produce a new thrust $\mathbf{T} = \mathbf{R} + \mathbf{E}$ and course for the vehicle as shown in the given diagram.

Example 2   The vector **v** in the given diagram represents the velocity of a plane flying northwest at 600 km/h. If the pilot reduces the plane's speed to 400 km/h and does not change its direction, the new velocity may be represented by a vector that has $\frac{2}{3}$ the length of the original vector.

These examples suggest the following definitions of **addition** of vectors and **multiplication** of vectors by scalars.

Let **a** and **b** be any vectors. Then:
To find $\mathbf{a} + \mathbf{b}$, first place the initial point of **b** at the terminal point of **a**. Then $\mathbf{a} + \mathbf{b}$ is the vector extending from the initial point of **a** to the terminal point of **b**.

This definition is equivalent to the **parallelogram rule** as is illustrated in Figure 6-2.

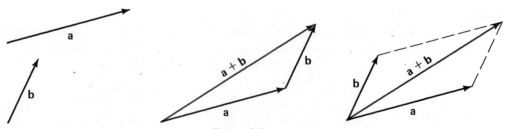

Figure 6-2

Let **a** be any vector and t be a scalar. Then:
To find t**a**, multiply the length of **a** by $|t|$ and reverse the direction if $t < 0$.

Figure 6-3 illustrates the definition for $t = 2$, $\frac{1}{2}$, -1.5, and -1.

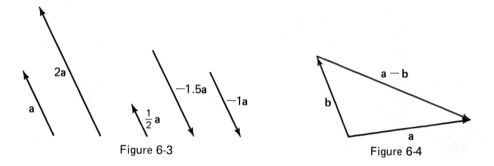

Figure 6-3   Figure 6-4

We write -1**a** as $-$**a** and define **a** $-$ **b** to be **a** $+ (-$**b**$)$. Thus, **a** $-$ **b** is the vector which must be added to **b** to give **a** as shown in Figure 6-4. The **zero vector 0** has magnitude 0 and thus its direction is not defined.

It can be shown that the operations defined above obey all the appropriate laws of algebra. For example,

$$(\mathbf{a} + \mathbf{b}) + \mathbf{c} = \mathbf{a} + (\mathbf{b} + \mathbf{c}) \text{ and } t(\mathbf{a} + \mathbf{b}) = t\mathbf{a} + t\mathbf{b}.$$

Expressions of the form

$$t_1\mathbf{a}_1 + t_2\mathbf{a}_2 + \cdots + t_n\mathbf{a}_n,$$

where $\mathbf{a}_1, \mathbf{a}_2, \ldots, \mathbf{a}_n$ are vectors and $t_1, t_2, \ldots, t_n$ are scalars, are called **linear combinations** of $\mathbf{a}_1, \mathbf{a}_2, \ldots, \mathbf{a}_n$.

Example 3 (a) Express the vectors **u** and **v** as linear combinations of the vectors **a** and **b**.
  (b) Express **a** and **b** as linear combinations of **u** and **v**.

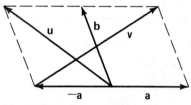

Solution   (a) From the figure **u** $= -$**a** $+$ **b** and **v** $=$ 2**a** $+$ **b**.
  (b) Set up a system of simultaneous equations from the linear combinations of part (a). Solve this system first for **a** and then for **b**.

$$
\begin{array}{ll}
-\mathbf{a} + \mathbf{b} = \mathbf{u} & \quad -2\mathbf{a} + 2\mathbf{b} = 2\mathbf{u} \\
\underline{\ \ 2\mathbf{a} + \mathbf{b} = \mathbf{v}} & \quad \underline{\ \ \ 2\mathbf{a} + \ \mathbf{b} = \ \mathbf{v}} \\
-3\mathbf{a} \qquad = \mathbf{u} - \mathbf{v} & \quad \ \ \ 3\mathbf{b} = 2\mathbf{u} + \mathbf{v} \\
\qquad \mathbf{a} = -\dfrac{1}{3}\mathbf{u} + \dfrac{1}{3}\mathbf{v} & \qquad \ \ \mathbf{b} = \dfrac{2}{3}\mathbf{u} + \dfrac{1}{3}\mathbf{v}
\end{array}
$$

Let us now consider vectors that lie in a fixed plane into which we have introduced a coordinate system. We denote by **i** and **j** the vectors having their initial points at the origin and their terminal points at $(1, 0)$ and $(0, 1)$, respectively. Any vector **a** now can be expressed as a linear combination of **i** and **j**. For example, if the initial point of **a** is placed at the origin, its terminal point will have coordinates (a_1, a_2). Thus,

$$\mathbf{a} = a_1\mathbf{i} + a_2\mathbf{j}.$$

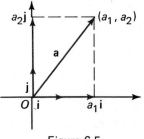

Figure 6-5

A vector, such as **a**, with its endpoint at the origin is said to be in **standard position.** The numbers a_1 and a_2 are called the **components** or **scalar components** of **a**. The algebraic properties of vectors can be used to prove the following.

If $\mathbf{a} = a_1\mathbf{i} + a_2\mathbf{j}$ and $\mathbf{b} = b_1\mathbf{i} + b_2\mathbf{j}$, then

$a_1\mathbf{i} + a_2\mathbf{j} = b_1\mathbf{i} + b_2\mathbf{j}$ if and only if
$a_1 = b_1$ and $a_2 = b_2$.   (1)

$\mathbf{a} + \mathbf{b} = (a_1 + b_1)\mathbf{i} + (a_2 + b_2)\mathbf{j}$ (2)

$t\mathbf{a} = ta_1\mathbf{i} + ta_2\mathbf{j}$ (3)

Example 4   Solve $(3\mathbf{i} + 2\mathbf{j}) + (s\mathbf{i} + t\mathbf{j}) = 7\mathbf{i} - 3\mathbf{j}$ for the scalars s and t.

Solution   By (2) above, the given equation is equivalent to

$$(3 + s)\mathbf{i} + (2 + t)\mathbf{j} = 7\mathbf{i} - 3\mathbf{j}.$$

Hence, by (1) above, we have

$$3 + s = 7 \text{ or } s = 4 \qquad \text{and} \qquad 2 + t = -3 \text{ or } t = -5.$$

Example 5   Let $\mathbf{a} = -2\mathbf{i} + 3\mathbf{j}$ and $\mathbf{b} = 2\mathbf{i} + \mathbf{j}$. Find $\mathbf{a} + 2\mathbf{b}$ and $\mathbf{a} - 2\mathbf{b}$ and draw all the vectors involved with their initial points at the origin.

Solution

$$\begin{aligned}
\mathbf{a} + 2\mathbf{b} &= (-2\mathbf{i} + 3\mathbf{j}) + 2(2\mathbf{i} + \mathbf{j}) \\
&= -2\mathbf{i} + 3\mathbf{j} + 4\mathbf{i} + 2\mathbf{j} \\
&= 2\mathbf{i} + 5\mathbf{j}
\end{aligned}$$

$$\begin{aligned}
\mathbf{a} - 2\mathbf{b} &= (-2\mathbf{i} + 3\mathbf{j}) - 2(2\mathbf{i} + \mathbf{j}) \\
&= -2\mathbf{i} + 3\mathbf{j} - 4\mathbf{i} - 2\mathbf{j} \\
&= -6\mathbf{i} + \mathbf{j}
\end{aligned}$$

The magnitude of a vector **a** is also called its **norm** and is denoted by ‖**a**‖. In terms of its components, the norm of **a** is given by:

> If $\mathbf{a} = a_1\mathbf{i} + a_2\mathbf{j}$, then $\|\mathbf{a}\| = \sqrt{a_1^2 + a_2^2}$.

Example 6 Let $\mathbf{a} = -9\mathbf{i} + 12\mathbf{j}$ and $\mathbf{b} = 4\mathbf{i} - 12\mathbf{j}$. Find $\|\mathbf{a}\|$, $\|\mathbf{b}\|$, and $\|\mathbf{a} + \mathbf{b}\|$.

Solution
$$\|\mathbf{a}\| = \sqrt{(-9)^2 + 12^2} = \sqrt{81 + 144} = \sqrt{225} = 15$$
$$\|\mathbf{b}\| = \sqrt{4^2 + (-12)^2} = \sqrt{16 + 144} = \sqrt{160} = 4\sqrt{10}$$
$$\|\mathbf{a} + \mathbf{b}\| = \|(-9\mathbf{i} + 12\mathbf{j}) + (4\mathbf{i} - 12\mathbf{j})\| = \|-5\mathbf{i}\| = 5$$

Two important properties of the norm are:

> $$\|\mathbf{a} + \mathbf{b}\| \le \|\mathbf{a}\| + \|\mathbf{b}\| \quad \text{and} \quad \|t\mathbf{a}\| = |t|\|\mathbf{a}\|$$

The first of these properties will be proved as Exercise 40 in Section 6-4. The second will be proved in Exercise 30 of this section.

A vector having norm 1 is called a **unit vector.** The unit vector in the direction of **a** ($\mathbf{a} \ne \mathbf{0}$) is

$$\frac{1}{\|\mathbf{a}\|}\mathbf{a}, \quad \text{or,} \quad \text{more simply,} \quad \frac{\mathbf{a}}{\|\mathbf{a}\|}.$$

We sometimes use the symbol $\hat{\mathbf{a}}$ to indicate the unit vector in the direction of **a.** Thus, if $\mathbf{a} = \mathbf{i} - \mathbf{j}$, then $\hat{\mathbf{a}} = \dfrac{\mathbf{i} - \mathbf{j}}{\sqrt{2}}$.

Exercises 6-1

In Exercises 1–6, draw the vectors **a**, **b**, **a** + **b**, and **a** − **b** with their initial points at the origin. Then find the norm of each of the four vectors.

A   1. $\mathbf{a} = \mathbf{i}$; $\mathbf{b} = \mathbf{j}$

2. $\mathbf{a} = \mathbf{i} + \mathbf{j}$; $\mathbf{b} = \mathbf{i} - \mathbf{j}$

3. $\mathbf{a} = 2\mathbf{i} + \mathbf{j}$; $\mathbf{b} = 2\mathbf{i} + 4\mathbf{j}$

4. $\mathbf{a} = 12\mathbf{i} + 5\mathbf{j}$; $\mathbf{b} = -5\mathbf{j}$

5. $\mathbf{a} = 4\mathbf{i} - 3\mathbf{j}$; $\mathbf{b} = -\mathbf{i} + 2\mathbf{j}$

6. $\mathbf{a} = \mathbf{i} - \sqrt{3}\,\mathbf{j}$; $\mathbf{b} = \mathbf{i} + \sqrt{3}\,\mathbf{j}$

Find unit vectors parallel to the four vectors in each of the following.

7. Exercise 3

8. Exercise 4

In Exercises 9–12, find the specified linear combinations of $\mathbf{u} = 3\mathbf{i} + \mathbf{j}$ and $\mathbf{v} = 2\mathbf{i} - 3\mathbf{j}$.

9. $3\mathbf{u} + \mathbf{v}$ 10. $2\mathbf{u} - 3\mathbf{v}$

11. $\dfrac{1}{2}\mathbf{u} + \dfrac{1}{2}\mathbf{v}$ 12. $\dfrac{1}{3}\mathbf{u} - \dfrac{2}{3}\mathbf{v}$

In Exercises 13–16, solve for the scalars s and t.

13. $(s\mathbf{i} + t\mathbf{j}) + (2\mathbf{i} - \mathbf{j}) = \mathbf{i}$

14. $(s\mathbf{i} + t\mathbf{j}) + (2\mathbf{i} - \mathbf{j}) = \mathbf{j}$

B 15. $(s\mathbf{i} - 3\mathbf{j}) + [(s - t)\mathbf{i} + t\mathbf{j}] = 5\mathbf{i} + 4\mathbf{j}$

16. $[(s + t)\mathbf{i} + 4\mathbf{j}] + [2\mathbf{i} + (s - t)\mathbf{j}] = 0$

In Exercises 17–20, (a) express $\mathbf{u}$ and $\mathbf{v}$ in terms of $\mathbf{a}$ and $\mathbf{b}$. Then (b) express $\mathbf{a}$ and $\mathbf{b}$ in terms of $\mathbf{u}$ and $\mathbf{v}$.

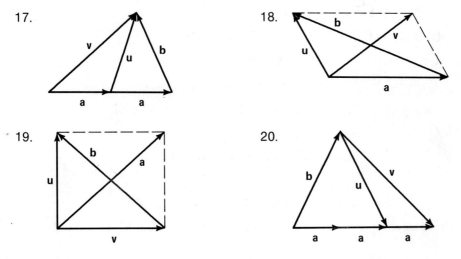

17.

18.

19.

20.

In Exercises 21–28, let $\mathbf{a} = a_1\mathbf{i} + a_2\mathbf{j}$, $\mathbf{b} = b_1\mathbf{i} + b_2\mathbf{j}$, and $\mathbf{c} = c_1\mathbf{i} + c_2\mathbf{j}$, and let s and t be scalars. Verify the following statements.

21. $(\mathbf{a} + \mathbf{b}) + \mathbf{c} = \mathbf{a} + (\mathbf{b} + \mathbf{c})$ 22. $\mathbf{a} + 0 = \mathbf{a}$

23. $\mathbf{a} + (-\mathbf{a}) = 0$ 24. $\mathbf{a} + \mathbf{b} = \mathbf{b} + \mathbf{a}$

25. $s(\mathbf{a} + \mathbf{b}) = s\mathbf{a} + s\mathbf{b}$ 26. $(s + t)\mathbf{a} = s\mathbf{a} + t\mathbf{a}$

27. $s(t\mathbf{a}) = (st)\mathbf{a}$ 28. $1\mathbf{a} = \mathbf{a}$

The statements in Exercises 21–28 are the *vector-space axioms,* and any system of objects $\{\mathbf{a}, \mathbf{b}, \mathbf{c}, \ldots\}$ satisfying them is called a **vector space.** For the kinds of vectors we are considering, the axioms can be verified geometrically, as illustrated in the next exercise.

29. The solid-line portions of the given diagrams are congruent. Explain how the figures can be used to prove that

$$(a + b) + c = a + (b + c).$$

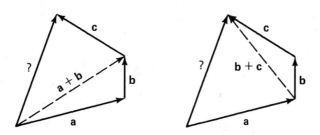

C 30. Prove algebraically that $\|t\mathbf{a}\| = |t|\|\mathbf{a}\|$.

In Exercises 31–34, find a linear combination of **a**, **b**, and **c** that equals **0**.

Example $\mathbf{a} = \mathbf{i} + 2\mathbf{j};\ \mathbf{b} = \mathbf{i} - 2\mathbf{j};\ \mathbf{c} = -2\mathbf{i} + \mathbf{j}.$

Solution   We wish to find scalars r, s, and t such that

$$r\mathbf{a} + s\mathbf{b} + t\mathbf{c} = 0.$$

Substituting for **a**, **b**, and **c**, we have

$$r(\mathbf{i} + 2\mathbf{j}) + s(\mathbf{i} - 2\mathbf{j}) + t(-2\mathbf{i} + \mathbf{j}) = 0\mathbf{i} + 0\mathbf{j},$$

or

$$(r + s - 2t)\mathbf{i} + (2r - 2s + t)\mathbf{j} = 0\mathbf{i} + 0\mathbf{j}.$$

Thus,

$$r + s - 2t = 0 \text{ and } 2r - 2s + t = 0.$$

Solve this system for r and s in terms of t.

$$r = \frac{3}{4}t,\ s = \frac{5}{4}t$$

In order to avoid fractions, set $t = 4$ and obtain

$$r = 3,\ s = 5,\ t = 4.$$

Therefore,

$$3\mathbf{a} + 5\mathbf{b} + 4\mathbf{c} = 0.$$

31. $\mathbf{a} = \mathbf{i} + \mathbf{j};\ \mathbf{b} = \mathbf{j};\ \mathbf{c} = \mathbf{i} - \mathbf{j}$

32. $\mathbf{a} = \mathbf{i} - \mathbf{j};\ \mathbf{b} = -\mathbf{i} + \mathbf{j};\ \mathbf{c} = \mathbf{i} + \mathbf{j}$

33. $\mathbf{a} = \mathbf{i} + \mathbf{j};\ \mathbf{b} = -2\mathbf{i} + \mathbf{j};\ \mathbf{c} = \mathbf{i} - 2\mathbf{j}$

34. $\mathbf{a} = \mathbf{i} + 2\mathbf{j};\ \mathbf{b} = 2\mathbf{i} - \mathbf{j};\ \mathbf{c} = 3\mathbf{i} - \mathbf{j}$

6-2   *Vectors and Navigation*

In certain applications it is convenient to denote the vector from point A to point B by $\overrightarrow{AB}$. For example, if an object moves from A to B, its **displacement** is given by the vector $\overrightarrow{AB}$. If it then moves from B to C, its resultant displacement from A to C is given by the sum of $\overrightarrow{AB}$ and $\overrightarrow{BC}$:

$$\overrightarrow{AC} = \overrightarrow{AB} + \overrightarrow{BC}$$

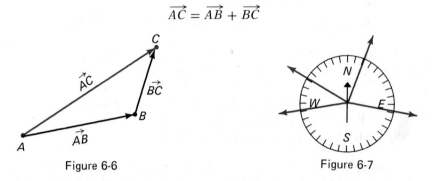

Figure 6-6   Figure 6-7

In navigation, directions and velocities are usually given by their bearings. The **bearing** of vector $\mathbf{v}$ is the angle θ, $0° \leq \theta < 360°$, measured clockwise, that $\mathbf{v}$ makes with a vector pointing due north. Figure 6-7 shows a compass face and vectors having bearings of 20°, 100°, 260°, and 300°. When describing the direction in which a craft is headed, we use the term **heading** instead of bearing.

Example 1   A ship leaves port A and sails 150 km with heading 115° to point B. It then changes its heading to 75° and sails 120 km to C. Find its distance and bearing from A.

Solution   The situation is illustrated in the figure.

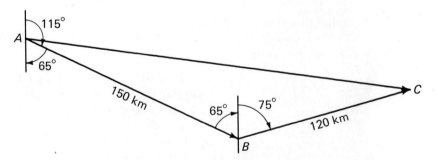

The angle B in triangle ABC is 140°, and the lengths of the adjacent sides are 150 km and 120 km. By the law of cosines,

$$\begin{aligned}(AC)^2 &= 150^2 + 120^2 - 2 \cdot 150 \cdot 120 \cos 140° \\ &= 22{,}500 + 14{,}400 - 36{,}000(-0.7660) \\ &= 36{,}900 + 27{,}576 = 64{,}476\end{aligned}$$

Therefore: $AC = 254$ km

Applying the law of sines to $\triangle ABC$, we have

$$\frac{\sin A}{120} = \frac{\sin 140°}{254}$$

$$\sin A = \frac{120(0.6428)}{254} = 0.3037$$

Therefore, $\angle A = 17.7°$ or $17°40'$.

Thus, the bearing of $\overrightarrow{AC} = 115° - 17.7° = 97.3°$ or $97°20'$.

A plane's **ground speed** (speed relative to the ground) may be different from its **air speed** (speed in still air) because of wind. Also, when a wind is blowing, the bearing of the plane's path relative to the ground, its **true course**, usually is different from its heading. The heading of the plane is the direction in which the plane is pointing, while the true course is the direction in which the engines and wind propel the plane.

Example 2   The air speed of a light plane is 200 km/h and its heading is 90°. A 40 km/h wind is blowing with a velocity vector having bearing 160°.
(a) Find the ground speed and the true course of the plane.
(b) What heading should the pilot use so that the true course will be 90°? What will the ground speed be then?

Solution   (a) In the figure, **u** and **v** are the velocities of the plane relative to the air and the ground, respectively, and **w** is the wind velocity.

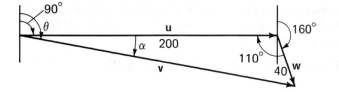

The ground speed is $\|\mathbf{v}\|$, and the true course is θ. By the law of cosines we have

$$\|\mathbf{v}\|^2 = 200^2 + 40^2 - 2 \cdot 200 \cdot 40 \cos 110°$$
$$= 40{,}000 + 1600 - 16{,}000(-0.3420)$$
$$= 41{,}600 + 5472 = 47{,}072$$
$$\|\mathbf{v}\| = 217$$

Therefore, the ground speed is 217 km/h.

Now applying the law of sines we have

$$\frac{\sin \alpha}{40} = \frac{\sin 110°}{217} \qquad \text{or} \qquad \sin \alpha = \frac{40(0.9397)}{217} = 0.1732$$

$$\alpha = 10°$$

Therefore, $\theta = 90° + 10° = 100°$, and the true course is 100°.

(Solution continued on next page.)

(b) In this case we denote the plane's air and ground velocities by $\mathbf{u'}$ and $\mathbf{v'}$, respectively.

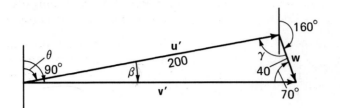

The required heading is θ. Apply the law of sines.

$$\frac{\sin \beta}{40} = \frac{\sin 70°}{200}$$

$$\sin \beta = \frac{40(0.9397)}{200}$$

$$= 0.1879$$

$$\beta = 10.8° \text{ or } 10°50'$$

Therefore, $\theta = 90° - 10.8° = 79.2°$, and the heading should be $79.2°$ or $79°10'$.

Noting that $\gamma = 180.0° - (10.8° + 70.0°) = 99.2°$, we apply the law of sines again.

$$\frac{\|\mathbf{v'}\|}{\sin 99.2°} = \frac{200}{\sin 70°}$$

$$\|\mathbf{v'}\| = \frac{200(0.9871)}{0.9397}$$

$$= 210$$

Therefore, the new ground speed would be 210 km/h.

Exercises 6-2

A 1. A plane flies due east for 500 km and then on a heading of $120°$ for 150 km. What are its distance and bearing from its starting point?

2. A ship leaves port and sails due west for 120 km, then due south for 40 km. What are the distance and bearing of the port from the ship?

3. A plane heads due east with an air speed of 300 km/h. A 45 km/h wind is blowing from 150°. Find the plane's true course and ground speed.

4. If the plane in Exercise 3 was heading due north while everything else remained the same, what would be its ground speed and true course?

5. A lake ferry leaves port A bound for port B, which is 80 km away and bears 340° from A. Because of shallow water the ferry first travels 20 km with heading 15°. What heading should it then use to proceed directly to B and how far must it travel?

6. Two planes take off in still air from the same airport at the same time. The speed of one plane is 420 km/h and its heading is 195°. The speed of the other plane is 500 km/h and its heading is 215°. How far apart are they at the end of one hour and what is the bearing of each from the other at that time?

7. The speed in still water of a power boat is 20 km/h. The boat heads directly west across an 80 m wide river that is flowing due south at 6 km/h. (a) How far downstream will the boat land? (b) What heading should the operator have used in order to land directly opposite the starting point?

8. A current is flowing so that if an outboard motor boat is pointed 21° upstream with a velocity of 12 km/h it will land directly opposite its starting point. What is the speed of the current?

B   9. A ship leaves Port Cod and sails due east for 200 km, then on a heading of 45° for 80 km, and then due north until the bearing of Port Cod from the ship is 225°. How far away is the ship from Port Cod?

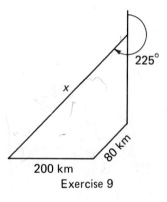

225°

x

80 km

200 km

Exercise 9

10. A search plane leaves its base and flies due north for 400 km. It then flies due east for 100 km and then on a heading of 225° for 200 km. At this point what are the plane's distance and bearing from its base?

C 11. Airport *SFO* bears 320° from airport *LAX* and is 540 km away. A pilot is planning a straight-line flight from *LAX* to *SFO* to leave at 1 P.M. The plane's air speed will be 650 km/h and there will be a 60 km/h wind blowing from 280°. What will the compass heading be and what is the plane's ETA (estimated time of arrival) to the nearest minute?

12. When the pilot of Exercise 11 plans the return trip to leave *SFO* at 6 P.M. conditions are the same as before. What now are the heading and ETA?

6-3   *Vectors and Force*

The basic unit of force is the **newton.** This is the force necessary to accelerate a 1 kg mass 1 m/s². The symbol for newton is N. Near the surface of the earth the force **G** exerted by gravity on a 1 kg mass is about 9.80 N. This force is directed toward the center of the earth and may be calculated by using the formula

$$\|\mathbf{G}\| = 9.8m$$

where m is the mass of the object in kilograms.

When working with forces we often use the following physical principle.

> When a body is at rest or moving with constant velocity, the sum of the forces acting on it is zero.

If the sum of the forces is not **0,** the object will accelerate and its velocity will vary.

Example 1   A loading ramp makes an angle of 24° with the horizontal. What is the frictional force that will keep a 150 kg crate from sliding down the ramp?

Solution   The figure shows the three forces acting on the crate. **G** is the gravitational force and has magnitude

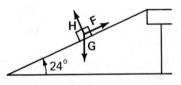

$$\|\mathbf{G}\| = 150 \times 9.80 = 1470 \, \text{N}.$$

The forces **F** and **H** are respectively parallel and perpendicular to the ramp. **F** is the frictional force we wish to find.

Since **F** prevents the crate from moving, the sum of **G**, **H**, and **F** is **0** as indicated in the given figure.

Then $\dfrac{\|\mathbf{F}\|}{\|\mathbf{G}\|} = \sin 24°$

$\|\mathbf{F}\| = \|\mathbf{G}\| \sin 24°$

$= 1470(0.4067) = 598.$

Therefore the frictional force is 598 N.

The **tension** in a rope or cable is the magnitude of the force it exerts.

Example 2 The figure shows a 250 kg piano suspended from a cable. The piano is being held away from the building by a rope attached to the cable at *P*. The tension in the rope is 900 N. Find the tension in the part of the cable above *P* and the angle that the cable makes with the building.

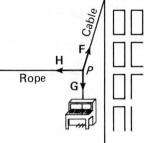

Solution The forces acting on *P* are shown at the right. **G** is the force of gravity acting on the piano. Its magnitude is

$$\|\mathbf{G}\| = 250 \times 9.80 = 2450 \text{ N}.$$

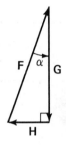

H is the force exerted by the rope. $\|\mathbf{H}\| = 900$ N. **F** is the tension we wish to find. Since *P* is at rest, **G**, **H**, and **F** form a triangle as shown. Since the triangle is a right triangle, we can use the Pythagorean theorem to find $\|\mathbf{F}\|$.

$$\|\mathbf{F}\|^2 = \|\mathbf{G}\|^2 + \|\mathbf{H}\|^2 = 2450^2 + 900^2$$
$$= 6{,}002{,}500 + 810{,}000 = 6{,}812{,}500$$
$$\|\mathbf{F}\| = 2610 \text{ N}$$

The angle that the cable makes with the building is given by

$$\alpha = \text{Tan}^{-1} \frac{\|\mathbf{H}\|}{\|\mathbf{G}\|} = \text{Tan}^{-1} \frac{900}{2450} = 20.2° \text{ or } 20°10'.$$

Let **v** be a nonzero vector having bearing θ. Introduce a coordinate system with the *x*- and *y*-axes having bearings 90° and 0°, respectively, as shown in Figure 6-8. The vector **v** can now be written in terms of its *x*- and *y*-components as

$$\mathbf{v} = [\|\mathbf{v}\| \sin \theta]\mathbf{i} + [\|\mathbf{v}\| \cos \theta]\mathbf{j}.$$

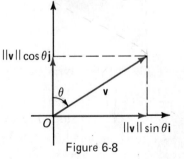

Figure 6-8

Now, suppose we are given **v** in the form

$$\mathbf{v} = a\mathbf{i} + b\mathbf{j}.$$

To express the bearing of **v** in terms of *a* and *b*, we compare the preceding two equations and obtain

$$\|\mathbf{v}\| \sin \theta = a, \text{ and } \|\mathbf{v}\| \cos \theta = b.$$

Since $\|\mathbf{v}\| = \sqrt{a^2 + b^2}$, we have

$$\sin \theta = \frac{a}{\sqrt{a^2 + b^2}} \text{ and } \cos \theta = \frac{b}{\sqrt{a^2 + b^2}}.$$

This method of finding the bearing of a vector will be illustrated in the following example.

Example 3 Three forces act on a particle P:

F_1: 12 N, bearing 50°
F_2: 10 N, bearing 155°
F_3: 20 N, bearing 250°

Find the magnitude to the nearest tenth and the bearing to the nearest tenth of a degree of the additional force F that will keep P stationary.

Solution   Since we must have $F + F_1 + F_2 + F_3 = 0$, we must have

$$F = -(F_1 + F_2 + F_3).$$

Introduce a coordinate system with P at the origin. Then:

$$F_1 = 12 \sin 50°i + 12 \cos 50°j = 9.1925i + 7.7135j$$
$$F_2 = 10 \sin 155°i + 10 \cos 155°j = 4.2262i - 9.0631j$$
$$F_3 = 20 \sin 250°i + 20 \cos 250°j = -18.7939i - 6.8404j$$

Therefore,

$$F = -[(9.1925i + 7.7135j) + (4.2262i - 9.0631j) +$$
$$(-18.7939i - 6.8404j)]$$
$$= 5.3752i + 8.1900j \approx 5.4i + 8.2j$$

The magnitude of F is

$$\|F\| = \sqrt{(5.4)^2 + 8.2^2} = \sqrt{96.4} = 9.8 \text{ N}.$$

The bearing, θ, of F is given by

$$\sin \theta = \frac{5.4}{\sqrt{96.4}} = 0.5500; \cos \theta = \frac{8.2}{\sqrt{96.4}} = 0.8352.$$

Therefore, $\theta = 33.4°$.

Exercises 6-3

A 1. A 30 kg object is on a plane inclined at 22° with the horizontal. Find the components of the gravitational force on the object parallel and perpendicular to the plane.

2. A 100 kg box slides with constant velocity down a ramp inclined at 20° with the horizontal. Find the frictional force acting on the box.

3. In Example 2, what force must the rope exert in order to have the cable make an angle of 25° with the building?

4. A 300 kg object is suspended by two cables each of which makes an angle of 45° with the horizontal. Find the tension in each cable.

5. A 500 kg motor is suspended by two cables which make angles of 40° and 60° with the horizontal. Find the tension in each cable.

6. Work Example 2 assuming that the rope, instead of being horizontal, makes an angle of 30° ~~with the horizontal.~~ *down from the*

In Exercises 7–9, the given forces act on a particle *P*. Find the magnitude and bearing of the additional force **F** that will keep *P* stationary.

7. F_1: 10 N, bearing 80°
   F_2: 15 N, bearing 200°
   F_3: 5 N, bearing 340°

8. F_1: 14 N, bearing 95°
   F_2: 20 N, bearing 200°
   F_3: 16 N, bearing 300°

B 9. F_1: 30 N, bearing 60°
   F_2: 50 N, bearing 90°
   F_3: 70 N, bearing 240°
   F_4: 80 N, bearing 330°

10. The figure represents a 50 kg sled on a 20° slope. What force making a 45° angle with the horizontal is necessary to keep the sled from sliding down? Assume that friction is negligible.

Exercise 10

C 11. A box is on a ramp making a 15° angle with the horizontal. A force of 480 N *up* the ramp moves the box with constant velocity and a force of 20 N *down* the ramp also moves it with constant velocity. Find the mass of the box.

12. Each side of a regular hexagon is one unit long. Five forces act at one vertex *A*, one directed toward each of the other vertices. The magnitude of each force equals the distance from *A* to its corresponding vertex. Describe the sum of the forces.

The Dot Product and Its Applications

6-4 *The Dot Product*

Let **a** and **b** be any two nonzero vectors. The angle between **a** and **b** is defined to be the angle θ, $0° \le \theta \le 180°$, that the vectors determine when their initial points are placed together.

We define the **dot product**, or **inner product**, **a · b** as

$$\mathbf{a} \cdot \mathbf{b} = \|\mathbf{a}\|\|\mathbf{b}\| \cos \theta,$$

where θ is the angle between **a** and **b**. If either **a** or **b** is the zero vector **0**, then $\mathbf{a} \cdot \mathbf{b} = 0$. A very useful formula expresses the dot product of two vectors in terms of their components.

Let $\mathbf{a} = a_1\mathbf{i} + a_2\mathbf{j}$ and $\mathbf{b} = b_1\mathbf{i} + b_2\mathbf{j}$. Then

$$\mathbf{a} \cdot \mathbf{b} = a_1b_1 + a_2b_2.$$

For example, if $\mathbf{a} = 2\mathbf{i} + 5\mathbf{j}$ and $\mathbf{b} = 3\mathbf{i} - 2\mathbf{j}$, then

$$\mathbf{a} \cdot \mathbf{b} = 2 \cdot 3 + 5(-2) = -4.$$

To prove the formula, let θ be the angle between $\mathbf{a}$ and $\mathbf{b}$ and apply the law of cosines to the triangle shown in Figure 6-9 having sides $\mathbf{a}$, $\mathbf{b}$, and $\mathbf{a} - \mathbf{b}$.

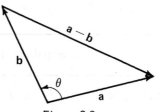

$$\|\mathbf{a} - \mathbf{b}\|^2 = \|\mathbf{a}\|^2 + \|\mathbf{b}\|^2 - 2\|\mathbf{a}\|\|\mathbf{b}\| \cos\theta$$

From this we have

$$\|\mathbf{a}\|\|\mathbf{b}\| \cos\theta = \frac{1}{2}[\|\mathbf{a}\|^2 + \|\mathbf{b}\|^2 - \|\mathbf{a} - \mathbf{b}\|^2].$$

Figure 6-9

Therefore, $\mathbf{a} \cdot \mathbf{b} = \dfrac{1}{2}[a_1^2 + a_2^2 + b_1^2 + b_2^2 - (a_1 - b_1)^2 - (a_2 - b_2)^2]$

$$= \frac{1}{2}[2a_1b_1 + 2a_2b_2] = a_1b_1 + a_2b_2$$

It is easy to see that

$$\mathbf{i} \cdot \mathbf{i} = \mathbf{j} \cdot \mathbf{j} = 1 \text{ and } \mathbf{i} \cdot \mathbf{j} = \mathbf{j} \cdot \mathbf{i} = 0$$

and that for any vector $\mathbf{a}$,

$$\mathbf{a} \cdot \mathbf{a} = \|\mathbf{a}\|^2.$$

Example 1   Prove the distributive law $\mathbf{a} \cdot (\mathbf{b} + \mathbf{c}) = \mathbf{a} \cdot \mathbf{b} + \mathbf{a} \cdot \mathbf{c}$.

Solution   Let $\mathbf{a} = a_1\mathbf{i} + a_2\mathbf{j}$, $\mathbf{b} = b_1\mathbf{i} + b_2\mathbf{j}$, and $\mathbf{c} = c_1\mathbf{i} + c_2\mathbf{j}$.

$$\begin{aligned}
\mathbf{a} \cdot (\mathbf{b} + \mathbf{c}) &= (a_1\mathbf{i} + a_2\mathbf{j}) \cdot [(b_1 + c_1)\mathbf{i} + (b_2 + c_2)\mathbf{j}] \\
&= a_1(b_1 + c_1) + a_2(b_2 + c_2) \\
&= (a_1b_1 + a_2b_2) + (a_1c_1 + a_2c_2) \\
&= \mathbf{a} \cdot \mathbf{b} + \mathbf{a} \cdot \mathbf{c}
\end{aligned}$$

The definition of dot product provides us with a formula for finding the angle θ between two nonzero vectors $\mathbf{a}$ and $\mathbf{b}$. We have

$$\cos\theta = \frac{\mathbf{a} \cdot \mathbf{b}}{\|\mathbf{a}\|\|\mathbf{b}\|}.$$

Therefore, $\theta = \text{Cos}^{-1}\dfrac{\mathbf{a} \cdot \mathbf{b}}{\|\mathbf{a}\|\|\mathbf{b}\|}$.

Example 2   Find the angle between $\mathbf{a} = 2\mathbf{i} - 3\mathbf{j}$ and $\mathbf{b} = 5\mathbf{i} + 7\mathbf{j}$ to the nearest tenth of a degree.

Solution
$$\cos\theta = \frac{2\cdot 5 + (-3)7}{\sqrt{13}\sqrt{74}} = \frac{-11}{\sqrt{13\cdot 74}}$$

$$\theta = \mathrm{Cos}^{-1}\frac{-11}{\sqrt{13\cdot 74}} = 110.8°$$

Two vectors are **orthogonal** if they are perpendicular, that is, if the angle between them is 90°. Two vectors are **parallel** if this angle is 0° or 180°. The zero vector, **0**, is considered to be both parallel and orthogonal to every other vector. Thus:

Two vectors are orthogonal if and only if their dot product is zero.

Example 3   Find a unit vector orthogonal to **c** = 3**i** + 5**j**.

Solution   First find *some* vector orthogonal to **c**. The vector **v** = *x***i** + *y***j** will be orthogonal to **c** if **c** · **v** = 0, that is, if

$$(3\mathbf{i} + 5\mathbf{j})\cdot(x\mathbf{i} + y\mathbf{j}) = 0, \text{ or } 3x + 5y = 0.$$

One solution of this equation is *x* = 5, *y* = −3. Thus **v** = 5**i** − 3**j** is orthogonal to **c** and the required unit vector in the direction of **v** is

$$\hat{\mathbf{v}} = \frac{\mathbf{v}}{\|\mathbf{v}\|} = \frac{5\mathbf{i} - 3\mathbf{j}}{\sqrt{34}}.$$

We also have the following.

Two vectors are parallel if and only if one is a scalar multiple of the other.

For example, **a** = −3**i** + 12**j** and **b** = **i** − 4**j** are parallel since **a** = − 3**b**.

Let θ be the angle between the vectors **v** and **a** and let **â** be the unit vector in the direction of **a**. The vector projection of **v** onto **a**, **v**$_a$, is the vector whose terminal point is found by dropping a perpendicular from the terminal point of **v** to the line on which **a** lies (Figure 6-10).

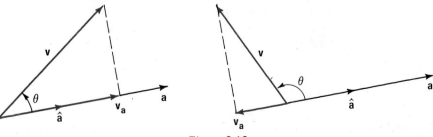

Figure 6-10

The norm of v_a is called the **scalar projection** of v onto a and is given by

$$\|v\| \cos \theta = \frac{\|a\|\|v\| \cos \theta}{\|a\|} = \frac{a \cdot v}{\|a\|}.$$

Thus, we can write v_a as

$$v_a = [\|v\| \cos \theta]\hat{a}.$$

Since $\hat{a} = \dfrac{a}{\|a\|}$, we have

$$v_a = \|v\| \cos \theta \frac{a}{\|a\|} = \frac{\|a\|\|v\| \cos \theta}{\|a\|^2} a.$$

Therefore, $v_a = \dfrac{a \cdot v}{\|a\|^2} a,$ or $v_a = \dfrac{a \cdot v}{a \cdot a} a.$

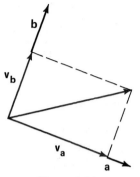

Figure 6-11

Let a and b be mutually orthogonal vectors and let v be an arbitrary vector (Figure 6-11). By the definition of vector addition we have

$$v = v_a + v_b,$$

or

$$v = \frac{a \cdot v}{a \cdot a} a + \frac{b \cdot v}{b \cdot b} b.$$

We call the vectors v_a and v_b the **components** of v parallel to a and b, respectively, and say that v has been **resolved** into these components.

Example 4   Resolve $v = -2i + 5j$ into components parallel to $a = 6i - 2j$ and $b = i + 3j$.

Solution   Since $a \cdot b = 6 \cdot 1 + (-2)3 = 0$, the vectors a and b are orthogonal, and we may apply the formula derived above.

$$a \cdot v = -12 - 10 = -22 \qquad b \cdot v = -2 + 15 = 13$$
$$a \cdot a = 36 + 4 = 40 \qquad b \cdot b = 1 + 9 = 10$$

Therefore, $v = -\dfrac{11}{20} a + \dfrac{13}{10} b.$

Exercises 6-4

Find the dot product of the given vectors.

A 1. $a = i + j;\ b = i - j$ 2. $a = 2i - j;\ b = 2i + j$

   3. $a = pi - qj;\ b = pi + qj$ 4. $a = pi + pj;\ b = qi - qj$

5. Which pairs of the given vectors are (a) parallel? (b) orthogonal?

$\mathbf{a} = 4\mathbf{i} + 6\mathbf{j}$ $\mathbf{b} = -4\mathbf{i} + 6\mathbf{j}$

$\mathbf{c} = 3\mathbf{i} + 2\mathbf{j}$ $\mathbf{d} = 2\mathbf{i} - 3\mathbf{j}$

6. Which pairs of the given vectors are (a) parallel? (b) orthogonal?

$\mathbf{a} = 2\mathbf{i} - 4\mathbf{j}$ $\mathbf{b} = \mathbf{i} + 2\mathbf{j}$

$\mathbf{c} = 8\mathbf{i} + 4\mathbf{j}$ $\mathbf{d} = 2\mathbf{i} + \mathbf{j}$

Use the distributive law to find $\mathbf{a} \cdot (\mathbf{b} + \mathbf{c})$ for the following vectors.

7. $\mathbf{a} = 2\mathbf{i} + 3\mathbf{j}$, $\mathbf{b} = 3\mathbf{i} + 8\mathbf{j}$, $\mathbf{c} = \mathbf{i} + 2\mathbf{j}$

8. $\mathbf{a} = 5\mathbf{i} + 2\mathbf{j}$, $\mathbf{b} = 2\mathbf{i} - 6\mathbf{j}$, $\mathbf{c} = 4\mathbf{i} + \mathbf{j}$

9. $\mathbf{a} = 3\mathbf{i} - 4\mathbf{j}$, $\mathbf{b} = 8\mathbf{i} - 4\mathbf{j}$, $\mathbf{c} = -3\mathbf{i} - 5\mathbf{j}$

10. $\mathbf{a} = -3\mathbf{i} + 2\mathbf{j}$, $\mathbf{b} = -7\mathbf{i} - 3\mathbf{j}$, $\mathbf{c} = 2\mathbf{i} + 5\mathbf{j}$

In Exercises 11–16, find the angle between the given vectors.

11. $\mathbf{i} + \sqrt{3}\mathbf{j}$; $-\mathbf{i} + \sqrt{3}\mathbf{j}$

12. $\sqrt{3}\mathbf{i} - \mathbf{j}$; $-\mathbf{i} + \sqrt{3}\mathbf{j}$

13. $\mathbf{i} + 2\mathbf{j}$; $4\mathbf{i} + 3\mathbf{j}$

14. $4\mathbf{i} + 3\mathbf{j}$; $4\mathbf{i} - 3\mathbf{j}$

15. $2\mathbf{i} + 5\mathbf{j}$; $3\mathbf{i} - 4\mathbf{j}$

16. $5\mathbf{i} + \mathbf{j}$; $\mathbf{i} - 2\mathbf{j}$

In Exercises 17–22, find unit vectors (a) parallel, and (b) perpendicular to the given vector.

17. $-\mathbf{i} + \mathbf{j}$

18. $2\mathbf{i} + \mathbf{j}$

19. $4\mathbf{i} - 3\mathbf{j}$

20. $5\mathbf{i} + 12\mathbf{j}$

21. $x\mathbf{i} + y\mathbf{j}$

22. $x\mathbf{i} - y\mathbf{j}$

In Exercises 23–26, find the scalar projection of $\mathbf{v}$ onto $\mathbf{a}$.

23. $\mathbf{v} = \mathbf{j}$; $\mathbf{a} = \mathbf{i} + \mathbf{j}$

24. $\mathbf{v} = \mathbf{i}$; $\mathbf{a} = -\mathbf{i} + \mathbf{j}$

25. $\mathbf{v} = \mathbf{i} + \mathbf{j}$; $\mathbf{a} = 4\mathbf{i} - 3\mathbf{j}$

26. $\mathbf{v} = -\mathbf{i} + \mathbf{j}$; $\mathbf{a} = 3\mathbf{i} + 4\mathbf{j}$

In Exercises 27–30, resolve $\mathbf{v}$ into components parallel to $\mathbf{a}$ and $\mathbf{b}$.

27. $\mathbf{v} = \mathbf{j}$; $\mathbf{a} = \mathbf{i} + \mathbf{j}$, $\mathbf{b} = \mathbf{i} - \mathbf{j}$

28. $\mathbf{v} = \mathbf{i} + 2\mathbf{j}$; $\mathbf{a} = \mathbf{i} + \mathbf{j}$, $\mathbf{b} = -\mathbf{i} + \mathbf{j}$

29. $\mathbf{v} = 2\mathbf{i} - 3\mathbf{j}$; $\mathbf{a} = 4\mathbf{i} + \mathbf{j}$, $\mathbf{b} = \mathbf{i} - 4\mathbf{j}$

30. $\mathbf{v} = \mathbf{i} + 5\mathbf{j}$; $\mathbf{a} = 3\mathbf{i} - 2\mathbf{j}$, $\mathbf{b} = 2\mathbf{i} + 3\mathbf{j}$

Give an algebraic proof for each of the following.

B 31. $\mathbf{a} \cdot \mathbf{b} = \mathbf{b} \cdot \mathbf{a}$

32. $r(\mathbf{a} \cdot \mathbf{b}) = (r\mathbf{a}) \cdot \mathbf{b}$

33. $(r\mathbf{a}) \cdot (s\mathbf{b}) = (rs)(\mathbf{a} \cdot \mathbf{b})$

34. $(\mathbf{a} + \mathbf{b}) \cdot (\mathbf{a} + \mathbf{b}) = \mathbf{a} \cdot \mathbf{a} + 2\mathbf{a} \cdot \mathbf{b} + \mathbf{b} \cdot \mathbf{b}$

35. $(\mathbf{a} + \mathbf{b}) \cdot (\mathbf{a} - \mathbf{b}) = \mathbf{a} \cdot \mathbf{a} - \mathbf{b} \cdot \mathbf{b}$

36. $\|\mathbf{a} + \mathbf{b}\|^2 = (\mathbf{a} + \mathbf{b}) \cdot (\mathbf{a} + \mathbf{b})$

37. $\|\mathbf{a} + \mathbf{b}\|^2 = \|\mathbf{a}\|^2 + 2\mathbf{a} \cdot \mathbf{b} + \|\mathbf{b}\|^2$

38. $(\mathbf{a} \cdot \mathbf{b})^2 \leq \|\mathbf{a}\|^2 \|\mathbf{b}\|^2$

C 39. Show that $\mathbf{a} + \mathbf{b}$ and $\mathbf{a} - \mathbf{b}$ are orthogonal if and only if $\|\mathbf{a}\| = \|\mathbf{b}\|$.

40. Use the results of Exercises 37 and 38 to prove that $\|\mathbf{a} + \mathbf{b}\| \leq \|\mathbf{a}\| + \|\mathbf{b}\|$.

6-5 *Vectors, Work, and Energy*

When a force acts to move an object, it does *work* and expends *energy*. The simplest case occurs when the force $\mathbf{F}$ moves the object from A to B along a straight-line path. $\mathbf{F}$ has the same direction as the displacement vector $\mathbf{d} = \overrightarrow{AB}$. In this case the **work** done by $\mathbf{F}$ is

$$W = \|\mathbf{F}\|\|\mathbf{d}\|.$$

This is also the **energy** expended by $\mathbf{F}$.

The basic unit of work and energy is the **joule** (J). A joule is the work done by a force of one newton in moving an object one meter. Another unit is the **kilowatt-hour** (kW $\cdot$ h). One kilowatt-hour equals 3.6×10^6 J.

Example 1 How much energy in kilowatt-hours is needed to lift a 2000 kg elevator 60 m?

Solution The motor lifting the elevator must overcome the force of gravity and therefore must exert an upward force, $\mathbf{F}$, of magnitude

$$2000 \times 9.8 = 1.96 \times 10^4 \text{ N}.$$

The displacement $\mathbf{d}$ is also upward and has a norm of 60 m. Thus, the work done is

$$W = \|\mathbf{F}\|\|\mathbf{d}\| = (1.96 \times 10^4) \times 60 = 1.176 \times 10^6 \text{ J}.$$

Therefore, the energy needed is

$$\frac{1.176 \times 10^6}{3.6 \times 10^6} = 0.3267 \text{ kW} \cdot \text{h}.$$

Let us now consider a constant force **F** which moves an object from A to B but does *not* have the same direction as $\mathbf{d} = \overrightarrow{AB}$. The situation is shown in Figure 6-12. In this case only the projection of **F** onto **d**, $\mathbf{F_d}$, is effective in doing work. Thus, the work done by **F** is

Figure 6-12

$$W = \|\mathbf{F}\| \cos \theta \, \|\mathbf{d}\|.$$

This formula can be written as a dot product.

$$W = \mathbf{F} \cdot \mathbf{d}$$

Notice that W may be 0 if $\theta = 90°$ or may be negative if $90° < \theta \le 180°$.

Example 2 Find the work done by the force $\mathbf{F} = 6\mathbf{i} - 7\mathbf{j}$ in moving an object from $A(-1, 0)$ to $B(2, 4)$. Assume that the force is given in newtons and the distance in meters.

Solution The displacement of the object is $\mathbf{d} = \overrightarrow{AB} = 3\mathbf{i} + 4\mathbf{j}.$
Thus the work done by **F** is

$$W = \mathbf{F} \cdot \mathbf{d} = (6\mathbf{i} - 7\mathbf{j}) \cdot (3\mathbf{i} + 4\mathbf{j})$$
$$= 18 + (-28) = -10 \, \text{J}.$$

Example 3 How much energy does the engine of a 1200 kg car expend in propelling the car up a 15% grade for a horizontal distance of 5 km? (Ignore friction and consider only the force of gravity.)

Solution Introduce a coordinate system as shown. Then the force of gravity acting on the car is

$$\mathbf{G} = -1200 \times 9.8\mathbf{j} = -11{,}760\mathbf{j}.$$

Since the top of the slope is 15% of 5000 m, or 750 m, higher than the bottom, the displacement vector is

$$\mathbf{d} = 5000\mathbf{i} + 750\mathbf{j}.$$

Thus, the work done by gravity on the car is

$$W = \mathbf{G} \cdot \mathbf{d} = (-11{,}760\mathbf{j}) \cdot (5000\mathbf{i} + 750\mathbf{j})$$
$$= -11{,}760 \times 750 = -8.82 \times 10^6 \, \text{J}.$$

Therefore, in overcoming gravity, the car's engine must expend

$$8.82 \times 10^6 \, \text{J, or } 2.45 \, \text{kW} \cdot \text{h of energy.}$$

Exercises 6-5

In Exercises 1–4, forces are measured in newtons and distances in meters. Give answers in joules.

A 1. Find the work done by the force $\mathbf{F} = 2\mathbf{i} + 3\mathbf{j}$ in moving a particle from $A(1, 0)$ to $B(0, 4)$.

2. Find the work done by the force $\mathbf{F} = 5\mathbf{j}$ in moving an object from $A(-1, 4)$ to $B(5, 2)$.

3. How much work is done by force $\mathbf{F} = 10\mathbf{i} - 5\mathbf{j}$ in moving an object (a) from $A(-2, 0)$ to $B(3, 7)$ to $C(5, 0)$? (b) from $A(-2, 0)$ to $C(5, 0)$?

4. What is the combined work done by forces $\mathbf{F} = 6\mathbf{i} + 5\mathbf{j}$ and $\mathbf{G} = 2\mathbf{i} - 7\mathbf{j}$ in moving an object from $A(-4, 1)$ to $B(1, -4)$? What is the work done by $\mathbf{F} + \mathbf{G}$ in moving an object from A to B?

5. The mass of a loaded helicopter is 1250 kg. How much energy does its engine expend in ascending vertically for 50 m?

6. How much energy does a crane expend by lifting 25 buckets of concrete from street level to a point 20 m above street level? The mass of a bucket of concrete is 1800 kg.

B 7. How much energy does a 90 kg man expend in climbing a 10 m ladder that makes an angle of 60° with the horizontal?

8. A rock is dragged for 60 m along a horizontal sidewalk by a force of 220 N exerted on a rope that is tied to the rock. If the rope makes an angle of 32° with the sidewalk, how much work is done?

9. A rope tow transports skiers up a 25° slope for a horizontal distance of 600 m. How much energy does the tow use in one day if it transports 235 skiers with an average mass of 62 kg?

10. A conveyor belt carries coal for a horizontal distance of five kilometers up a 20% slope. How many kilowatt-hours of energy are used in transporting 100 metric tons of coal? (1 metric ton = 1000 kilograms.)

11. Work Example 3 assuming that there is a frictional force of magnitude 500 N. Remember that frictional forces always act in direct opposition to motion.

In Exercises 12 and 13 prove the stated facts.

C 12. The work done by a constant force $\mathbf{F}$ in moving an object along $\overrightarrow{AB}$ and then along $\overrightarrow{BC}$ is the same as the work $\mathbf{F}$ does in moving the object along $\overrightarrow{AC}$.

13. Let $\mathbf{F}$ and $\mathbf{G}$ be constant forces. The work done by $\mathbf{F} + \mathbf{G}$ in moving an object along $\overrightarrow{AB}$ equals the combined work done by $\mathbf{F}$ and by $\mathbf{G}$ in moving the object along $\overrightarrow{AB}$.

Challenge

The *coefficient of friction* of two surfaces is

$$k = \frac{\|\mathbf{M}\|}{\|\mathbf{H}\|},$$

where **H** is a force holding the surfaces together and **M** is the force just sufficient to make one surface move relative to the other. (In Example 1, page 158, $\mathbf{M} = -\mathbf{F}$.) The *angle of repose* of the two surfaces is $\tan^{-1} k$. Give a physical interpretation of this angle. Describe a way to find coefficients of friction without measuring forces.

CHAPTER SUMMARY

1. Quantities having both magnitude and direction are called **vectors.** Quantities having magnitude, but without direction, are called **scalars.** Vectors are represented by arrows with parallel arrows of the same length representing equal vectors. Vectors may be added by using the parallelogram rule, or may be multiplied by scalars. If a vector **a** is expressed in terms of its components, $\mathbf{a} = a_1\mathbf{i} + a_2\mathbf{j}$, then the **norm** or magnitude of **a** can be defined as

$$\|\mathbf{a}\| = \sqrt{a_1^2 + a_2^2}.$$

2. Vector directions used in navigation are called **bearings** and are measured in a clockwise direction from a vector pointing due north.

3. Vectors may be used to solve force problems by using the fact that the vector sum of all the forces on an object will be zero if the object is at rest or if it is moving with a constant velocity. It is usually helpful to write a vector in terms of its x and y components. For example if

$$\mathbf{v} = a\mathbf{i} + b\mathbf{j}, \text{ then } a = \|\mathbf{v}\| \sin\theta \text{ and } b = \|\mathbf{v}\| \cos\theta.$$

4. The **dot product** of two vectors **a** and **b** is defined as

$$\mathbf{a} \cdot \mathbf{b} = \|\mathbf{a}\|\|\mathbf{b}\| \cos\theta,$$

where θ is the angle between the two vectors. If **a** and **b** are given in terms of their components, $\mathbf{a} = a_1\mathbf{i} + a_2\mathbf{j}$, $\mathbf{b} = b_1\mathbf{i} + b_2\mathbf{j}$, then $\mathbf{a} \cdot \mathbf{b} = a_1b_1 + a_2b_2$. The angle θ between two vectors may be found by using the formula

$$\theta = \text{Cos}^{-1} \frac{\mathbf{a} \cdot \mathbf{b}}{\|\mathbf{a}\|\|\mathbf{b}\|}.$$

The dot product of two perpendicular vectors is zero.

5. When a force **F** moves an object a distance **d**, the work done or energy expended is given by the dot product $W = \mathbf{F} \cdot \mathbf{d}$.

CHAPTER TEST

6-1 1. Let $\mathbf{u} = 2\mathbf{i} - 5\mathbf{j}$ and $\mathbf{v} = \mathbf{i} + 2\mathbf{j}$.

  (a) Draw an arrow with initial point at the origin representing $\mathbf{u} + 2\mathbf{v}$.

  (b) Find $\|\mathbf{u} - \mathbf{v}\|$.

  (c) Express $\mathbf{w} = 3\mathbf{i} + 15\mathbf{j}$ as a linear combination of $\mathbf{u}$ and $\mathbf{v}$.

6-2 2. A ship sails due west for 150 km, then on a heading of 195° for 70 km. How far is the ship from its initial position and what heading should it take to return to its starting point?

6-3 3. Suppose the following forces are acting on a particle P:

  $\mathbf{F_1}$: 30 N, bearing 20°
  $\mathbf{F_2}$: 10 N, bearing 160°
  $\mathbf{F_3}$: 20 N, bearing 260°

  Find the magnitude and bearing of the additional force that will keep P stationary.

For Exercises 4–8, let $\mathbf{a} = \mathbf{i} + 3\mathbf{j}$, $\mathbf{b} = 2\mathbf{i} - 5\mathbf{j}$, and $\mathbf{c} = n\mathbf{i} + 6\mathbf{j}$.

6-4 4. Find $\mathbf{a} \cdot \mathbf{b}$.

  5. Find the angle between $\mathbf{a}$ and $\mathbf{b}$.

  6. For what value of n is $\mathbf{c}$ perpendicular to $\mathbf{b}$?

  7. Find a unit vector perpendicular to $\mathbf{a}$.

  8. Using the value for n obtained in Exercise 6, resolve $\mathbf{a}$ into components that are parallel to $\mathbf{b}$ and $\mathbf{c}$.

6-5 9. Find the work done in moving a 50 kg box 10 m up a ramp with a 30° incline. (Ignore friction and assume that the force is applied parallel to the ramp.)

  10. What force is needed to hold the box in Exercise 9 stationary on the ramp if the force is applied parallel to the ramp?

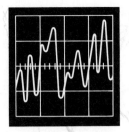

FINDING THE BEST SKI SLOPE

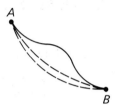

A famous question in the history of mathematics deals with moving from one point to another in the shortest time. Simply stated, the question asks what mathematical curve passing through two given points A and B, not on the same vertical line nor in the same horizontal plane, will enable an object sliding without friction to move from point A to point B in the shortest time.

   This question is called the brachistochrone or "shortest time" problem. It was first posed as a challenge to the scientific world in 1696 by Johann Bernoulli, a member of the Swiss family of distinguished scientists and mathematicians. Bernoulli, of course, knew the answer, but after several months no one else had solved the problem. When it came to the attention of Sir Isaac Newton, one of the greatest mathematicians of all time, he solved the problem, along with another one Bernoulli had posed, in one evening and anonymously communicated his solutions to the Royal Society of England. When Bernoulli saw the solutions, he wrote, "I know the lion by his claw," immediately recognizing the work of Newton.

   The answer to the question is, surprisingly, not a straight line, but an inverted **cycloid**, the path traced by a fixed point P on a circle as it rolls along the underside of a line. The curve is shown in the diagram below. If the cycloid is placed so that its lowest point is at the origin, then it may be represented by the following equations.

$$x = r(\theta + \sin \theta)$$
$$y = r(1 - \cos \theta)$$

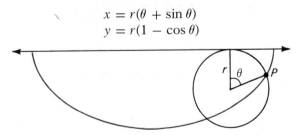

The cycloid is also the answer to another question: What shape has the property that an object sliding without friction will reach its lowest point in the same amount of time, no matter where on the curve it starts? This is called the tautochrone, or "equal time," problem. The fact that a cycloid also possesses this property enabled Christian Huygens, a Dutch scientist known for his important work in optics, to design the first accurate pendulum clock in 1673.

Two interesting curves related to the cycloid are the epicycloid and the hypocycloid. An **epicycloid** is the path traced by a fixed point P on a circle as it rolls around the outside of the circumference of a second circle. If the circles have the same radius, the curve traced is also referred to as a **cardioid.**

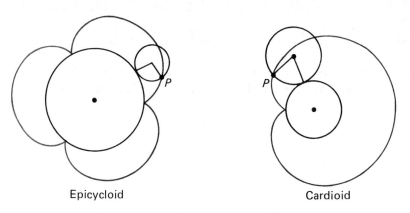

Epicycloid   Cardioid

A **hypocycloid** is the path traced by a fixed point on a circle as it rolls around the inside of the circumference of a second circle. If the radius, r, of the small circle is $\frac{1}{4}$ the radius, R, of the large circle, the curve traced is called an **astroid** and can be defined by the equations

$$x = R \cos^3\theta \quad ; \quad y = R \sin^3\theta$$

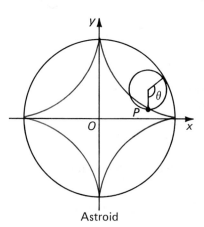

Astroid

Both the cardioid and the astroid are tautochrones for objects moving under the influence of physical forces different from gravity.

FINDING NORMS OF VECTORS AND NORMS OF SUMS

The following program first finds the norms of two separate vectors. The program then adds the two vectors and finds the norm of their sum.

Exercises

1. Copy the following program and RUN it for the values given in Example 6 on page 151.

```
10   PRINT "TO FIND THE NORMS OF TWO VECTORS"
20   PRINT "        AND THEIR SUM:"
30   PRINT "    A1I+A2J, B1I+B2J"
40   PRINT "INPUT A1, A2";
50   INPUT A1,A2
60   PRINT "INPUT B1, B2";
70   INPUT B1,B2
80   LET S1=A1+B1
90   LET S2=A2+B2
100  PRINT "NORM A=";SQR(A1*A1+A2*A2)
110  PRINT "NORM B=";SQR(B1*B1+B2*B2)
120  PRINT "NORM SUM=";SQR(S1*S1+S2*S2)
130  END
```

2. Change line 20 and insert lines to produce a program that will also find the norm of the difference of the two vectors.

3. RUN the revised program to check your work in Exercises 1–6 on page 151.

4. Use the program on pages 120–121 for solving a triangle when two sides and the included angle are given to solve Example 1 on page 154.

5. Use the program on pages 120–121 to illustrate the solution of Example 2 part (a) on page 155.

6. Alter the given program so that it will find the dot product of two vectors.

Examples of spirals, such as that seen in the shell of the chambered nautilus, are often found in nature. Spirals can be graphed by using polar coordinates.

COMPLEX NUMBERS

OBJECTIVES

1. To define polar coordinates and complex numbers.

2. To express complex numbers in polar form.

3. To use De Moivre's theorem to find products, quotients, and roots of complex numbers.

Polar Coordinates

7-1 *The Polar Coordinate System*

If we know how far away and in what direction a point is from a given point, we can locate that point. This is the principle of the polar coordinate system.

The reference system for polar coordinates in the plane consists of a point O, called the **pole,** or origin, and a ray called the **polar axis,** having O as its endpoint. We can now describe the position of a point P other than O by giving its **polar coordinates,** (r, θ), where r is the distance from O to P and θ is the measure of an angle from the polar axis to the segment $\overline{OP}$ (Figure 7-1). The pole, O, has polar coordinates $(0, \theta)$, where θ is arbitrary.

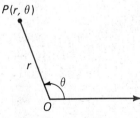

Figure 7-1

Each point has an unlimited number of pairs of polar coordinates. For example, $(4, 210°)$, $(4, 570°)$, and $(4, -150°)$ are all polar coordinates of the same point P since the angles are coterminal. Moreover, it will be convenient to let a negative value of r denote the *negative* of the distance OP. In this case P is on the extension of the terminal side of θ through O, and is $|r|$ units from O (Figure 7-2). Therefore, $(-4, 30°)$ and $(-4, -330°)$ are additional pairs of polar coordinates of P.

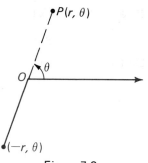

Figure 7-2

Example 1   Graph the points whose polar coordinates are

(a) $(3, 120°)$ (b) $(-4, 60°)$ and   (c) $\left(4, -\dfrac{\pi}{4}\right)$.

Solution·   First find the ray from the pole making angle θ with the polar axis. Then measure r units either along this ray or, in (b), along the extension of the ray through the pole. Note that in (c) the absence of the ° symbol indicates *radian* measure.

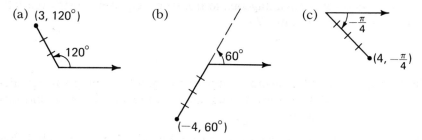

(a) $(3, 120°)$ (b)   (c)

When polar coordinates are used with rectangular coordinates, the polar axis is taken to coincide with the non-negative x-axis as shown in Figure 7-3. The equations in the table below can be derived from Figure 7-3. They enable us to change from one coordinate system to the other.

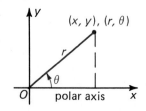

Figure 7-3

Coordinate Changes	
From polar to rectangular	From rectangular to polar
$x = r \cos \theta$ $y = r \sin \theta$	$r = \pm \sqrt{x^2 + y^2}$ $\cos \theta = \dfrac{x}{r}, \sin \theta = \dfrac{y}{r}$

Example 2 Convert (a) $(3, -3)$ to polar coordinates and (b) $\left(-2, \frac{2\pi}{3}\right)$ to rectangular coordinates.

Solution (a) Since $r = \pm\sqrt{3^2 + (-3)^2}$, we may take $r = 3\sqrt{2}$. Then, since

$$\cos\theta = \frac{3}{3\sqrt{2}} = \frac{\sqrt{2}}{2} \text{ and } \sin\theta = -\frac{3}{3\sqrt{2}} = -\frac{\sqrt{2}}{2}, \text{ we may take}$$

$\theta = -\frac{\pi}{4}$ to obtain $\left(3\sqrt{2}, -\frac{\pi}{4}\right)$ as a pair of polar coordinates.

(b) Since $x = -2\cos\frac{2\pi}{3} = -2\left(-\frac{1}{2}\right) = 1$ and

$$y = -2\sin\frac{2\pi}{3} = -2\left(\frac{\sqrt{3}}{2}\right) = -\sqrt{3}, \text{ we obtain } (1, -\sqrt{3}).$$

Some curves have very simple polar-coordinate equations. Consider, for example, the circle C of radius a with center at the origin. A point is on C if and only if it has polar coordinates of the form (r, θ), where $r = a$. Thus, a polar equation of C is $r = a$. Similar reasoning shows that the line through the origin making angle α with the polar axis has polar equation $\theta = \alpha$.

In some cases it is advantageous to transform equations using the relations in the table on page 176.

Example 3 (a) Plot several points of the graph of the polar equation $r = 2\cos\theta$.
(b) Transform the equation to rectangular form and identify the graph.

Solution (a) When constructing a table of values, it usually is best to assign values to θ and compute the corresponding values of r.

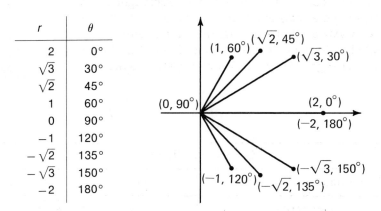

r	θ
2	0°
$\sqrt{3}$	30°
$\sqrt{2}$	45°
1	60°
0	90°
-1	120°
$-\sqrt{2}$	135°
$-\sqrt{3}$	150°
-2	180°

Continuation of the table in multiples of 30° and 45° produces no new points.

(Solution continued on next page.)

(b) Transforming $r = 2 \cos \theta$ to rectangular coordinates is somewhat easier if we multiply both sides by r.

$$r^2 = 2r \cos \theta$$

(Since the origin is a point of the graph of $r = 2 \cos \theta$, we add no new points in doing this.) Using the equations
$r = \pm \sqrt{x^2 + y^2}$ and $x = r \cos \theta$, we have

$$x^2 + y^2 = 2x.$$

This can be put into the form $(x - 1)^2 + y^2 = 1$ to show that the graph is the circle of radius 1 with center at $(1, 0)$.

Exercises 7-1

Plot the point whose polar coordinates are given and find its rectangular coordinates.

A 1. $(2, 30°)$ 2. $(3, 150°)$ 3. $(-2, 30°)$ 4. $(-3, 150°)$

5. $\left(2, -\dfrac{\pi}{6}\right)$ 6. $\left(3, -\dfrac{2\pi}{3}\right)$ 7. $(-2, -30°)$ 8. $(-3, -150°)$

Graph each of the given pairs of rectangular coordinates in the plane and find a pair of polar coordinates of the point.

9. $(\sqrt{3}, 1)$ 10. $(-1, -\sqrt{3})$ 11. $(-2, 2\sqrt{3})$ 12. $(-\sqrt{2}, -\sqrt{2})$

13. $(-\sqrt{2}, \sqrt{2})$ 14. $(\sqrt{3}, -\sqrt{3})$ 15. $(\sqrt{6}, \sqrt{2})$ 16. $(\sqrt{2}, -\sqrt{6})$

Find a polar equation for each of the following.

17. $y = -2$ 18. $x = 4$ 19. $x + 2 = 0$

20. $3 - y = 0$ 21. $y = x$ 22. $x + y = 0$

23. $x^2 + y^2 = 4$ 24. $x^2 + y^2 = 3$ 25. $x + \sqrt{3}y = 0$

26. $y = x\sqrt{3}$ 27. $x + y = 4$ 28. $x - y = 2$

Graph the following polar equations.

29. $r = 1$ 30. $\theta = 90°$ 31. $r \cos \theta = 2$

32. $r \sin \theta = 3$ 33. $\theta = \dfrac{3\pi}{4}$ 34. $\theta + \dfrac{\pi}{6} = 0$

(a) Plot several points of the graph of the given polar equation. Then (b) transform the equation to rectangular form and identify the graph.

B 35. $r = 2 \sin \theta$ 36. $r = -4 \cos \theta$ 37. $r = \cos \theta + \sin \theta$ 38. $r = \tan \theta \sec \theta$

39. Prove the following **distance formula**.

    The distance between the points $P_1(r_1, \theta_1)$ and $P_2(r_2, \theta_2)$ is
    $$P_1P_2 = \sqrt{r_1^2 + r_2^2 - 2r_1r_2 \cos(\theta_2 - \theta_1)}.$$

40. Use Exercise 39 to show that
    (a) if $\theta_1 = \theta_2$, then $P_1P_2 = |r_1 - r_2|$
    (b) if $r_1 = r_2 = r_0$, then $P_1P_2 = 2\left|r_0 \sin\dfrac{1}{2}(\theta_2 - \theta_1)\right|.$

41. Find a polar equation of the circle of radius a with center at the point (r_0, θ_0).

C 42. Let l be a line not passing through the origin O. Let (p, ω) be the foot of the perpendicular dropped from O to l. Show that a polar equation of l is
    $$r \cos(\omega - \theta) = p.$$

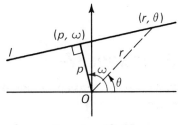

Exercises 42, 43

43. (a) Let l, p, and ω be as defined in Exercise 42. Show that a rectangular-coordinate equation of l is
    $$(\cos \omega)x + (\sin \omega)y = p.$$

    (b) Show that when the equation $Ax + By = C$ is put into the form given in (a), the result is
    $$\frac{A}{\pm\sqrt{A^2 + B^2}}x + \frac{B}{\pm\sqrt{A^2 + B^2}}y = \frac{C}{\pm\sqrt{A^2 + B^2}},$$

    where the sign of the radical is chosen so that the right-hand member is positive. $\left(\text{In other words, } p = \dfrac{C}{\pm\sqrt{A^2 + B^2}} \text{ is the distance from the origin to the line } Ax + By = C.\right)$

7-2 Graphs of Polar Equations

We often graph polar equations by observing how the trigonometric functions vary.

Example 1 Graph $r = 2 \cos 3\theta$.

Solution   We set up a table that shows how r varies as θ increases. Notice that we have chosen intervals of θ so that $\cos 3\theta$ is restricted to no more than a quarter of a period.

(Solution continued on next page.)

θ	3θ	$\cos 3\theta$	$r = 2\cos 3\theta$
$0° \rightarrow 30°$	$0° \rightarrow 90°$	$1 \rightarrow 0$	$2 \rightarrow 0$
$30° \rightarrow 60°$	$90° \rightarrow 180°$	$0 \rightarrow -1$	$0 \rightarrow -2$
$60° \rightarrow 90°$	$180° \rightarrow 270°$	$-1 \rightarrow 0$	$-2 \rightarrow 0$
$90° \rightarrow 120°$	$270° \rightarrow 360°$	$0 \rightarrow 1$	$0 \rightarrow 2$
$120° \rightarrow 150°$	$360° \rightarrow 450°$	$1 \rightarrow 0$	$2 \rightarrow 0$
$150° \rightarrow 180°$	$450° \rightarrow 540°$	$0 \rightarrow -1$	$0 \rightarrow -2$

In the first entry of the table, we see that as θ increases from 0° to 30°, r decreases from 2 to 0. Then as θ continues to increase from 30° to 60°, r becomes negative and decreases from 0 to -2. The corresponding portions of the graph are shown at the left below. The complete graph, called a **three-leaved rose,** appears at the right below. (As θ increases from 180°, the curve is repeated.)

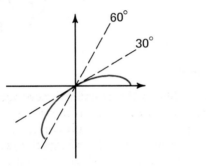
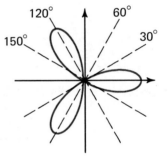

Several tests for symmetry of polar curves may be deduced from Figure 7-4. For example, a curve is symmetric with respect to the polar axis if replacing θ by $-\theta$ leaves its equation unchanged. This was the case in Example 1.

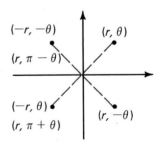

Figure 7-4

Example 2   Graph $r^2 = 4\sin 2\theta$.

Solution   Replacing r by $-r$ leaves the equation unchanged. The graph, therefore, is symmetric with respect to the origin. We first draw the part of the curve above the polar axis. In doing this, it is sufficient to graph $r = \sqrt{4\sin 2\theta} = 2\sqrt{\sin 2\theta}$.

θ	2θ	$\sin 2\theta$	$4\sin 2\theta$	$r = 2\sqrt{\sin 2\theta}$
$0° \rightarrow 45°$	$0° \rightarrow 90°$	$0 \rightarrow 1$	$0 \rightarrow 4$	$0 \rightarrow 2$
$45° \rightarrow 90°$	$90° \rightarrow 180°$	$1 \rightarrow 0$	$4 \rightarrow 0$	$2 \rightarrow 0$
$90° \rightarrow 180°$	$180° \rightarrow 360°$	negative	negative	—

From the table, we have the portion of the curve shown at the left below. The complete graph, called a **lemniscate,** is obtained by using symmetry with respect to the origin and is shown at the right.

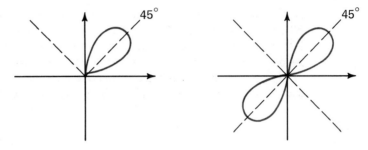

Example 3   Graph $r = 1 - 2\sin\theta$.

Solution   Replacing θ by $\pi - \theta$ leaves the equation unchanged. The graph, therefore, is symmetric with respect to the vertical line $\theta = 90°$. We shall first draw the portion of the graph corresponding to $-90° \le \theta \le 90°$ and then obtain the rest of the graph by using symmetry. Note that r is sometimes positive and sometimes negative. To find when $r = 0$, we solve $1 - 2\sin\theta = 0$, obtaining $\theta = 30°$.

θ	$\sin\theta$	$-2\sin\theta$	$r = 1 - 2\sin\theta$
$-90° \rightarrow 0°$	$-1 \rightarrow 0$	$2 \rightarrow 0$	$3 \rightarrow 1$
$0° \rightarrow 30°$	$0 \rightarrow \dfrac{1}{2}$	$0 \rightarrow -1$	$1 \rightarrow 0$
$30° \rightarrow 90°$	$\dfrac{1}{2} \rightarrow 1$	$-1 \rightarrow -2$	$0 \rightarrow -1$

From the table, we obtain the curve at the left below. The complete graph, called a **limaçon,** is shown at the right.

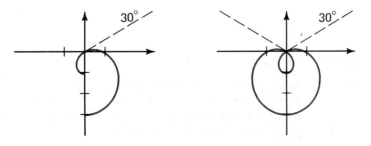

Sometimes, as in the next example, the variable θ appears independently of a trigonometric function. In such a case, θ denotes *radian* measure.

Example 4   Draw the spiral $r = \dfrac{1}{\theta}$, $\theta > 0$.

Solution   We see that the larger θ is, the smaller r is. A table of values is helpful.

$\theta =$	$\dfrac{\pi}{6}$	$\dfrac{\pi}{3}$	$\dfrac{\pi}{2}$	$\dfrac{3\pi}{4}$	π	$\dfrac{3\pi}{2}$	2π	$\dfrac{5\pi}{2}$	3π	4π
$r =$	$\dfrac{6}{\pi}$	$\dfrac{3}{\pi}$	$\dfrac{2}{\pi}$	$\dfrac{4}{3\pi}$	$\dfrac{1}{\pi}$	$\dfrac{2}{3\pi}$	$\dfrac{1}{2\pi}$	$\dfrac{2}{5\pi}$	$\dfrac{1}{3\pi}$	$\dfrac{1}{4\pi}$
$r =$	1.91	0.95	0.64	0.42	0.32	0.21	0.16	0.13	0.11	0.08

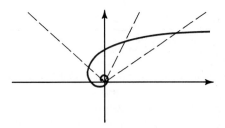

The spiral winds around the origin infinitely many times.

Exercises 7-2

Graph the following polar equations.

A   1. $r = 1 - \sin \theta$ (**cardioid**–a heart-shaped curve)

2. $r = 1 - \cos \theta$ (cardioid)

3. $r = 1 + \cos \theta$ (cardioid)

4. $r = 1 + \sin \theta$ (cardioid)

5. $r = 2 \sin 3\theta$ (three-leaved rose)

6. $r = 2 \cos 2\theta$ (four-leaved rose)

7. $r = 6 \sin 2\theta$ (four-leaved rose)

8. $r = 2 \cos 5\theta$ (five-leaved rose)

9. $r^2 = 4\cos 2\theta$ (lemniscate)

10. $r = 1 + 2\cos\theta$ (limaçon with small loop)

11. $r = 1 + 2\sin\theta$ (limaçon with small loop)

12. $r = 2 - 4\sin\theta$ (limaçon with small loop)

13. $r = \theta$, $\theta \geq 0$ (spiral)

14. $r = \dfrac{2}{\theta}$, $\theta > 0$ (spiral)

B 15. $r = |\cos\theta|$ (two circles)

16. $r = |\sin\theta|$ (two circles)

17. $r^2 - 3r + 2 = 0$ (two circles)

18. $r^2 - r - 2 = 0$ (two circles)

19. $r^2 - 3r = 0$ (circle and a point)

20. $r = \cos\theta + \sin\theta$ (circle)

21. $r = 2\cos\theta + 2\sin\theta$ (circle)

22. $r = 2\cos\dfrac{1}{2}\theta$

23. $r = 2\sin\dfrac{1}{2}\theta$

C 24. A line segment l of length 2 moves so that one endpoint is on the x-axis and the other is on the y-axis. Let P be the foot of the perpendicular dropped from the origin to l. Show algebraically that P traces out a four-leaved rose. (Hint: Let a and b be the x- and y-intercepts of the line containing l. Express these in terms of r and θ, and derive a relationship between r and θ from a relationship between a and b.)

25. Let F_1 and F_2 be the points $(1, 0°)$ and $(1, 180°)$, respectively. A point P moves so that $PF_1 \cdot PF_2 = 1$. Show that P moves on a "horizontal" lemniscate (see Exercise 9).

7-3 *Conic Sections (Optional)*

You may already have studied parabolas, ellipses, and hyperbolas. These curves, formed when a plane intersects a cone, are called *conic sections*. They share a common property that can be used as a definition.

In a plane, let D be a fixed line and let F be a fixed point not on D. Let e be a positive number. Then the set of all points P such that

$$\frac{PF}{PD} = e$$

where PD is the perpendicular distance from the point P to the line D, is called a **conic section** with **eccentricity** e. F is the **focus** of the conic section and D is the corresponding **directrix**. We often abbreviate *conic section* to *conic*.

If $e = 1$, the conic section is a parabola. In this case P moves so that $PF = PD$. A conic is an ellipse if $0 < e < 1$ and a hyperbola if $e > 1$.

Example 1   Find a polar equation of the parabola having the pole as focus and the line $x = -4$ as directrix.

Solution   For any point $P(r, \theta)$, we see from the diagram that $PF = r$ and $PD = 4 + r \cos \theta$. Thus, P is on the parabola if and only if

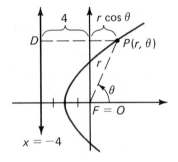

$$\frac{PF}{PD} = \frac{r}{4 + r \cos \theta} = 1.$$

When we solve this equation for r, we obtain

$$r = \frac{4}{1 - \cos \theta}.$$

We can use the method of Example 1 to derive equations of conics having the pole as a focus and either a horizontal or a vertical directrix. For example, if the directrix is the line $y = p \ (p > 0)$, we have (Figure 7-5)

$$PF = r \text{ and } PD = p - r \sin \theta.$$

If the eccentricity of the conic is e, then

$$\frac{PF}{PD} = \frac{r}{p - r \sin \theta} = e.$$

When we solve this equation for r, we obtain

$$r = \frac{ep}{1 + e \sin \theta}.$$

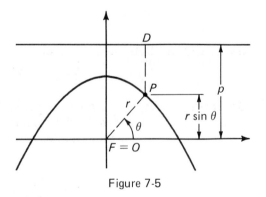

Figure 7-5

The various cases are summarized in Figure 7-6. In each case, p is the distance from the focus to the directrix, and e is the eccentricity. Each of these conics has one of the lines $\theta = 0°$ or $\theta = 90°$ as an axis of symmetry.

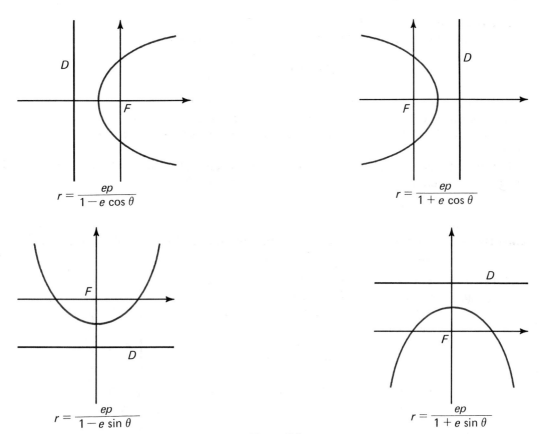

$$r = \frac{ep}{1 - e\cos\theta}$$

$$r = \frac{ep}{1 + e\cos\theta}$$

$$r = \frac{ep}{1 - e\sin\theta}$$

$$r = \frac{ep}{1 + e\sin\theta}$$

Figure 7-6

Example 2 Determine the type of conic and sketch its graph.

$$\text{(a) } r = \frac{6}{2 + \cos\theta} \qquad \text{(b) } r = \frac{10}{2 - 3\sin\theta}$$

Solution

(a) To compare the given equation with those shown in Figure 7-6, we make the constant term of the denominator equal to 1 by dividing numerator and denominator by 2.

$$r = \frac{6}{2 + \cos\theta} = \frac{3}{1 + \frac{1}{2}\cos\theta}, \quad \text{or} \quad r = \frac{\frac{1}{2}\cdot 6}{1 + \frac{1}{2}\cos\theta}.$$

Comparing this equation with $r = \dfrac{ep}{1 + e\cos\theta}$, we see that

$e = \dfrac{1}{2} < 1$, so that the conic is an *ellipse*. We can set up the following table from which we draw the graph.

(Solution continued on next page.)

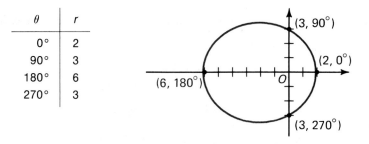

θ	r
0°	2
90°	3
180°	6
270°	3

(b) Again, we first put the equation into one of the forms shown in Figure 7-6.

$$r = \frac{10}{2 - 3\sin\theta} = \frac{5}{1 - \frac{3}{2}\sin\theta}, \text{ or } r = \frac{\frac{3}{2}\cdot\frac{10}{3}}{1 - \frac{3}{2}\sin\theta}.$$

Since $e = \frac{3}{2} > 1$, the conic is a *hyperbola*. We complete the exercise as in part (a).

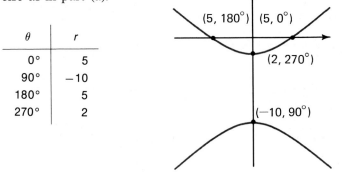

θ	r
0°	5
90°	−10
180°	5
270°	2

You may be familiar with the following definition of an ellipse.

An **ellipse** is a set of points in a plane such that, for each point of the set, the sum of its distances to two fixed points is a constant.

We can show that the ellipse

$$r = \frac{6}{2 + \cos\theta}$$

from Example 2(a) satisfies this condition. For the ellipse, the fixed points are $F_1(0, 0°)$ and $F_2(4, 180°)$. (See Figure 7-7.) Let $P(r, \theta)$ be any point of the curve above the polar axis. Note first that $2 \leq r \leq 6$. Clearly, $PF_1 = r$. Now we apply the law of cosines to $\triangle F_1F_2P$.

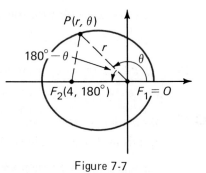

Figure 7-7

$$PF_2^2 = r^2 + 4^2 - 2\cdot r\cdot 4\cos(180° - \theta) = r^2 + 16 + 8r\cos\theta$$

From the equation $r = \dfrac{6}{2 + \cos \theta}$, we have $2r + r \cos \theta = 6$, and hence $r \cos \theta = 6 - 2r$. Therefore,

$$PF_2^2 = r^2 + 16 + 8(6 - 2r)$$
$$= r^2 - 16r + 64 = (r - 8)^2.$$

Since $r < 8$, $8 - r > 0$, so that $PF_2 = \sqrt{(r - 8)^2} = 8 - r$. Therefore,

$$PF_1 + PF_2 = r + (8 - r) = 8, \text{ a constant.}$$

The case where P is below the polar axis is similar.

Exercises 7-3

Determine the type of each conic.

A 1. $r = \dfrac{4}{1 - \sin \theta}$ 2. $r = \dfrac{3}{1 - \dfrac{1}{2} \sin \theta}$

3. $r = \dfrac{6}{2 - \cos \theta}$ 4. $r = \dfrac{4}{1 - \cos \theta}$

5. $r = \dfrac{8}{1 + 3 \cos \theta}$ 6. $r = \dfrac{6}{1 + 2 \sin \theta}$

7. $r = \dfrac{15}{3 + 2 \sin \theta}$ 8. $r = \dfrac{3}{2 + 2 \cos \theta}$

9. $r = \dfrac{5}{2 - 2 \sin \theta}$ 10. $r = \dfrac{10}{2 + 3 \cos \theta}$

11-20. Sketch the graphs of the conics in Exercises 1-10.

B 21. Identify the curve $r^2 = \sec 2\theta$.

22. Identify the curve $r^2 = \csc 2\theta$.

23. Let P be any point on the conic $r = \dfrac{6}{2 - \cos \theta}$. Let F_1 and F_2 be the points $(0, 0°)$ and $(4, 0°)$, respectively. Show that $PF_1 + PF_2 = 8$.

24. Let P be any point on the conic $r = \dfrac{12}{3 - \sin \theta}$. Let F_1 and F_2 be the points $(0, 0°)$ and $(3, 90°)$, respectively. Show that $PF_1 + PF_2 = 9$.

C 25. Every ellipse (and every hyperbola) has two foci, each with its corresponding directrix. Show that for the ellipse $r = \dfrac{6}{2 + \cos \theta}$, $(0, 0°)$ is one focus with a corresponding directrix of $x = 6$, while $(4, 180°)$ is the other focus with a corresponding directrix of $x = -10$.

Representing Complex Numbers

7-4   *Complex Numbers*

From algebra we know that quadratic equations such as $x^2 = -1$ have no solutions over the set of real numbers. In order to solve such equations, mathematicians introduced the number i with the property that:

$$i^2 = -1, \text{ or } i = \sqrt{-1}$$

We can see that

$$i^3 = i^2 \cdot i = (-1)i = -i$$
$$i^4 = i^2 \cdot i^2 = (-1)(-1) = 1$$
$$i^5 = i^4 \cdot i = 1 \cdot i = i, \text{ and so on.}$$

Thus, if n is a nonnegative integer,

$$i^{4n+1} = i, \quad i^{4n+2} = -1, \quad i^{4n+3} = -i, \quad \text{and} \quad i^{4n+4} = 1.$$

A **complex number** is a number of the form

$$x + yi, \text{ or } x + iy,$$

where x and y are real numbers and i is the **imaginary unit.** The number x is called the **real part** of $x + yi$, and y (not yi) is called the **imaginary part.** Thus,

$$2 - i, 4i, \text{ and } -5$$

have real parts 2, 0, and -5, respectively, and imaginary parts -1, 4, and 0, respectively.

The complex numbers $x + iy$ and $u + iv$ are **equal** if and only if $x = u$ and $y = v$.

To add or subtract complex numbers, we treat them as ordinary binomials. For example,

$$(4 + i) + (3 - 2i) = (4 + 3) + (1 - 2)i = 7 - i$$
$$(4 + i) - (3 - 2i) = (4 - 3) + (1 + 2)i = 1 + 3i.$$

To multiply complex numbers, we first multiply them as ordinary binomials and then use the fact that $i^2 = -1$. For example,

$$(4 + i)(3 - 2i) = 12 - 8i + 3i - 2i^2$$
$$= 12 - 5i - 2(-1)$$
$$= 14 - 5i.$$

The general definitions of the sum and product of complex numbers are:

Let $w = u + vi$ and $z = x + yi$. Then

$$w + z = (u + x) + (v + y)i$$
$$wz = (ux - vy) + (uy + vx)i.$$

The definition of multiplication is suggested by the following.

$$wz = (u + vi)(x + yi) = ux + uyi + vxi + vyi^2$$
$$= ux + (uy + vx)i + vy(-1) = (ux - vy) + (uy + vx)i$$

The difference and quotient of complex numbers are defined as usual in terms of addition and multiplication, respectively. Example 1 illustrates how to find the quotient of two complex numbers.

Example 1 Let $w = 2 - i$ and $z = 3 - 4i$. Find $\dfrac{w}{z}$.

Solution
$$\frac{w}{z} = \frac{2 - i}{3 - 4i} = \frac{2 - i}{3 - 4i} \cdot \frac{3 + 4i}{3 + 4i}$$
$$= \frac{6 + 8i - 3i - 4i^2}{9 - 16i^2} = \frac{6 + 5i - 4(-1)}{9 - 16(-1)} = \frac{10 + 5i}{25}$$

Therefore, $\dfrac{w}{z} = \dfrac{2 + i}{5}$, or $\dfrac{2}{5} + \dfrac{1}{5}i$.

Notice that in Example 1 we multiplied both numerator and denominator by the *conjugate* of the denominator, that is, the number $3 + 4i$, obtained from $3 - 4i$ by changing the sign of the imaginary part. In general the **conjugate,** $\bar{z}$, of a complex number $z = x + yi$ is $x - yi$. Furthermore, since

$$z\bar{z} = (x + yi)(x - yi) = x^2 - y^2 i^2 = x^2 - y^2(-1) = x^2 + y^2,$$

we see that the product of two conjugate complex numbers is a nonnegative real number.

The **modulus,** or **absolute value,** of $z = x + iy$ is

$$|z| = \sqrt{x^2 + y^2}.$$

From the preceding paragraph we see that $z\bar{z} = |z|^2$.

Example 2 Find the reciprocal of $z = 3 + i$.

Solution
$$\frac{1}{z} = \frac{\bar{z}}{z\bar{z}} = \frac{\bar{z}}{|z|^2}$$

Since $\bar{z} = 3 - i$ and $|z| = \sqrt{3^2 + 1^2} = \sqrt{10}$, we have

$$\frac{1}{z} = \frac{3 - i}{10}, \text{ or } \frac{1}{z} = \frac{3}{10} - \frac{1}{10}i.$$

Exercises 7-4

Put all complex-number answers into the form $x + yi$.

Find (a) $w + z$,   (b) wz,   and   (c) $\dfrac{w}{z}$.

A 1. $w = 3 + i,\ z = 1 - i$ 2. $w = 1 - 3i,\ z = 2 - i$

3. $w = 5i,\ z = 3 + 4i$ 4. $w = i,\ z = 2 + 2i$

5. $w = -1 + i\sqrt{3},\ z = -1 - i\sqrt{3}$ 6. $w = 3 + 4i,\ z = 3 - 4i$

Find (a) the conjugate,   (b) the modulus,   and   (c) the reciprocal of z.

7. $z = 2 + i$ 8. $z = \dfrac{1}{2} + \dfrac{1}{2}i$

9. $z = \dfrac{\sqrt{3}}{2} - \dfrac{1}{2}i$ 10. $z = \dfrac{3}{5} - \dfrac{4}{5}i$

Find z^2 and z^3.

11. $z = -1 + i$ 12. $z = -1 - i$ 13. $z = -1 + i\sqrt{3}$ 14. $z = 1 + i\sqrt{3}$

Write each of the following as i, -1, $-i$, or 1.

B 15. (a) i^{25} (b) i^{34} (c) i^{100} 16. (a) i^{55} (b) i^{80} (c) i^{123}

17. $i + i^2 + i^3 + \cdots + i^{21}$ 18. $i + i^2 + i^3 + \cdots + i^{101}$

19. (a) $\dfrac{1}{i}$ (b) $\dfrac{1}{i^2}$ (c) $\dfrac{1}{i^3}$ (d) $\dfrac{1}{i^4}$ 20. i^{4n+1} for a negative integer n

21. i^{4n+2} for a negative integer n 22. i^{4n+3} for a negative integer n

23. i^{4n+4} for a negative integer n

A **field** is an algebraic system that consists of a set F together with two binary operations that satisfy the following axioms.

The Field Axioms		
Let $a, b, c \in F.$		
ADDITION AXIOMS		MULTIPLICATION AXIOMS
$a + b \in F$	Closure property	$ab \in F$
$(a + b) + c = a + (b + c)$	Associative property	$(ab)c = a(bc)$
$a + 0 = 0 + a = a$	Existence of identity	$a \cdot 1 = 1 \cdot a = a$
$a + (-a) = (-a) + a = 0$	Existence of inverse	$a \cdot \dfrac{1}{a} = \dfrac{1}{a} \cdot a = 1\ (a \neq 0)$
$a + b = b + a$	Commutative property	$ab = ba$
DISTRIBUTIVE AXIOM	$a(b + c) = ab + ac,\ (b + c)a = ba + ca$	

It can be shown that the set of complex numbers together with addition and multiplication forms a field. Show that the following axioms hold.

C 24. Associative property of addition

25. Associative property of multiplication

26. Existence of multiplicative inverse

27. Distributive axiom

28. Show that if z_1 and z_2 are nonreal complex numbers such that $z_1 + z_2$ and $z_1 z_2$ are both real, then $z_2 = \bar{z}_1$.

7-5 *The Complex Plane and the Polar Form*

We can represent complex numbers as points in the Cartesian plane by letting $x + yi$ correspond to the point (x, y), as indicated in Figure 7-8. When the plane is used for this purpose, it is called the **complex plane,** or the **Argand plane.** The x-axis is called the **real axis,** and the y-axis is called the **imaginary axis.**

Let $z = x + yi$ be any complex number. Graphically, then, $|z|$ is the distance from O to the point z. This is indicated in Figure 7-9, which also shows the graphical relationship between z, $\bar{z}$, and $-z$.

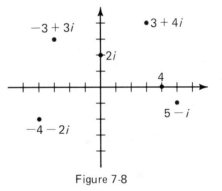

Figure 7-8

The sum of any two complex numbers $z = x + yi$ and $w = u + vi$ can be represented graphically by adding the vectors drawn from O to each point (Figure 7-10). The product of any two complex numbers is represented graphically on page 195.

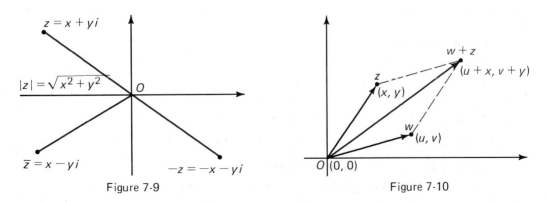

Figure 7-9

Figure 7-10

Example 1   Let $w = 4 + i$ and $z = -3 + 2i$. Draw a graphical representation of
(a) $w + z$ (b) $w - z$ (c) $z + \bar{z}$.

Solution   (a)

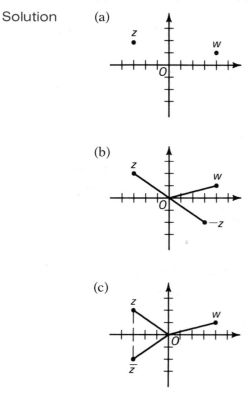

(b)

(c)

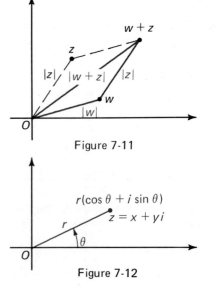

The following property, known as the **triangle inequality,** can also be illustrated graphically as shown in Figure 7-11:

$$|w + z| \leq |w| + |z|$$

This fact is proved algebraically in Exercises 15–20 on page 194.

Recall that each point (x, y) in the plane has polar coordinates (r, θ) with $r \geq 0$, and

$$x = r \cos \theta, \; y = r \sin \theta.$$

Thus, any complex number $z = x + iy$ can be written as

$$z = (r \cos \theta) + i(r \sin \theta), \text{ or}$$
$$z = r(\cos \theta + i \sin \theta) \text{ for } r \geq 0.$$

This form is called the **polar form** of the complex number z (Figure 7-12).

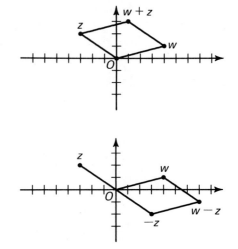

Figure 7-11

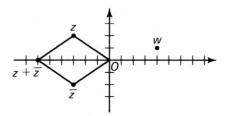

Figure 7-12

It is easy to see that $r = |z|$. The angle θ is called the **argument** of z, written **arg** z, and r is called the **modulus** of z.

Example 2   Write each complex number in polar form. If necessary, express angles to the nearest tenth of a degree.
(a) $z = -3 + 3i$ (b) $w = 4 - 3i$

Solution   (a) $r = |z| = \sqrt{(-3)^2 + 3^2} = 3\sqrt{2}$. Factor $3\sqrt{2}$ out of $-3 + 3i$, obtaining

$$z = 3\sqrt{2}\left(-\frac{1}{\sqrt{2}} + \frac{1}{\sqrt{2}}i\right).$$

Comparing this equation with $z = r(\cos\theta + i\sin\theta)$, we see that
$\cos\theta = -\dfrac{1}{\sqrt{2}}$ and $\sin\theta = \dfrac{1}{\sqrt{2}}$, so that $\theta = 135°$. Therefore,
$z = 3\sqrt{2}(\cos 135° + i\sin 135°)$.

(b) $r = |w| = \sqrt{(4)^2 + (-3)^2} = \sqrt{25} = 5$.

We have $w = 5\left(\dfrac{4}{5} - \dfrac{3}{5}i\right)$, so that $\cos\theta = \dfrac{4}{5} = 0.8$,

$\sin\theta = -\dfrac{3}{5} = -0.6$, and $\theta \approx 323.1°$. Therefore,

$w = 5(\cos 323.1° + i\sin 323.1°)$.

Exercises 7-5

For each pair of complex numbers, draw a graphical representation of the given combinations.

A 1. Let $w = 3 - i$, $z = 2 + 3i$. Draw $w + z$ and $w - z$ on the same plane.

2. Let $w = -4 + 2i$, $z = 4i$. Draw $w + z$ and $w - z$ on the same plane.

3. Let $w = 5 + 2i$, $z = -3 + 3i$. Draw $\overline{w + z}$ and $\overline{w} + \overline{z}$ on the same plane.

4. Let $w = 4 - 3i$, $z = -2 - i$. Draw $\overline{w + z}$ and $\overline{w} + \overline{z}$ on the same plane.

Graph the given complex number.

5. $2(\cos 30° + i\sin 30°)$ 6. $\sqrt{2}(\cos 135° + i\sin 135°)$

7. $3(\cos 270° + i\sin 270°)$ 8. $4(\cos 300° + i\sin 300°)$

Write each complex number in polar form. If necessary, express angles to the nearest tenth of a degree.

9. $2 - 2i$ 10. $-2 + i\sqrt{3}$ 11. $3i$

12. -2 13. $-4 - 2i\sqrt{3}$ 14. $1 - i$

The following sequence of Exercises 15–20 leads to a proof of the triangle inequality. We denote the real part of the complex number z by Re z and the imaginary part by Im z.

B 15. Show that Re $z \leq |z|$ and Im $z \leq |z|$.

16. Show that $z + \bar{z} = 2$ Re z and $z - \bar{z} = 2i$ Im z.

17. Show that $|\bar{z}| = |z|$ and $|-z| = |z|$.

18. Show that $\overline{w + z} = \bar{w} + \bar{z}$ and $\overline{wz} = \bar{w}\bar{z}$.

19. Show that $w\bar{z} + \bar{w}z \leq 2|w||z|$ by justifying the following steps:

$$w\bar{z} + \bar{w}z = w\bar{z} + \overline{w\bar{z}} = 2 \text{ Re } w\bar{z}$$
$$\leq 2|w||\bar{z}| = 2|w||z|$$

20. Prove the triangle inequality by justifying the following steps.

$$|w + z|^2 = (w + z)\overline{(w + z)} = (w + z)(\bar{w} + \bar{z})$$
$$= w\bar{w} + w\bar{z} + \bar{w}z + z\bar{z}$$
$$= |w|^2 + (w\bar{z} + \bar{w}z) + |z|^2$$
$$\leq |w|^2 + 2|w||z| + |z|^2$$

Therefore, $|w + z|^2 \leq (|w| + |z|)^2$.

Since both $|w + z|$ and $|w| + |z|$ are nonnegative, the last inequality implies

$$|w + z| \leq |w| + |z|.$$

C 21. Under what conditions is it true that $|w + z| = |w| + |z|$?

22. Prove that $|w| - |z| \leq |w + z|$. (Hint: Apply the triangle inequality to $|(w + z) - z|$.)

23. Show that the distance between the graphs of any two complex numbers w and z is $|w - z|$.

Using the Polar Form of Complex Numbers

7-6   *The Polar Form of Products and Quotients*

In this section we shall investigate how the polar form can be used to multiply and divide complex numbers.

Let $z_1 = r_1(\cos \theta_1 + i \sin \theta_1)$ and $z_2 = r_2(\cos \theta_2 + i \sin \theta_2)$ be any two complex numbers. To express the product $z_1 z_2$ in polar form we have

$$z_1 z_2 = r_1(\cos \theta_1 + i \sin \theta_1) \cdot r_2(\cos \theta_2 + i \sin \theta_2)$$
$$= r_1 r_2(\cos \theta_1 \cos \theta_2 + i \sin \theta_2 \cos \theta_1 + i \sin \theta_1 \cos \theta_2 - \sin \theta_1 \sin \theta_2)$$
$$= r_1 r_2[(\cos \theta_1 \cos \theta_2 - \sin \theta_1 \sin \theta_2) + i(\sin \theta_1 \cos \theta_2 + \cos \theta_1 \sin \theta_2)].$$

Using the sum formulas for sine and cosine, we have

$$z_1 z_2 = r_1 r_2[\cos(\theta_1 + \theta_2) + i \sin (\theta_1 + \theta_2)]$$

Example 1 Find the product of $z_1 = 3(\cos 30° + i \sin 30°)$ and $z_2 = 2(\cos 150° + i \sin 150°)$.

Solution
$$z_1 z_2 = 3 \cdot 2[\cos (30° + 150°) + i \sin (30° + 150°)]$$
$$= 6(\cos 180° + i \sin 180°) = 6(-1 + 0 \cdot i)$$
$$= -6$$

The polar form of the product of two complex numbers z_1 and z_2 can be used to represent $z_1 z_2$ graphically. Since

$$z_1 z_2 = r_1 r_2[(\cos (\theta_1 + \theta_2) + i \sin (\theta_1 + \theta_2)],$$

the graph of $z_1 z_2$ will be on the terminal side $\theta = \theta_1 + \theta_2$ (Figure 7-13). The distance $r_1 r_2$ can be found as follows. Let P_1 be the graph of z_1, P_2 be the graph of z_2, and A be the point $(1, 0)$. Choose the point P on the terminal side $\theta = \theta_1 + \theta_2$ so that $\triangle OP_2P$ and $\triangle OAP_1$ are similar. Let $r = OP$. Then

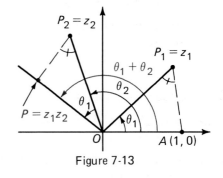

Figure 7-13

$$\frac{OP}{OP_2} = \frac{OP_1}{OA} \quad \text{or} \quad \frac{r}{r_2} = \frac{r_1}{1}$$

and $r = r_1 r_2$. The point P is therefore the graph of $z_1 z_2$.

The reciprocal of $z = r(\cos \theta + i \sin \theta)$, $z \neq 0$, can be expressed in polar form as follows.

$$\frac{1}{z} = \frac{\bar{z}}{z\bar{z}} = \frac{\bar{z}}{|z|^2} = \frac{r(\cos \theta - i \sin \theta)}{r^2}$$

Therefore:

$$\frac{1}{z} = \frac{1}{r}[\cos (-\theta) + i \sin (-\theta)]$$

We can now find the polar form of the quotient $\dfrac{z_1}{z_2}$ by writing it as $z_1 \cdot \dfrac{1}{z_2}$, $z_2 \neq 0$, and by using the reciprocal and product formulas. The result is:

$$\frac{z_1}{z_2} = \frac{r_1}{r_2}[\cos(\theta_1 - \theta_2) + i\sin(\theta_1 - \theta_2)]$$

Example 2   Let $z = 3(\cos 120° + i\sin 120°)$. Find $\dfrac{z}{i}$.

Solution   The modulus of i is 1 and its argument is 90°. Hence in polar form

$$i = 1(\cos 90° + i\sin 90°).$$

Therefore, $\dfrac{z}{i} = \dfrac{3}{1}[\cos(120° - 90°) + i\sin(120° - 90°)]$
$$= 3(\cos 30° + i\sin 30°).$$

The graphical effect of multiplying or dividing a complex number z by i is indicated in Figure 7-14. (See also Exercises 13–16 on page 197.) Note that

$$|iz| = |z| = \left|\frac{z}{i}\right|.$$

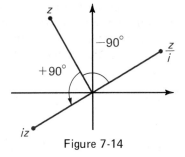

Figure 7-14

Exercises 7-6

For each of the following find $z_1 z_2$ in polar form.

A 1. $z_1 = \sqrt{2}(\cos 135° + i\sin 135°)$, $z_2 = 2(\cos 90° + i\sin 90°)$

2. $z_1 = \sqrt{2}(\cos 45° + i\sin 45°)$, $z_2 = \sqrt{2}(\cos 135° + i\sin 135°)$

3. $z_1 = 2(\cos 60° + i\sin 60°)$, $z_2 = 4(\cos 120° + i\sin 120°)$

4. $z_1 = 2\sqrt{3}(\cos 300° + i\sin 300°)$, $z_2 = 2(\cos 30° + i\sin 30°)$

Find the reciprocal of z in (a) polar form and (b) the form $x + iy$.

5. $z = 2(\cos 120° + i\sin 120°)$ 6. $z = \cos 135° + i\sin 135°$

7. $z = -1 + i$ 8. $z = -\sqrt{3} - i$

9–12. In Exercises 1–4 find $\dfrac{z_1}{z_2}$ in polar form.

Find the polar forms of (a) iz and (b) $\frac{z}{i}$. Also graph z, iz, and $\frac{z}{i}$ in the same complex plane.

B 13. $z = 2(\cos 100° + i \sin 100°)$ 14. $z = \cos(-40°) + i \sin(-40°)$

15. $z = -2 - 2i$ 16. $z = -\sqrt{3} + i$

17. Show that if $z = r(\cos \theta + i \sin \theta)$ then $z^2 = r^2(\cos 2\theta + i \sin 2\theta)$.

18. Show that if $z = r(\cos \theta + i \sin \theta)$ then $z^3 = r^3(\cos 3\theta + i \sin 3\theta)$.

Let z, z_1, and z_2 be complex numbers. Use the polar form to represent each of the following graphically. (See Figure 7-13.)

C 19. z^2 20. $\frac{1}{z}$, $z \neq 0$ 21. $\frac{z_1}{z_2}$, $z_2 \neq 0$

Challenge

We can regard complex numbers as vectors by letting $x + yi$ correspond to $x\mathbf{i} + y\mathbf{j}$ (in the notation of Chapter 6). Show that the product z_1z_2 of two complex numbers corresponds to the sum of two orthogonal vectors. (Hint: Consider the vector that is the product of z_1 and the real part of z_2.)

7-7 De Moivre's Theorem

In Section 7-6 we derived the formula

$$z_1z_2 = r_1r_2[\cos(\theta_1 + \theta_2) + i \sin(\theta_1 + \theta_2)].$$

Setting $z_1 = z_2 = z = r(\cos \theta + i \sin \theta)$, we obtain

$$z \cdot z = r \cdot r[\cos(\theta + \theta) + i \sin(\theta + \theta)].$$

Therefore, $z^2 = r^2(\cos 2\theta + i \sin 2\theta)$.
Also,

$$z^2 \cdot z = r^2(\cos 2\theta + i \sin 2\theta) \cdot r(\cos \theta + i \sin \theta)$$
$$z^3 = r^3[\cos(2\theta + \theta) + i \sin(2\theta + \theta)].$$

Therefore, $z^3 = r^3(\cos 3\theta + i \sin 3\theta)$.
These results suggest that

$$z^n = r^n(\cos n\theta + i \sin n\theta), \text{ for all positive integers } n.$$

This is indeed true. Furthermore, the formula, called De Moivre's theorem, holds for *every* integer n.

De Moivre's Theorem
For every integer n,
$$[r(\cos \theta + i \sin \theta)]^n = r^n(\cos n\theta + i \sin n\theta)$$

When $r = 1$, we can express De Moivre's theorem in the following simplified form:
$$(\cos \theta + i \sin \theta)^n = \cos n\theta + i \sin n\theta$$

Example 1   Express $(-1 + i\sqrt{3})^6$ in the form $x + iy$.

Solution   First express $-1 + i\sqrt{3}$ in polar form.
$$-1 + i\sqrt{3} = 2\left(-\frac{1}{2} + i\frac{\sqrt{3}}{2}\right) = 2(\cos 120° + i \sin 120°)$$

Therefore $(-1 + i\sqrt{3})^6 = [2(\cos 120° + i \sin 120°)]^6$
$$= 2^6[\cos (6 \cdot 120°) + i \sin (6 \cdot 120°)]$$
$$= 64(\cos 720° + i \sin 720°)$$
$$= 64(\cos 0° + i \sin 0°)$$
$$= 64(1 + i \cdot 0) = 64$$

De Moivre's theorem can be proved for positive integers n by using mathematical induction (Exercise 25 on page 200). Let us assume that this has been done and prove it for *negative* integers n.

Let $n = -m$, where m is a *positive* integer. Then
$$(\cos \theta + i \sin \theta)^n = (\cos \theta + i \sin \theta)^{-m}$$

$$= \frac{1}{(\cos \theta + i \sin \theta)^m}$$

$$= \frac{1}{\cos m\theta + i \sin m\theta} \quad \begin{array}{l}\text{by De Moivre's theorem for} \\ \text{positive integers}\end{array}$$

$$= \cos (-m\theta) + i \sin (-m\theta) \quad \begin{array}{l}\text{by the reciprocal} \\ \text{formula on page 195}\end{array}$$

$$= \cos n\theta + i \sin n\theta \quad \text{since } -m = n.$$

Example 2   Express $(1 + i)^{-5}$ in the form $x + iy$.

Solution   First express $1 + i$ in polar form.
$$1 + i = \sqrt{2}\left(\frac{1}{\sqrt{2}} + i\frac{1}{\sqrt{2}}\right) = \sqrt{2}(\cos 45° + i \sin 45°)$$

Then $(1 + i)^{-5} = [\sqrt{2}(\cos 45° + i \sin 45°)]^{-5}$
$$= (\sqrt{2})^{-5}[\cos (-5 \cdot 45°) + i \sin (-5 \cdot 45°)]$$
$$= \frac{1}{4\sqrt{2}}[\cos (-225°) + i \sin (-225°)]$$
$$= \frac{1}{4\sqrt{2}}\left(-\frac{1}{\sqrt{2}} + i\frac{1}{\sqrt{2}}\right) = \frac{1}{8}(-1 + i) = -\frac{1}{8} + \frac{1}{8}i.$$

De Moivre's theorem can be used to derive certain identities; for example, identities which express cos $n\theta$ and sin $n\theta$ in terms of functions of θ for any integer n.

Example 3 Use De Moivre's theorem to derive identities for cos 3θ and sin 3θ in terms of cos θ and sin θ.

Solution   By De Moivre's theorem,

$$\cos 3\theta + i \sin 3\theta = (\cos \theta + i \sin \theta)^3.$$

Expanding the cube of the binomial, we obtain

$$\cos^3 \theta + 3 \cos^2 \theta \cdot i \sin \theta + 3 \cos \theta \cdot i^2 \sin^2 \theta + i^3 \sin^3 \theta,$$

which can be written as

$$(\cos^3 \theta - 3 \cos \theta \sin^2 \theta) + i(3 \cos^2 \theta \sin \theta - \sin^3 \theta).$$

Therefore,

$$\cos 3\theta + i \sin 3\theta = (\cos^3 \theta - 3 \cos \theta \sin^2 \theta) + i(3 \cos^2 \theta \sin \theta - \sin^3 \theta).$$

By applying the definition of equality of complex numbers given on page 188, we have

$$\cos 3\theta = \cos^3 \theta - 3 \cos \theta \sin^2 \theta \quad \text{and}$$
$$\sin 3\theta = 3 \cos^2 \theta \sin \theta - \sin^3 \theta.$$

Using the identity $\sin^2 \theta + \cos^2 \theta = 1$, we can write these formulas as

$$\cos 3\theta = 4 \cos^3 \theta - 3 \cos \theta \quad \text{and}$$
$$\sin 3\theta = 3 \sin \theta - 4 \sin^3 \theta.$$

Exercises 7-7

Use De Moivre's theorem to express each of the following in the form $x + iy$.

A 1. $[2(\cos 60° + i \sin 60°)]^6$

2. $[2(\cos 30° + i \sin 30°)]^6$

3. $\left[\frac{1}{2}(\cos 135° + i \sin 135°)\right]^4$

4. $(\cos 36° + i \sin 36°)^{100}$

5. $[\cos (-60°) + i \sin (-60°)]^9$

6. $\left[\frac{1}{2}(\cos 45° + i \sin 45°)\right]^7$

7. $(1 + i\sqrt{3})^3$

8. $(\sqrt{3} - i)^6$

9. $(1 + i)^{10}$

10. $(-1 + i)^{12}$

Express each in the form $x + iy$.

11. $[3(\cos 12° + i \sin 12°)]^5 \left[\frac{1}{3}(\cos 10° + i \sin 10°)\right]^3$

12. $[5(\cos 25° + i \sin 25°)]^5 \left[\frac{2}{3}(\cos 70° + i \sin 70°)\right]^4$

13. $\dfrac{[2(\cos 23° + i \sin 23°)]^6}{[2(\cos 6° + i \sin 6°)]^8}$

14. $\dfrac{(\cos 25° + i \sin 25°)^5}{(\cos 10° + i \sin 10°)^8}$

15. $\left[\frac{1}{2}(1 + i\sqrt{3})\right]^5 \left[\frac{1}{2}(\sqrt{3} - i)\right]^4$

16. $\left[\frac{1}{2}(-\sqrt{3} - i)\right]^3 \left[\frac{1}{2}(-1 + i\sqrt{3})\right]^4$

17. $\dfrac{(1 + i\sqrt{3})^2}{(-\sqrt{3} + i)^3}$

18. $\dfrac{(-\sqrt{3} + i)^3}{(1 - i)^6}$

B 19. Use De Moivre's theorem to prove $\cos 2\theta = \cos^2 \theta - \sin^2 \theta$ and $\sin 2\theta = 2 \sin \theta \cos \theta$.

20. Use De Moivre's theorem to derive identities for $\cos 4\theta$ and $\sin 4\theta$ in terms of $\cos \theta$ and $\sin \theta$.

21. Use De Moivre's theorem to derive identities for $\cos 5\theta$ and $\sin 5\theta$ in terms of $\cos \theta$ and $\sin \theta$.

22. Define z^0 by $z^0 = 1$. Verify that De Moivre's theorem holds for the case $n = 0$.

C 23. Verify that De Moivre's theorem holds for the case $n = \frac{1}{2}$ by expanding $\left(\cos \frac{1}{2}\theta + i \sin \frac{1}{2}\theta\right)^2$ and using the double angle formula.

24. Verify that De Moivre's theorem holds for the case $n = \frac{1}{3}$ by expanding $\left(\cos \frac{1}{3}\theta + i \sin \frac{1}{3}\theta\right)^3$ and using the results of Example 3 on page 199.

25. Prove De Moivre's theorem for positive integers n by using mathematical induction. (Hint: First show that the theorem is true for $n = 1$. Next, show that if the theorem holds for any given positive integer k, then it holds for $k + 1$.)

7-8 Roots of Complex Numbers

The complex number w is an **nth root** of z if $w^n = z$. Example 1 illustrates a method of finding roots.

Example 1   Find the cube roots of -8.

Solution   We wish to find the solutions of

$$w^3 = -8.$$

Using the polar form, we have

$$w = r(\cos \phi + i \sin \phi),$$
$$-8 = 8(\cos 180° + i \sin 180°).$$

By De Moivre's theorem,

$$w^3 = r^3(\cos 3\phi + i \sin 3\phi).$$

Thus $w^3 = -8$ can be written as

$$r^3(\cos 3\phi + i \sin 3\phi) = 8(\cos 180° + i \sin 180°).$$

Hence $r^3 = 8$, and since r is a positive real number,

$$r = 8^{\frac{1}{3}} = 2.$$

Also

$$\cos 3\phi = \cos 180°, \text{ and } \sin 3\phi = \sin 180°.$$

Thus 3ϕ and $180°$ are coterminal, and

$$3\phi = 180° + k \cdot 360°,$$

where k is an integer. Hence

$$\phi = 60° + k \cdot 120°,$$

and

$$w = 2[\cos (60° + k \cdot 120°) + i \sin (60° + k \cdot 120°)].$$

When $k = 0$,

$$w_1 = 2(\cos 60° + i \sin 60°) = 2 \left(\frac{1}{2} + i \frac{\sqrt{3}}{2} \right) = 1 + i\sqrt{3}.$$

When $k = 1$,

$$w_2 = 2(\cos 180° + i \sin 180°) = 2(-1 + i \cdot 0) = -2.$$

When $k = 2$,

$$w_3 = 2 \left[\frac{1}{2} + i \left(-\frac{\sqrt{3}}{2} \right) \right] = 1 - i\sqrt{3}.$$

Any other value of k will produce a root of $w^3 = -8$ equal to one of the three found above. Thus -8 has exactly three cube roots, namely, $1 + i\sqrt{3}$, -2, and $1 - i\sqrt{3}$.

Using the method of Example 1 we can show that every nonzero complex number has exactly n nth roots. Moreover:

The n nth roots of $r(\cos\theta + i\sin\theta)$ are given by

$$r^{\frac{1}{n}}\left(\cos\frac{\theta + k\cdot 360°}{n} + i\sin\frac{\theta + k\cdot 360°}{n}\right),$$

where $k = 0, 1, 2, \ldots, n - 1$.

Example 2   Find the five fifth roots of $-1 + i$. Leave your answer in polar form.

Solution   In polar form, $-1 + i = \sqrt{2}(\cos 135° + i\sin 135°)$.

Using $r = \sqrt{2}$ and $\theta = 135°$ in the formula displayed above, we have

$$w = (\sqrt{2})^{\frac{1}{5}}\left(\cos\frac{135° + k\cdot 360°}{5} + i\sin\frac{135° + k\cdot 360°}{5}\right)$$

or

$$w = 2^{\frac{1}{10}}[\cos(27° + k\cdot 72°) + i\sin(27° + k\cdot 72°)].$$

Setting k equal, in turn, to 0, 1, 2, 3, and 4, we obtain the five fifth roots of $-1 + i$:

$$w_1 = 2^{\frac{1}{10}}(\cos 27° + i\sin 27°)$$
$$w_2 = 2^{\frac{1}{10}}(\cos 99° + i\sin 99°)$$
$$w_3 = 2^{\frac{1}{10}}(\cos 171° + i\sin 171°)$$
$$w_4 = 2^{\frac{1}{10}}(\cos 243° + i\sin 243°)$$
$$w_5 = 2^{\frac{1}{10}}(\cos 315° + i\sin 315°)$$

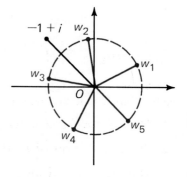

Figure 7-15 shows $-1 + i$ and its five fifth roots. Notice that the roots are equally spaced around the circle of radius $2^{\frac{1}{10}} \approx 1.07$, centered at O. In general, the nth roots of $r(\cos\theta + i\sin\theta)$ are equally spaced around the circle of radius $r^{\frac{1}{n}}$ centered at O.

The nth roots of the complex number 1, called the nth roots of unity, are easily found. Since

$$1 = 1(\cos 0° + i\sin 0°),$$

the n nth roots of unity are given by

$$\cos\frac{k\cdot 360°}{n} + i\sin\frac{k\cdot 360°}{n}, \quad k = 0, 1, \ldots, n - 1,$$

or, if radian measure is used, by

$$\cos \frac{2k\pi}{n} + i \sin \frac{2k\pi}{n}, \; k = 0, 1, \ldots, n-1.$$

It is easy to prove that if

$$w = \cos \frac{2\pi}{n} + i \sin \frac{2\pi}{n},$$

then the n nth roots of unity are

$$w, w^2, \ldots, w^{n-1}, w^n = 1.$$

Example 3 Find and graph the sixth roots of unity, labeling each as a power of one of them.

Solution The sixth roots of unity are given by $\cos \dfrac{2k\pi}{6} + i \sin \dfrac{2k\pi}{6} =$

$\cos \dfrac{k\pi}{3} + i \sin \dfrac{k\pi}{3}$, for $k = 0, 1, 2, 3, 4, 5$.

Hence the roots are

$\cos 0 + i \sin 0 = 1$

$\cos \dfrac{\pi}{3} + i \sin \dfrac{\pi}{3} = \dfrac{1}{2} + i \dfrac{\sqrt{3}}{2}$

$\cos \dfrac{2\pi}{3} + i \sin \dfrac{2\pi}{3} = -\dfrac{1}{2} + i \dfrac{\sqrt{3}}{2}$

$\cos \pi + i \sin \pi = -1$

$\cos \dfrac{4\pi}{3} + i \sin \dfrac{4\pi}{3} = -\dfrac{1}{2} - i \dfrac{\sqrt{3}}{2}$

$\cos \dfrac{5\pi}{3} + i \sin \dfrac{5\pi}{3} = \dfrac{1}{2} - i \dfrac{\sqrt{3}}{2}.$

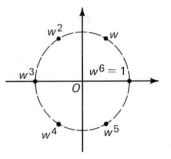

In the diagram,

$$w = \cos \frac{\pi}{3} + i \sin \frac{\pi}{3} = \frac{1}{2} + i \frac{\sqrt{3}}{2}.$$

Exercises 7-8

In Exercises 1–4, find and graph the indicated roots of unity in the form $x + iy$, labeling each root as a power of one of them as in Example 3.

A 1. The cube roots of unity

2. The fourth roots of unity

3. The fifth roots of unity

4. The eighth roots of unity

Find the indicated roots in the form specified, and sketch their graphs along with that of the given number.

5. The cube roots of i in the form $x + iy$

6. The fourth roots of -1 in the form $x + iy$

7. The fifth roots of $-1 - i$ in polar form

8. The tenth roots of $-\sqrt{3} - i$ in polar form

9. The sixth roots of $-\sqrt{3} + i$ in polar form

10. The fifth roots of $-i$ in polar form

B 11. Without actually substituting, explain why the cube roots of unity other than 1 satisfy

$$z^2 + z + 1 = 0.$$

(Hint: $z^3 - 1 = (z - 1)(z^2 + z + 1)$.)

12. Without actually substituting, explain why the sixth roots of unity other than 1 satisfy

$$z^5 + z^4 + z^3 + z^2 + z + 1 = 0.$$

(Hint: $z^6 - 1 = (z - 1)(?)$.)

13. Generalize Exercises 11 and 12.

14. Show that if k is an integer between 0 and $\frac{n}{2}$, then

$$\cos \frac{2(n - k)\pi}{n} = \cos \frac{2k\pi}{n} \text{ and } \sin \frac{2(n - k)\pi}{n} = -\sin \frac{2k\pi}{n}.$$

Explain why this implies that the imaginary nth roots of unity occur in conjugate pairs.

15. Find the roots of the quadratic equation $z^2 + 2z + 1 + 2i = 0$ in the form $x + yi$. (Hint: Use the quadratic formula and find the necessary square roots by the method of this section.)

16. Find the roots of the quadratic equation $z^2 + 2iz - 1 - 8i = 0$ in the form $x + yi$.

An nth root of unity, w, is said to be **primitive** if every nth root of unity is an integral power of w.

C 17. Show that $\cos \frac{2\pi}{n} + i \sin \frac{2\pi}{n}$ is a primitive nth root of unity.

18. Let $w = \cos \dfrac{2\pi}{8} + i \sin \dfrac{2\pi}{8} = \cos \dfrac{\pi}{4} + i \sin \dfrac{\pi}{4}$. By Exercise 17,

$$w, w^2, w^3, \ldots, w^7, w^8 = 1$$

are all the eighth roots of unity. (a) Show that w^3 is primitive. (Consider $(w^3)^2 = w^6$, $(w^3)^3 = w^9 = w$, etc.) (b) Show that w^2 is *not* primitive. (c) Which of the eighth roots of unity are primitive and which are not?

19. Find the primitive twelfth roots of unity.

20. Explain why, if n is not a prime number, at least one nth root of unity, besides 1, is not primitive.

CHAPTER SUMMARY

1. A point may be located in a plane by using the **polar coordinates** (r, θ) where r is the distance from the point to the **pole** and θ is the angle formed by the segment from the point to the pole and the **polar axis.** Polar coordinates may be related to rectangular coordinates by using the equations $x = r \cos \theta$, $y = r \sin \theta$, and $r = \sqrt{x^2 + y^2}$.

2. To graph a polar equation first set up a table of values for r and θ. Some useful tests for symmetry of polar equation graphs are:

If the equation is unchanged when	then the graph is symmetric with respect to
θ is replaced with $-\theta$	the polar axis
r is replaced with $-r$	the origin
θ is replaced with $\pi - \theta$	the vertical line $\theta = \dfrac{\pi}{2}$

3. A **conic section** is the set of all points whose distance from a fixed point, called the **focus,** and a fixed line, called the **directrix,** have a constant ratio e, called the **eccentricity.** When the focus is the pole and the directrix is a horizontal or vertical line p units from the pole, the conic has a polar equation of the form $r = \dfrac{ep}{1 \pm e \cos \theta}$ for a vertical directrix or $r = \dfrac{ep}{1 \pm e \sin \theta}$ for a horizontal directrix.

4. A **complex number** is a number that may be written in the form $a + bi$ where $i = \sqrt{-1}$. To add or subtract complex numbers, add or subtract the corresponding real and imaginary parts. To multiply complex numbers, multiply as binomials and use $i^2 = -1$. The **conjugate** of $z = a + bi$ is $a - bi$. To find the quotient of two complex numbers, multiply both numerator and denominator by the conjugate of the denominator. The **modulus** or **absolute value** of $z = a + bi$ is $|z| = \sqrt{a^2 + b^2}$.

5. Complex numbers may be represented as points in a plane. Addition of complex numbers may be represented as vector addition. The polar form of a complex number z is $r(\cos \theta + i \sin \theta)$ where $r \geq 0$. θ is called the **argument** of z; r is called the **modulus** of z.

6. To multiply complex numbers in polar form, multiply their moduli and add their arguments. To divide complex numbers, divide their moduli and subtract their arguments.

7. Integral powers of complex numbers may be found by using De Moivre's theorem:

$$z = [r(\cos \theta + i \sin \theta)]^n = r^n(\cos n\theta + i \sin n\theta).$$

The theorem is also useful in deriving identities for $\cos n\theta$ or $\sin n\theta$.

8. The nth roots of a complex number $z = r(\cos \theta + i \sin \theta)$ can be found using $z^{\frac{1}{n}} = r^{\frac{1}{n}} \left(\cos \dfrac{\theta + k \cdot 360°}{n} + i \sin \dfrac{\theta + k \cdot 360°}{n} \right)$

for $k = 0, 1, 2, \ldots, n - 1$.

CHAPTER TEST

7-1   1. Give three different polar coordinates for $(\sqrt{3}, -3)$ with at least one having $r < 0$.

2. Give rectangular coordinates for $(-8, 300°)$.

7-2   3. Graph $r = \dfrac{1}{2} - \sin \theta$.

4. Graph $r = \dfrac{1}{\theta^2}$.

7-3   5. (*Optional*) Graph $r = \dfrac{3}{\dfrac{1}{2} - \cos \theta}$ and identify the resulting

conic.

7-4   6. If $c = 5 + 2i$, $d = 3 - 4i$, and $e = 3 + i$, find

$$cd + \frac{d}{e} + |d|.$$

7-5   7. Write $8 - 6i$ in polar form.

7-6   8. Write the complex number z in the form $x + yi$ if $z =$
$$\frac{[2(\cos 85° + i \sin 85°)] \cdot [12(\cos 75° + i \sin 75°)]}{8(\cos 60° + i \sin 60°)}.$$

7-7   9. Use De Moivre's theorem to find $(-50\sqrt{2} + 50\,i\sqrt{2})^{-8}$.

7-8   10. Give the cube roots of $4 - 4i\sqrt{3}$ in polar form.

CUMULATIVE REVIEW   *Chapters 6 and 7*

Chapter 6

1. If $\mathbf{a} = 3\mathbf{i} + 2\mathbf{j}$ and $\mathbf{b} = \mathbf{i} + \mathbf{j}$, graph the vector $\mathbf{c} = \mathbf{a} - 2\mathbf{b}$.

2. Find the norm of $\mathbf{d} = 6\mathbf{i} - 8\mathbf{j}$.

3. A pilot wishes to fly a plane at 500 km/h on a bearing of 280°. However, there is a 40 km/h wind, blowing from 65°. What should be the heading and engine speed of the plane?

4. A 150 kg crate is resting on a ramp inclined at an angle of 25° with the horizontal. The crate is held in place by a rope parallel to the ramp as shown in the diagram. Find the tension in the rope.

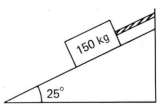

5. Resolve $\mathbf{v} = 3\mathbf{i} - 2\mathbf{j}$ into components parallel to $\mathbf{a} = \mathbf{i} + 4\mathbf{j}$ and $\mathbf{b} = 4\mathbf{i} - \mathbf{j}$.

6. Find the angle θ between **c** and **d** if **c** $=$ **i** $+$ 5**j** and **d** $=$ 2**i** $-$ 3**j**.

7. Use a specific example to test the following properties for a scalar s and vectors **a** and **b**. State which property is true.
   (1) $s(\mathbf{a} \cdot \mathbf{b}) = (s\mathbf{a}) \cdot (s\mathbf{b})$
   (2) $s(\mathbf{a} \cdot \mathbf{b}) = (s\mathbf{a}) \cdot \mathbf{b}$

8. Suppose the crate in Exercise 4 is moved 10 m up the ramp. Ignoring friction, how much energy is expended?

Chapter 7

1. Give rectangular coordinates for $(-7, 300°)$.

2. Give polar coordinates for $(-3, -3)$.

3. Graph $r = |1 + \theta|$ for $-\pi \le \theta \le \pi$.

4. (*Optional*) Graph $r = \dfrac{2}{1 + \cos \theta}$ and identify the resulting conic.

5. If $z = 6 - 6i$ and $w = 4(\cos 90° + i \sin 90°)$,
   (a) find $z + w$ using vector addition
   (b) find $z \cdot w$ using polar form.

6. If $w = 4 - 3i$ and $z = 3 + i$, find $\dfrac{w}{z}$,

   (a) using rectangular coordinates
   (b) using polar form.

Graph the given complex number.

7. $2(\cos 60° + i \sin 60°)$

8. $4(\cos 330° + i \sin 330°)$

Use De Moivre's theorem to express each of the following in the form $x + iy$.

9. $(2 + i)^6$

10. $(-\sqrt{3} + i)^6$

11. Find the fourth roots of i.

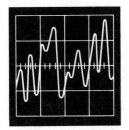

AN APPLICATION OF COMPLEX NUMBERS

Since ancient times, mathematicians have investigated the possibilities of geometric construction using only a compass and a straightedge. The early Greeks had discovered ways to construct regular polygons of 3, 4, 5, 6, and 8 sides, but were unable to construct regular polygons of 7, 9, or 11 sides. It was not until 1796 that Karl Friedrich Gauss proved that such constructions are mathematically impossible.

Gauss' method of proof relied on the properties of complex numbers. Consider, for example, a regular polygon of 9 sides. Such a polygon could be drawn with the aid of a protractor, by inscribing nine 40° angles with a common vertex inside a unit circle, as shown at the left below. If the polygon is drawn in the complex plane with its center at the origin (at the right below), we can let $(a + bi)$ be that complex

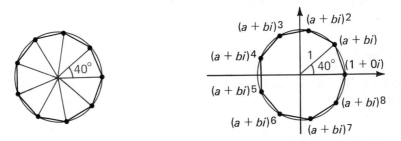

number on the unit circle whose angle is 40°. Then, by the definition of complex multiplication (recall De Moivre's theorem on page 198), $1 + 0i$, $(a + bi)$, $(a + bi)^2$, $(a + bi)^3$, . . . , $(a + bi)^8$ are the vertices of the polygon.

Gauss matched the complex numbers representing the vertices of a regular polygon, other than $1 + 0i$, with the roots of the equation

$$z^{n-1} + z^{n-2} + \cdots + z^2 + z + 1 = 0,$$

where n is the number of sides of the polygon. He then characterized the values of n for which these complex roots can be constructed. Gauss found that a regular polygon of n sides is constructible if and only if
(a) n is a prime number of the form $2^{2^k} + 1$ for $k = 0, 1, 2, \ldots$, or
(b) when n is factored into primes, no prime other than 2 appears more than once.

POLAR COORDINATES

The program below will find a pair of polar coordinates in terms of $R \geq 0$ and $0° \leq \theta < 360°$, for the rectangular coordinates (X, Y).

The only inverse trigonometric function in BASIC is the inverse tangent function (ATN). From page 59, we know that if $\text{Tan } \theta \geq 0$, then $0° \leq \text{Tan}^{-1} \theta < 90°$, and if $\text{Tan } \theta < 0$, then $-90° < \text{Tan}^{-1} \theta < 0°$. But we want $0° \leq \theta < 360°$. Therefore, if $T = \text{ATN}(Y/X)$, $X \neq 0$, we have:

for (X, Y) in the first quadrant, $\theta = T$ radians
for (X, Y) in the second quadrant, $\theta = (P1 + T)$ radians
for (X, Y) in the third quadrant, $\theta = (P1 + T)$ radians
for (X, Y) in the fourth quadrant, $\theta = (2P1 + T)$ radians.

The program also provides for (X, Y) on an axis. A sample RUN follows the program.

```
10    PRINT "TO FIND POLAR COORDINATES FOR (X, Y):"
20    PRINT "INPUT X, Y";
30    INPUT X,Y
40    LET P1=3.14159
45    REM******FIND R.
50    LET R=SQR(X*X+Y*Y)
60    LET R=INT(R*100+.5)/100
65    REM******FIND THETA.
70    IF X=0 THEN 150
80    LET T=ATN(Y/X)
90    IF X>0 THEN 120
95    REM*********X<0
100   LET T=P1+T
110   GOTO 190
115   REM*********X>0
120   IF Y>=0 THEN 190
130   LET T=2*P1+T
140   GOTO 190
145   REM*********X=0
150   IF Y>=0 THEN 180
160   LET T=3*P1/2
170   GOTO 190
```

(Program continued on next page.)

```
180   LET T=P1/2
185   REM******PRINT R AND THETA.
190   PRINT "R=";R;
200   IF R=0 THEN 250
210   LET T=T*180/P1
220   LET T=INT(T*100+.5)/100
230   PRINT "    THETA=";T; "DEG."
240   STOP
250   PRINT "     THETA IS ARBITRARY."
260   END

RUN
TO FIND POLAR COORDINATES FOR (X, Y):
INPUT X, Y?-1,0
R=1     THETA=180 DEG.
```

Exercises

1. RUN the program for Example 2 (a) on page 177.

2. RUN the program for Exercises 9–16 on page 178.

3. Write a program that will convert (r, θ) to (x, y).

4. Write a program that will give the sum and product of two complex numbers.

5. Use the program to find r and θ for Example 2(a) and (b) on page 193.

6. Insert the following lines into the program to find the sum of two vectors when r and θ are given. RUN the revised program for $(10, 30°)$ and $(10, 90°)$.

```
10   PRINT "TO FIND THE SUM OF TWO VECTORS:"
15   PRINT "INPUT R1, THETA 1 (DEG.)";
20   INPUT R1,T1
25   PRINT "INPUT R2, THETA 2 (DEG.)";
30   INPUT R2,T2
31   LET P1=3.14159
32   REM******FIND (X, Y) FOR SUM.
33   LET T1=T1*P1/180
34   LET T2=T2*P1/180
35   LET X1=R1*COS(T1)
36   LET Y1=R1*SIN(T1)
37   LET X2=R2*COS(T2)
38   LET Y2=R2*SIN(T2)
39   LET X=X1+X2
40   LET Y=Y1+Y2
```

The transformation of a tennis player's position during a
serve can be analyzed by means of a computer, as
shown above.

TRANSFORMATIONS

OBJECTIVES

1. To define and use transformations.
2. To apply transformations to the conic sections.

8-1 *Translations*

Sometimes the equation of a given curve can be simplified by introducing a new coordinate system. For example, consider a line with a slope of 2 and a y-intercept of -3. The equation of this line is

$$y = 2x - 3.$$

If we introduce a new coordinate system (the (x', y')-system) whose origin O' is at $(0, -3)$ as shown in Figure 8-1, the equation takes the simpler form

$$y' = 2x'.$$

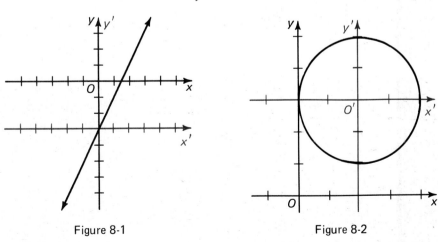

Figure 8-1   Figure 8-2

As a second example consider the equation

$$x^2 + y^2 - 4x - 6y + 9 = 0.$$

By completing squares, we can put the equation into the form

$$(x - 2)^2 + (y - 3)^2 = 4,$$

which represents a circle with center at (2, 3). If now we let

$$x' = x - 2 \text{ and } y' = y - 3,$$

the equation of the circle takes the simpler form

$$(x')^2 + (y')^2 = 4.$$

What we have done in this case is to introduce a new coordinate system with origin O' at the point whose old (x, y)-coordinates are (2, 3), as shown in Figure 8-2 on the previous page.

We can think of what we did above as moving the (x, y)-coordinate system so that the new positions of the coordinate axes are parallel to, or coincident with, their original positions. Such a motion is called a **translation.**

The general translation moves the origin to the point having old coordinates (h, k). The equations that relate the new coordinates (x', y') of a point P and its old coordinates (x, y) are:

$$x = x' + h \qquad y = y' + k$$
$$\text{or} \quad x' = x - h \qquad y' = y - k$$

Example 1 Draw the graph of $x^2 - 6x - 2y + 5 = 0$ after first translating the origin to the point (3, −2).

Solution   The equations of the translation are

$$x = x' + 3, \quad y = y' - 2.$$

Substituting in the given equation, we obtain

$$(x' + 3)^2 - 6(x' + 3) - 2(y' - 2) + 5 = 0.$$

This equation simplifies to

$$(x')^2 = 2y', \text{ or } y' = \frac{1}{2}(x')^2.$$

We see from this simpler equation that the graph is the parabola shown in the figure.

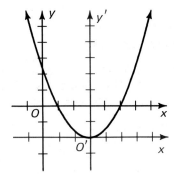

Usually we must decide for ourselves what translation to use. In the example of the circle with center at (2, 3), we decided after completing squares.

Example 2   Graph $y = \cos\left(2x + \dfrac{\pi}{3}\right)$, $x \geq 0$.

Solution   The given equation can be written as $y = \cos 2\left(x + \dfrac{\pi}{6}\right)$.

This equation suggests that we let

$$x' = x + \frac{\pi}{6} \quad \text{and} \quad y' = y.$$

These are the equations of the translation that moves the origin to the point $\left(-\dfrac{\pi}{6}, 0\right)$. In the new coordinate system the curve has the equation

$$y' = \cos 2x'.$$

The graph appears in the figure below.

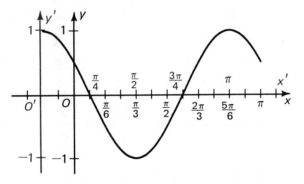

Exercises 8-1

(a) Find the equations of the translation that moves the origin to the specified point, O'. (b) Then find the new equation of the curve C, and (c) graph the curve, showing both sets of axes.

A   1. $O'(2, -4)$ $C: y = x^2 - 4x$ (parabola)

2. $O'(1, 1)$ $C: y = 2x - x^2$ (parabola)

3. $O'(3, 1)$ $C: 3x - 4y - 5 = 0$ (line)

4. $O'(0, -3)$ $C: 3x - y - 3 = 0$ (line)

5. $O'(2, -1)$ $C: xy + x - 2y - 3 = 0$ (hyperbola)

6. $O'(-4, 1)$ $C: xy - x + 4y - 5 = 0$ (hyperbola)

Find the translation that moves the origin to the center of the given circle. Then draw both sets of axes and the circle.

7. $x^2 + y^2 - 4x + 6y + 9 = 0$

8. $x^2 + y^2 + 8x - 6y = 0$

9. $x^2 + y^2 = 4y$

10. $x^2 + y^2 + 4x + 4y + 4 = 0$

By means of a translation, write the equation in the form $y' = a \sin \omega x'$ or $y' = a \cos \omega x'$. Then draw the curve and both sets of axes.

11. $y = \cos\left(2x - \dfrac{\pi}{3}\right)$

12. $y = 2 \sin\left(3x + \dfrac{3\pi}{4}\right)$

13. $y = \cos\left(3x + \dfrac{3\pi}{4}\right) - 1$

14. $y = \sin\left(2x - \dfrac{\pi}{3}\right) + 1$

Choose a translation that changes the given equation into a simpler form. Then graph the equation, showing both sets of axes.

15. $y - 3 = (x + 1)^2$ (parabola)

16. $x + 4 = (y - 2)^2$ (parabola)

17. $\dfrac{(x - 1)^2}{9} + \dfrac{(y - 2)^2}{4} = 1$ (ellipse)

18. $\dfrac{(x + 2)^2}{4} + \dfrac{(y - 3)^2}{9} = 1$ (ellipse)

B 19. $x^2 + 3y^2 + 6x + 6 = 0$

20. $9x^2 - 16y^2 - 18x + 96y = 279$

21. $4x + y^2 + 4y - 4 = 0$

22. $4x^2 + 4y^2 - 12x + 8y = 3$

23. $x^2 - 4y^2 + 4x + 32y = 64$

24. $4x^2 + 8y^2 + 4x - 24y + 1 = 0$

C 25. Under the general translation $x = x' + h$ and $y = y' + k$, points $A(x_1, y_1)$ and $B(x_2, y_2)$ are given the coordinates $A'(x_1', y_1')$ and $B'(x_2', y_2')$, respectively. Prove algebraically that the distance from A to B is the same as the distance from A' to B'.

8-2 *Rotations*

A translation is only one type of *transformation of coordinates*. Another is a **rotation** of the coordinate axes through an angle ϕ about the origin (Figure 8-3). All rotations discussed in this chapter will have the origin as the center of rotation. Equations that relate the old (x, y)-coordinates of a point P and its new (x', y')-coordinates are:

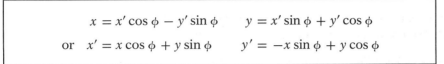

$$x = x' \cos \phi - y' \sin \phi \qquad y = x' \sin \phi + y' \cos \phi$$

$$\text{or} \quad x' = x \cos \phi + y \sin \phi \qquad y' = -x \sin \phi + y \cos \phi$$

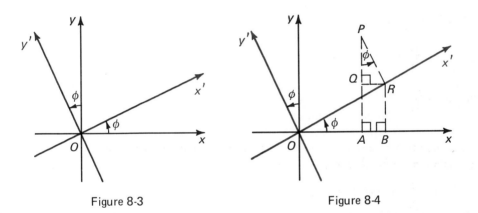

Figure 8-3   Figure 8-4

To derive these equations suppose that P is in the first quadrant of both systems as shown in Figure 8-4. From the figure we see that $OR = x'$, $RP = y'$, $OA = x$, and $AP = y$. Also,

$$OA = OB - AB = OB - QR$$

and $$AP = AQ + QP = BR + QP$$

since $AB = QR$ and $AQ = BR$. Using the definition of sine and the fact that triangles OBR and PQR are right triangles, we obtain the following:

$$OB = OR \cos \phi = x' \cos \phi$$
$$QR = PR \sin \phi = y' \sin \phi$$
$$BR = OR \sin \phi = x' \sin \phi$$
$$QP = PR \cos \phi = y' \cos \phi$$

Therefore, $$x = x' \cos \phi - y' \sin \phi, \quad \text{and}$$
$$y = x' \sin \phi + y' \cos \phi.$$

The second set of equations can be obtained from the first by solving for x' and y'. A complete proof using polar coordinates is outlined in Exercise 14.

Example 1   Transform the equation

$$2x^2 + \sqrt{3}xy + y^2 = 10$$

by rotating the axes through a 30° angle.

Solution   Since $\cos 30° = \dfrac{\sqrt{3}}{2}$ and $\sin 30° = \dfrac{1}{2}$, the equations of the rotation

are

$$x = \frac{\sqrt{3}}{2}x' - \frac{1}{2}y' = \frac{\sqrt{3}x' - y'}{2}$$

and

$$y = \frac{1}{2}x' + \frac{\sqrt{3}}{2}y' = \frac{x' + \sqrt{3}y'}{2}.$$

Substituting in the given equation, we have

$$2\left(\frac{\sqrt{3}x' - y'}{2}\right)^2 + \sqrt{3}\left(\frac{\sqrt{3}x' - y'}{2}\right)\left(\frac{x' + \sqrt{3}y'}{2}\right) + \left(\frac{x' + \sqrt{3}y'}{2}\right)^2 = 10,$$

or $2\dfrac{3(x')^2 - 2\sqrt{3}x'y' + (y')^2}{4} + \sqrt{3}\dfrac{\sqrt{3}(x')^2 + 2x'y' - \sqrt{3}(y')^2}{4} +$

$$\frac{(x')^2 + 2\sqrt{3}x'y' + 3(y')^2}{4} = 10.$$

This equation simplifies to $5(x')^2 + (y')^2 = 20$.
The last equation can be written as

$$\frac{(x')^2}{4} + \frac{(y')^2}{20} = 1.$$

Example 2   Transform the equation

$$4xy + 3y^2 + 4 = 0$$

by rotating the axes through the angle $\text{Cos}^{-1}\left(\dfrac{1}{\sqrt{5}}\right)$.

Solution   Let $\phi = \text{Cos}^{-1}\left(\dfrac{1}{\sqrt{5}}\right)$. Then

$$\cos \phi = \frac{1}{\sqrt{5}} \quad \text{and} \quad \sin \phi = \sqrt{1 - \cos^2 \phi} = \frac{2}{\sqrt{5}}.$$

The equations of the rotation are

$$x = \frac{1}{\sqrt{5}}x' - \frac{2}{\sqrt{5}}y' = \frac{x' - 2y'}{\sqrt{5}} \quad \text{and} \quad y = \frac{2}{\sqrt{5}}x' + \frac{1}{\sqrt{5}}y' = \frac{2x' + y'}{\sqrt{5}}.$$

Substituting in the given equation, we have

$$4\left(\frac{x' - 2y'}{\sqrt{5}}\right)\left(\frac{2x' + y'}{\sqrt{5}}\right) + 3\left(\frac{2x' + y'}{\sqrt{5}}\right)^2 + 4 = 0.$$

This equation simplifies to

$$4(x')^2 - (y')^2 + 4 = 0, \text{ or } \frac{(y')^2}{4} - \frac{(x')^2}{1} = 1.$$

From your studies of conic sections, you will recall that the graphs of equations of the form $\frac{x^2}{a^2} + \frac{y^2}{b^2} = 1$ are ellipses. The graph of the equation in Example 1 is therefore an ellipse. It is shown in Figure 8-5.

The graph of the equation in Example 2 is the hyperbola shown in Figure 8-6.

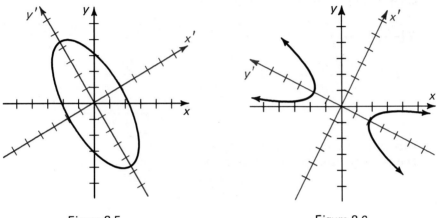

Figure 8-5   Figure 8-6

Exercises 8-2

In Exercises 1–12, simplify the equation by rotating the axes through the specified angle. Then draw its graph.

A 1. $4x - 3y = 15, \phi = \text{Cos}^{-1}\left(\frac{3}{5}\right)$ 2. $4x - 3y = 15, \phi = \text{Tan}^{-1}\left(-\frac{3}{4}\right)$

3. $x^2 - y^2 = 2, \phi = 45°$ 4. $x^2 + y^2 - \sqrt{2}x - \sqrt{2}y = 0, \phi = 45°$

5. $x^2 + 2xy + y^2 - \sqrt{2}x - \sqrt{2}y = 0, \phi = 45°$
   (Hint: Write the first three terms as the square of a binomial.)

6. $x^2 - 2xy + y^2 + x + y = 0, \phi = 45°$

7. $4x^2 + 4xy + y^2 + \sqrt{5}x - 2\sqrt{5}y = 0, \phi = \text{Cos}^{-1}\left(\frac{2}{\sqrt{5}}\right)$

8. $3x^2 - 4xy + 20 = 0$, $\phi = \text{Cos}^{-1}\left(\dfrac{1}{\sqrt{5}}\right)$

9. $\sqrt{3}xy - y^2 = 18$, $\phi = 30°$ 10. $x^2 + \sqrt{3}xy + 2y^2 - 10 = 0$, $\phi = 60°$

11. $2x^2 + 4xy - y^2 = 10$, $\phi = \text{Sin}^{-1}\left(-\dfrac{2}{\sqrt{5}}\right)$ 12. $2x^2 + 3xy + 2y^2 - 7 = 0$, $\phi = 45°$

B 13. Solve the system $\begin{aligned} x &= x' \cos\theta - y' \sin\theta \\ y &= x' \sin\theta + y' \cos\theta \end{aligned}$ for x' and y'.

14. In this exercise we derive the equations of the rotation through the angle ϕ. Let the arbitrary point P have polar coordinates (ρ, θ) in the (x, y)-system and (ρ', θ') in the (x', y')-system. Note that

$$\rho = \rho' \text{ and } \theta = \theta' + \phi.$$

Justify the following steps:

$$\begin{aligned} x &= \rho \cos\theta = \rho' \cos(\theta' + \phi) \\ &= \rho'(\cos\theta' \cos\phi - \sin\theta' \sin\phi) \\ &= (\rho' \cos\theta') \cos\phi - (\rho' \sin\theta') \sin\phi \\ &= x' \cos\phi - y' \sin\phi. \end{aligned}$$

Derive the equation $y = x' \sin\phi + y' \cos\phi$ in a similar way.

In Exercises 15–18 rotate the x- and y-axes so that the given line has a slope of 0 relative to the x'- and y'-axes. Then find the x'y'-equation of the line. (Hint: The slope of the line $Ax + By = C$ is $m = -\dfrac{A}{B}$.)

15. $x + y = -4$ 16. $2x - y = 3$

17. $3y + x = 6$ 18. $x - y = 2$

C 19. Under the general rotation $x = x' \cos\phi - y' \sin\phi$ and $y = x' \sin\phi + y' \cos\phi$, the points $A(x_1, y_1)$ and $B(x_2, y_2)$ are given the coordinates $A'(x_1', y_1')$ and $B'(x_2', y_2')$, respectively. Prove algebraically that the distance from A to B is the same as the distance from A' to B'.

20. Show algebraically that every line through the origin is rotated onto itself if $\phi = 180°$.

8-3   *Alias versus Alibi (Optional)*

In Sections 8-1 and 8-2 we took the point of view that we moved the coordinate system while points and curves in the plane remained fixed. Points acquired new coordinates, and curves acquired new equations; that is, these objects acquired *new names*. This is the **alias** (other name) interpretation of transformations.

The **alibi** (other place) interpretation takes a different point of view. In it, we think of the coordinate system as remaining fixed, while the points and curves move into new positions.

For example, consider the action of the translation

$$x = x' - 3 \qquad y = y' + 1$$
$$\text{or} \quad x' = x + 3 \qquad y' = y - 1$$

on the point P whose original coordinates are $(-1, 4)$. In the alias interpretation of the transformation, the axes are moved as indicated in Figure 8-7a, and the stationary point P acquires new coordinates $(2, 3)$. In the alibi interpretation, the point P moves from $(-1, 4)$ to the new position $(2, 3)$, as shown in Figure 8-7b.

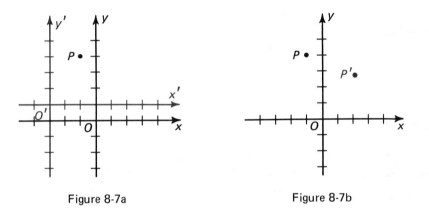

Figure 8-7a   Figure 8-7b

As a second example, let us consider Example 1 of Section 8-2. We used the alias interpretation and rotated the axes 30°, while the ellipse remained motionless (Figure 8-8a). In the alibi interpretation of the same transformation, we leave the axes fixed and rotate the ellipse $-30°$ (Figure 8-8b).

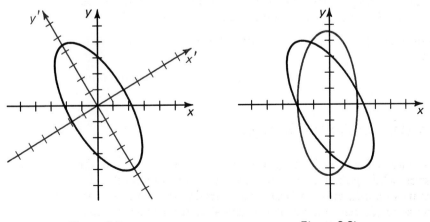

Figure 8-8a   Figure 8-8b

In general, the transformation

$$x = x' \cos \phi - y' \sin \phi, \; y = x' \sin \phi + y' \cos \phi$$

can be interpreted either as a rotation of the coordinate axes through the angle ϕ or as a rotation of the curve in question through the angle $-\phi$.

Example   Find an equation of the curve obtained by rotating the curve $4x^2 - 10xy + 4y^2 - 9 = 0$ through the angle $45°$.

Solution   Using $-\phi = 45°$, or $\phi = -45°$, the transformation equations become

$$x = x' \cos(-45°) - y' \sin(-45°) = \frac{1}{\sqrt{2}} x' + \frac{1}{\sqrt{2}} y'$$

$$y = x' \sin(-45°) + y' \cos(-45°) = -\frac{1}{\sqrt{2}} x' + \frac{1}{\sqrt{2}} y' \quad \text{or}$$

$$x = \frac{x' + y'}{\sqrt{2}}, \quad y = \frac{-x' + y'}{\sqrt{2}}.$$

Substituting in the given equation, we have

$$4 \left(\frac{x' + y'}{\sqrt{2}} \right)^2 - 10 \left(\frac{x' + y'}{\sqrt{2}} \right)\left(\frac{-x' + y'}{\sqrt{2}} \right) + 4 \left(\frac{-x' + y'}{\sqrt{2}} \right)^2 - 9 = 0$$

This equation simplifies to

$$9(x')^2 - (y')^2 - 9 = 0,$$

or

$$\frac{(x')^2}{1} - \frac{(y')^2}{9} = 1.$$

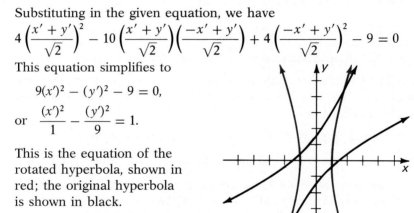

This is the equation of the rotated hyperbola, shown in red; the original hyperbola is shown in black.

Exercises 8-3

Draw figures illustrating the alias and the alibi interpretations of the given transformation's action on the given point or curve.

A   1. $x = x' - 3, \; y = y' + 1$; the point $(1, 0)$

2. $x = x' + 2, \; y = y' - 5$; the point $(-1, 3)$

3. $x = \dfrac{3x' - 4y'}{5}, \; y = \dfrac{4x' + 3y'}{5}$; the point $(6, 8)$

4. $x = \dfrac{3x' - 4y'}{5}, \; y = \dfrac{4x' + 3y'}{5}$; the line $3x + 4y = 10$

5. $x = \dfrac{x' - y'}{\sqrt{2}}$, $y = \dfrac{x' + y'}{\sqrt{2}}$; the parabola $4y = x^2$

6. $x = \dfrac{x' - y'}{\sqrt{2}}$, $y = \dfrac{x' + y'}{\sqrt{2}}$; the hyperbola $xy = 1$

Transform the given equation by means of the given translation or rotation.

B   7. $2x^2 + 8x - y + 3 = 0$; translate $(-2, -5)$ to the origin

8. $x^2 - 8x + y^2 + 4y = -11$; translate $(4, -2)$ to the origin

9. $3x^2 - 6\sqrt{3}xy + 9y^2 - 2\sqrt{3}x - 2y = 0$; rotate the curve $60°$

10. $13x^2 - 10xy + 13y^2 = 72$; rotate the curve $-45°$

8-4 *Second-Degree Equations (Optional)*

The general form of a second-degree equation in two variables is

$$ax^2 + bxy + cy^2 + dx + ey + f = 0.$$

By means of a rotation we can transform such an equation into an equation in x' and y' that has no $x'y'$-term. To determine what angle of rotation to use, let us substitute

$$x = x' \cos \phi - y' \sin \phi \quad \text{and} \quad y = x' \sin \phi + y' \cos \phi$$

in the general equation to find the coefficient of the resulting $x'y'$-term:

ax^2 contributes $a(-2 \sin \phi \cos \phi)x'y'$, or $-a(\sin 2\phi)x'y'$
bxy contributes $b(\cos^2 \phi - \sin^2 \phi)x'y'$, or $b(\cos 2\phi)x'y'$
cy^2 contributes $c(2 \sin \phi \cos \phi)x'y'$, or $c(\sin 2\phi)x'y'$

Therefore, the coefficient of $x'y'$ is $(c - a) \sin 2\phi + b \cos 2\phi$. We can always choose ϕ so that

$$(c - a) \sin 2\phi + b \cos 2\phi = 0.$$

Specifically:

To transform

$$ax^2 + bxy + cy^2 + dx + ey + f = 0$$

into an equation in x' and y' that has no $x'y'$-term, rotate the axes through the angle ϕ, where

$$\tan 2\phi = \frac{b}{a - c} \quad \text{if } a \neq c$$
$$\text{or } \phi = 45° \quad \text{if } a = c.$$

In Example 1 of Section 8-2, the given equation was

$$2x^2 + \sqrt{3}xy + y^2 = 10.$$

Here $a = 2$, $b = \sqrt{3}$, and $c = 1$, so that

$$\tan 2\phi = \frac{b}{a-c} = \frac{\sqrt{3}}{2-1} = \sqrt{3}.$$

Thus, $2\phi = 60°$, and $\phi = 30°$, the angle of rotation used in the example.

Example 1   Find a rotation that will transform

$$9x^2 + 24xy + 16y^2 + 20x - 15y = 0$$

into an equation having no $x'y'$-term. Then graph the equation.

Solution   Here $a = 9$, $b = 24$, and $c = 16$.

$$\tan 2\phi = \frac{24}{9-16} = -\frac{24}{7}$$

From the sketch we see that

$$\cos 2\phi = -\frac{7}{25}.$$

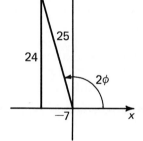

Therefore $\cos \phi = \sqrt{\dfrac{1+\cos 2\phi}{2}} =$

$$\sqrt{\dfrac{1-\dfrac{7}{25}}{2}} = \sqrt{\dfrac{9}{25}} = \dfrac{3}{5}$$

and $\sin \phi = \sqrt{1 - \cos^2 \phi} = \dfrac{4}{5}.$

Therefore, the equations of the rotation are

$$x = \frac{3}{5}x' - \frac{4}{5}y' = \frac{3x' - 4y'}{5}$$

and

$$y = \frac{4}{5}x' + \frac{3}{5}y' = \frac{4x' + 3y'}{5}.$$

The task of transforming the given equation is made easier when we notice that the first three terms form a perfect square:

$$9x^2 + 24xy + 16y^2 = (3x + 4y)^2$$

Since $3x + 4y = 3\left(\dfrac{3x' - 4y'}{5}\right) + 4\left(\dfrac{4x' + 3y'}{5}\right) = 5x'$,   the first three terms reduce to $25(x')^2$, and the transformed equation is

$$25(x')^2 + 20\left(\frac{3x' - 4y'}{5}\right) - 15\left(\frac{4x' + 3y'}{5}\right) = 0.$$

This reduces to $25(x')^2 - 25y' = 0$, or $y' = (x')^2$.

The graph is the parabola shown in the figure.

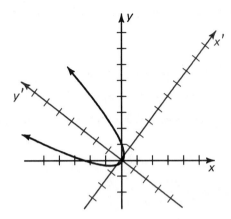

Example 2   Use a rotation followed by a translation to simplify the equation

$$x^2 - xy + y^2 - 3\sqrt{2}x = 0.$$

Then draw its graph.

Solution   Since the coefficients of x^2 and y^2 are equal, we use $\phi = 45°$. Since $\cos 45° = \sin 45° = \dfrac{1}{\sqrt{2}}$, equations of the rotation are

$$x = \frac{1}{\sqrt{2}}x' - \frac{1}{\sqrt{2}}y' = \frac{x' - y'}{\sqrt{2}}$$

and

$$y = \frac{1}{\sqrt{2}}x' + \frac{1}{\sqrt{2}}y' = \frac{x' + y'}{\sqrt{2}}.$$

Substituting in the given equation, we have

$$\frac{(x')^2 - 2x'y' + (y')^2}{2} - \frac{(x')^2 - (y')^2}{2} + \frac{(x')^2 + 2x'y' + (y')^2}{2} -$$

$$3\sqrt{2}\left(\frac{x' - y'}{\sqrt{2}}\right) = 0,$$

or $(x')^2 + 3(y')^2 - 6x' + 6y' = 0.$

(Solution continued on next page.)

Now, completing two squares, we obtain

$$((x')^2 - 6x' + 9) + 3((y')^2 + 2y' + 1) = 9 + 3,$$

or
$$(x' - 3)^2 + 3(y' + 1)^2 = 12.$$

Next use the translation

$$x'' = x' - 3, \quad y'' = y' + 1$$

to obtain the following equation of an ellipse:

$$(x'')^2 + 3(y'')^2 = 12, \quad \text{or} \quad \frac{(x'')^2}{12} + \frac{(y'')^2}{4} = 1$$

The graph of this ellipse is shown in the figure.

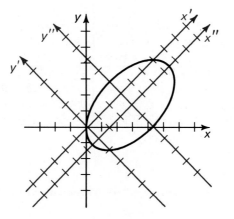

Exercises 8-4

In Exercises 1–10, find the equations of a rotation that will transform the equation into an equation in x' and y' without an $x'y'$-term.

A 1. $x^2 + 2xy + y^2 - \sqrt{2}x + \sqrt{2}y = 0$

2. $3x^2 - 2xy + 3y^2 = 4$

3. $x^2 - \sqrt{3}xy + 2y^2 = 25$

4. $\sqrt{3}xy + y^2 - 18 = 0$

5. $7x^2 - 4xy + 4y^2 = 240$

6. $24xy - 7y^2 = 144$

7. $4xy + 3y^2 + 4 = 0$

8. $3x^2 + 4xy + 4\sqrt{5}x + 4 = 0$

9. $x^2 + xy + y^2 - 3\sqrt{2}y - 12 = 0$

10. $x^2 + 4xy + 4y^2 - 5\sqrt{5}y = 0$

B 11–20. In Exercises 1–10, (a) find the transformed equation, (b) use a translation to simplify it further, if necessary, and (c) draw the graph.

C 21. If an arbitrary rotation is applied to the expression $ax^2 + bxy + cy^2$, the result will be an equation of the form $a'(x')^2 + b'x'y' + c'(y')^2$. Show that $a' + c' = a + c$.

Challenge

The graph of $ax^2 + bxy + cy^2 + dx + ey + f = 0$ contains the origin if the constant term f is 0. An equation for the line tangent at the origin to the graph of $ax^2 + bxy + cy^2 + dx + ey = 0$, where at least one of d and e is not 0, is $dx + ey = 0$. Find an equation for the line m that is tangent at the point $(4, -3)$ to the circle with the equation $x^2 + y^2 = 25$. (Hint: Translate the axes so that $(4, -3)$ is the new origin and find an equation for the tangent at the new origin. Then translate back to the original axes to find the required equation of the tangent.)

CHAPTER SUMMARY

1. Sometimes the equation of a given curve can be simplified by using a translation to slide the coordinate axes to a new position. If the new origin is the point with old coordinates (h, k), the new equation may be obtained by replacing

$$x \text{ with } (x' + h) \quad \text{and} \quad y \text{ with } (y' + k).$$

2. Rotation of the coordinate axes is another procedure that is helpful in simplifying equations. To rotate the axes through an angle θ about the origin, replace

$$x \text{ with } (x' \cos \theta - y' \sin \theta) \quad \text{and} \quad y \text{ with } (x' \sin \theta + y' \cos \theta).$$

3. The alias interpretation of transformations considers the points of a curve as fixed, but permits the coordinate axes to move, thus obtaining new coordinates and equations. The alibi interpretation considers the coordinate axes as fixed and the points of the curve as being moved onto a congruent curve.

4. The general form of a second-degree equation in two variables is

$$ax^2 + bxy + cy^2 + dx + ey + f = 0.$$

The shape of this curve is more easily identifiable if the xy-term is eliminated. To transform the equation into an equation in x' and y' with no $x'y'$-term, rotate the coordinate axes through the angle ϕ about the origin where

$$\phi = 45° \text{ if } a = c, \quad \text{or} \quad \tan 2\phi = \frac{b}{a - c} \text{ if } a \neq c.$$

CHAPTER TEST

8-1   1. Find the new equation of the line $2x - y - 1 = 0$ after translating the origin to the point $(1, 2)$. Write the equations of the translation.

Use a translation to simplify the equation. Then graph the equation, showing both sets of axes.

2. $y = \cos\left(2x - \frac{\pi}{4}\right) - 2$

3. $9x^2 + 4y^2 + 18x - 16y - 11 = 0$

Simplify the equation by rotating the axes through the specified angle. Then draw its graph.

8-2   4. $x^2 + xy + y^2 = 6; \ \theta = \frac{\pi}{4}$

5. $(x + \sqrt{3}y)^2 + 8\sqrt{3}x - 8y - 32 = 0; \ \theta = 60°$

6. Use the equations for the rotation of the coordinate axes through an angle ϕ about the origin to show that the equation of the unit circle $x^2 + y^2 = 1$ is unchanged by a rotation of axes.

8-3   7. (*Optional*) Consider the transformation $x = x' + 1, y = y' - 1$. Draw two figures, one illustrating the alias interpretation and the other the alibi interpretation of the transformation of the ellipse $x^2 - 2x + 4y^2 = 3$.

8-4   8. (*Optional*) Transform the following equation by a rotation which will eliminate the xy-term, and draw its graph.

$$16x^2 + 24xy + 9y^2 - 15x + 20y - 125 = 0$$

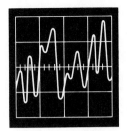

MATRICES AND LINEAR TRANSFORMATIONS

A transformation of coordinates that can be defined by equations of the form

$$x = ax' + by' \quad \text{and} \quad y = cx' + dy',$$

for some fixed real numbers a, b, c, and d, is called a **linear transformation.** Linear transformations can be represented by using matrix notation. The **matrix** (plural: matrices) of a linear transformation is simply a rectangular (often a square) array consisting of the constants that uniquely determine the linear transformation. For example, the transformation above can be written in matrix notation as

$$\begin{bmatrix} a & b \\ c & d \end{bmatrix}.$$

We have already considered one important class of linear transformations; namely, rotations. As we have seen, a rotation through an angle θ can be defined by the equations

$$x = x' \cos \theta - y' \sin \theta \quad \text{and} \quad y = x' \sin \theta + y' \cos \theta.$$

Thus the matrix of such a rotation is

$$\begin{bmatrix} \cos \theta & -\sin \theta \\ \sin \theta & \cos \theta \end{bmatrix}.$$

Note that from the definition of a linear transformation, any linear transformation of coordinates leaves the location of the origin unchanged. Therefore, a translation is *not* a linear transformation.

Another kind of linear transformation is called a **dilation** (or magnification or shrinking). This is a transformation that "expands" or "contracts" the coordinate system radially along the lines through the origin. A dilation can be defined by equations of the form

$$x = kx' \quad \text{and} \quad y = ky',$$

and thus has a matrix of the form

$$\begin{bmatrix} k & 0 \\ 0 & k \end{bmatrix}.$$

A diagram of the effect of a dilation where $k = 2$ is shown below.

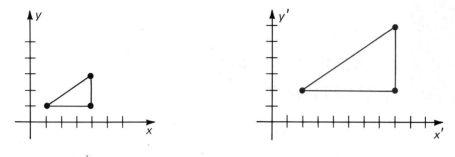

One advantage of matrix notation is that it allows us to find the **composition** of two linear transformations easily. That is, it enables us to find the single transformation that has the same effect as would be obtained by applying two given transformations *in succession*. We first define the **product** of two matrices as follows:

$$\begin{bmatrix} a & b \\ c & d \end{bmatrix} \cdot \begin{bmatrix} e & f \\ g & h \end{bmatrix} = \begin{bmatrix} ae + bg & af + bh \\ ce + dg & cf + dh \end{bmatrix}$$

The matrix for the composition of two linear transformations is the product of the matrices of the two factors. That is, if

$$\begin{aligned} x &= ax' + by' \\ y &= cx' + dy' \end{aligned} \quad \text{and} \quad \begin{aligned} x' &= ex'' + fy'' \\ y' &= gx'' + hy'' \end{aligned}$$

then the matrix associated with the transformation that changes the *xy*-coordinates of a point into the $x''y''$-coordinates is given by

$$\begin{bmatrix} ae + bg & af + bh \\ ce + dg & cf + dh \end{bmatrix}.$$

That is, we have

$$x = (ae + bg)x'' + (af + bh)y'' \quad \text{and} \quad y = (ce + dg)x'' + (cf + dh)y''.$$

Consider, for example, the composition of two rotations, one through an angle α and the other through an angle β. Clearly the composition of these two rotations is a rotation through the angle $\alpha + \beta$, shown in the diagram. This fact is verified as follows by matrix multiplication.

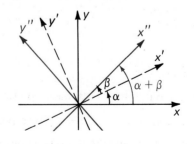

$$\begin{bmatrix} \cos \alpha & -\sin \alpha \\ \sin \alpha & \cos \alpha \end{bmatrix} \cdot \begin{bmatrix} \cos \beta & -\sin \beta \\ \sin \beta & \cos \beta \end{bmatrix} =$$

$$\begin{bmatrix} \cos (\alpha + \beta) & -\sin (\alpha + \beta) \\ \sin (\alpha + \beta) & \cos (\alpha + \beta) \end{bmatrix}$$

TRANSFORMING EQUATIONS

The following program uses the formulas $\tan \phi = \dfrac{b}{a - c}$ if $a \neq c$ and $\phi = 45°$ if $a = c$, to find the angle for a rotation that will produce a transformed equation without an $x'y'$–term. The program also gives the equations that relate the (x, y)–coordinates to the (x', y')–coordinates.

```
10    PRINT "TO ELIMINATE THE X'Y'-TERM:"
20    PRINT "INPUT A, B(<>0), C";
30    INPUT A, B, C
40    LET P1=3.14159
50    IF A=C THEN 110
60    LET P=ATN(B/(A−C))
70    IF P>0 THEN 90
80    LET P=P1+P
90    LET P=P/2
100   GOTO 120
110   LET P=P1/4
120   LET P2=INT(100*P*180/P1+.5)/100
130   PRINT "PHI=";P2; "DEG."
140   PRINT
150   LET S1=INT(100*SIN(P)+.5)/100
160   LET C1=INT(100*COS(P)+.5)/100
170   PRINT "X=";C1;"X'−";S1;"Y'"
180   PRINT "Y=";S1;"X'+";C1;"Y'"
190   END
```

Exercises

RUN the program for the values given.

1. $A = 1$, $B = -1$, $C = 1$ (Example 2 on page 225)

2. $A = 2$, $B = 1.732$, $C = 1$ (Example 1 on page 218)

3. $A = 0$, $B = 4$, $C = 3$ (Example 2 on page 218)

4. $A = 4$, $B = 4$, $C = 1$ (Exercise 7 on page 219)

The hydrostatic forces acting on a dam can be explained by using three-dimensional vectors.

VECTORS IN SPACE

OBJECTIVES

1. To define a coordinate system in space.
2. To study vectors, lines, and planes under this coordinate system.

Coordinates and Vectors

9-1 *Rectangular Coordinates in Space*

Since we live in three-dimensional space it is not surprising that we need a coordinate system for it. The most commonly used system is obtained by adding a z-axis to an xy-system in the manner shown in Figure 9-1.

The position of each point in space can now be described by an ordered triple (x, y, z) of numbers. Some examples are shown in Figure 9-2.

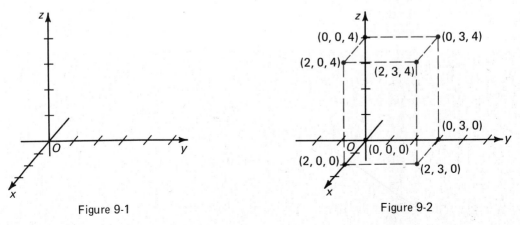

Figure 9-1   Figure 9-2

The plane determined by the x-axis and the z-axis is called the **xz-plane,** while the planes determined by the x- and y-axes and the y- and z-axes are called the xy- and yz-planes, respectively. These three **coordinate planes** divide space into eight regions, called **octants.** The octant in which points have all positive coordinates is the **first octant.** The other seven octants are not named.

In space the *distance* between the points $P_1(x_1, y_1, z_1)$ and $P_2(x_2, y_2, z_2)$ is given by:

$$P_1P_2 = \sqrt{(x_2 - x_1)^2 + (y_2 - y_1)^2 + (z_2 - z_1)^2}$$

We can prove this distance formula by applying the Pythagorean theorem to the right triangles P_1QR and P_1RP_2 shown in Figure 9-3 to obtain

$$(P_1R)^2 = |x_2 - x_1|^2 + |y_2 - y_1|^2$$

and

$$(P_1P_2)^2 = (P_1R)^2 + |z_2 - z_1|^2 =$$
$$(x_2 - x_1)^2 + (y_2 - y_1)^2 + (z_2 - z_1)^2.$$

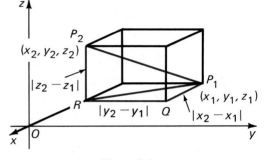

Figure 9-3

Example 1   Show that $A(1, 0, 2)$, $B(3, -1, 4)$, and $C(2, -2, 0)$ are the vertices of an isosceles right triangle.

Solution   Use the distance formula to find the lengths of the sides of $\triangle ABC$.

$$AB = \sqrt{(3 - 1)^2 + (-1 - 0)^2 + (4 - 2)^2} = \sqrt{4 + 1 + 4} = \sqrt{9} = 3$$
$$BC = \sqrt{(2 - 3)^2 + (-2 - (-1))^2 + (0 - 4)^2} = \sqrt{1 + 1 + 16} = 3\sqrt{2}$$
$$AC = \sqrt{(2 - 1)^2 + (-2 - 0)^2 + (0 - 2)^2} = \sqrt{1 + 4 + 4} = \sqrt{9} = 3$$

Since $AB = AC$, $\triangle ABC$ is isosceles, and since $(BC)^2 = (AB)^2 + (AC)^2$, it is also a right triangle.

Spheres are among the most important surfaces in space. The **sphere** of radius r centered at the point $C(h, k, l)$ is the set of all points $P(x, y, z)$ such that

$$CP = r, \text{ or } (CP)^2 = r^2.$$

Using the distance formula, we see that an equation of this sphere is

$$(x - h)^2 + (y - k)^2 + (z - l)^2 = r^2.$$

Spheres with center at the origin have equations of the form

$$x^2 + y^2 + z^2 = r^2.$$

Example 2 Find an equation of the sphere that has center at $(3, -1, 2)$ and is tangent to the *xy*-plane.

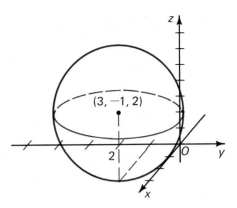

Solution To fulfill the tangency condition, the radius of the sphere must equal the distance from its center to the *xy*-plane. Thus $r = 2$, and, since $(h, k, l) = (3, -1, 2)$, we have

$$(x - 3)^2 + (y + 1)^2 + (z - 2)^2 = 2^2.$$

The equation $(x - h)^2 + (y - k)^2 + (z - l)^2 = r^2$ may be expanded and simplified, resulting in an equation of the form

$$x^2 + y^2 + z^2 + ax + by + cz + d = 0.$$

Thus, in Example 2, the equation of the sphere may also be given as $x^2 + y^2 + z^2 - 6x + 2y - 4z + 10 = 0$.

When an equation of a sphere is given in this expanded form, we can find its center and radius by completing squares.

Example 3 Find the center and radius of the sphere

$$x^2 + y^2 + z^2 - 2x + 6y + 4z + 5 = 0.$$

Solution We complete three squares:

$$(x^2 - 2x +) + (y^2 + 6y +) + (z^2 + 4z +) = -5$$
$$(x^2 - 2x + 1) + (y^2 + 6y + 9) + (z^2 + 4z + 4) = -5 + 1 + 9 + 4$$
$$(x - 1)^2 + (y + 3)^2 + (z + 2)^2 = 9$$

Therefore, the sphere has center $(1, -3, -2)$ and radius 3.

Exercises 9-1

In Exercises 1–6, (a) plot the given points and draw the triangle that they determine. (b) Use the distance formula to check whether the triangle is isosceles, right, both, or neither.

A 1. $(0, 0, 0)$, $(2, 3, 4)$, $(0, -2, 5)$

2. $(2, 1, 3)$, $(4, -1, 4)$, $(0, 0, 5)$

3. $(1, 0, 0)$, $(0, 2, -2)$, $(3, 2, 2)$

4. $(0, 1, -1)$, $(1, 1, 2)$, $(3, 4, -2)$

5. $(0, 1, 1)$, $(1, 0, 1)$, $(1, 1, 0)$

6. $(1, 0, 0)$, $(1, 1, 1)$, $(2, 2, -2)$

In Exercises 7–10, find an equation of the sphere having the given center and radius. Give your answer in the form $x^2 + y^2 + z^2 + ax + by + cz + d = 0$.

7. Center $(0, 0, 0)$, radius 5

8. Center $(0, 0, 1)$, radius 1

9. Center $(2, -1, -2)$, radius 3

10. Center $(1, -2, 3)$, radius 4

In Exercises 11–14, find the center and radius of the given sphere.

11. $x^2 + y^2 + z^2 = 2x$

12. $x^2 + y^2 + z^2 + 6x - 16z = 0$

13. $x^2 + y^2 + z^2 - 4x + 6y + 2z + 5 = 0$

14. $x^2 + y^2 + z^2 = x + y + z$

In Exercises 15 and 16 the given points are diagonally opposite vertices of a rectangular solid (box) that has its edges parallel to the coordinate axes. Find the other six vertices.

B 15. $(2, 0, 3)$, $(5, 2, -1)$

16. $(1, 3, 5)$, $(2, 4, 6)$

In Exercises 17 and 18 three vertices of a rectangle are given. Find the fourth vertex.

17. $(0, 2, 4)$, $(3, 2, 1)$, $(3, 0, 1)$

18. $(1, 2, 3)$, $(1, 5, 3)$, $(-1, 2, 5)$

In Exercises 19 and 20 find an equation for the set of points that are equidistant from the given points. (Hint: Let $P(x, y, z)$ be any point equidistant from A and B. Then $AP = BP$.)

19. $A(3, 0, 1)$, $B(0, 3, 1)$

20. $A(0, 0, 0)$, $B(2, 2, 2)$

In Exercises 21–24, find an equation of the sphere having the stated properties.

21. Center at $(3, 0, -4)$, passes through the origin.

22. Center at $(2, 3, -3)$, tangent to the xy-plane.

23. Radius 2, tangent to the xz-plane at $(2, 0, 3)$. (Two answers)

24. Radius 3, center in the first octant, tangent to all three coordinate planes.

C 25. A point P moves so that it is always twice as far from $(5, 0, 0)$ as from $(2, 0, 0)$. Show that P is always the same distance from some point. State the distance and find the coordinates of that point.

26. Show that the midpoint of the segment with endpoints (x_1, y_1, z_1) and (x_2, y_2, z_2) is the point $\left(\dfrac{x_1 + x_2}{2}, \dfrac{y_1 + y_2}{2}, \dfrac{z_1 + z_2}{2} \right)$.

27. Find an equation whose graph is the set of all points the sum of whose distances from $(0, 2, 0)$ and $(0, -2, 0)$ is 5 units.

9-2 *Vectors in Space*

In space we represent vector quantities by arrows and denote these vectors by boldface letters, just as we did in the plane. At this point it is a good idea to reread Sections 6-1 and 6-4. As you will see in this section, the definitions given for vectors in the plane can be readily extended to vectors in space.

Let us denote by **i**, **j**, and **k** the unit vectors having their initial points at the origin and their terminal points at $(1, 0, 0)$, $(0, 1, 0)$, and $(0, 0, 1)$, respectively. It can be seen from Figure 9-4 that any vector **a** is a linear combination of these unit vectors. That is,

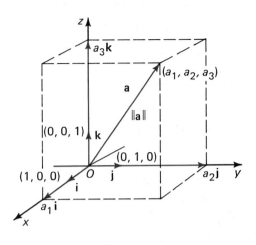

Figure 9-4

$$\mathbf{a} = a_1\mathbf{i} + a_2\mathbf{j} + a_3\mathbf{k}.$$

In terms of its components, a_1, a_2, and a_3, the norm of **a** is given by

$$\|\mathbf{a}\| = \sqrt{a_1^2 + a_2^2 + a_3^2}.$$

Algebraic properties of vectors can be used to show that if $\mathbf{a} = a_1\mathbf{i} + a_2\mathbf{j} + a_3\mathbf{k}$, $\mathbf{b} = b_1\mathbf{i} + b_2\mathbf{j} + b_3\mathbf{k}$, and t is a scalar, then:

$\mathbf{a} = \mathbf{b}$ if and only if $a_1 = b_1$, $a_2 = b_2$, and $a_3 = b_3$	(1)
$\mathbf{a} + \mathbf{b} = (a_1 + b_1)\mathbf{i} + (a_2 + b_2)\mathbf{j} + (a_3 + b_3)\mathbf{k}$	(2)
$t\mathbf{a} = ta_1\mathbf{i} + ta_2\mathbf{j} + ta_3\mathbf{k}$	(3)

Example 1 Let $\mathbf{a} = \mathbf{i} + 2\mathbf{j}$, $\mathbf{b} = \mathbf{j} + 2\mathbf{k}$, and $\mathbf{c} = 2\mathbf{i} + \mathbf{k}$.
(a) Find $2\mathbf{a} - 3\mathbf{b} + \mathbf{c}$. (b) Express $\mathbf{v} = 3\mathbf{i} - \mathbf{j} + 2\mathbf{k}$ as a linear combination of **a**, **b**, and **c**.

Solution (a) $2\mathbf{a} - 3\mathbf{b} + \mathbf{c} = 2(\mathbf{i} + 2\mathbf{j}) - 3(\mathbf{j} + 2\mathbf{k}) + (2\mathbf{i} - \mathbf{k})$
$= 4\mathbf{i} + \mathbf{j} - 5\mathbf{k}$

(b) We wish to find scalars r, s, and t such that

$$\mathbf{v} = r\mathbf{a} + s\mathbf{b} + t\mathbf{c},$$

or $3\mathbf{i} - \mathbf{j} + 2\mathbf{k} = r(\mathbf{i} + 2\mathbf{j}) + s(\mathbf{j} + 2\mathbf{k}) + t(2\mathbf{i} + \mathbf{k}).$

Multiplying and rearranging terms, we have

$$3\mathbf{i} - \mathbf{j} + 2\mathbf{k} = (r + 2t)\mathbf{i} + (2r + s)\mathbf{j} + (2s + t)\mathbf{k}.$$

(Solution continued on next page.)

This vector equation is equivalent to the following system of scalar equations.

$$
\begin{aligned}
r \quad\quad + 2t &= 3 \\
2r + s \quad\quad &= -1 \\
2s + t &= 2
\end{aligned}
$$

Solving this system gives $r = -\dfrac{5}{9}$, $s = \dfrac{1}{9}$, and $t = \dfrac{16}{9}$.

Therefore, $\mathbf{v} = -\dfrac{5}{9}\mathbf{a} + \dfrac{1}{9}\mathbf{b} + \dfrac{16}{9}\mathbf{c}.$

Recall that the dot product of $\mathbf{a}$ and $\mathbf{b}$ is defined by

$$\mathbf{a} \cdot \mathbf{b} = \|\mathbf{a}\|\|\mathbf{b}\| \cos\theta,$$

where θ is the angle between $\mathbf{a}$ and $\mathbf{b}$. Using a slight modification of the discussion given on page 162, we can show that if $\mathbf{a} = a_1\mathbf{i} + a_2\mathbf{j} + a_3\mathbf{k}$ and $\mathbf{b} = b_1\mathbf{i} + b_2\mathbf{j} + b_3\mathbf{k}$, then:

$$\mathbf{a} \cdot \mathbf{b} = a_1b_1 + a_2b_2 + a_3b_3$$

We note that the formula

$$\cos\theta = \frac{\mathbf{a} \cdot \mathbf{b}}{\|\mathbf{a}\|\|\mathbf{b}\|}$$

enables us to find the angle between two nonzero vectors. In particular, $\mathbf{a}$ and $\mathbf{b}$ are orthogonal if and only if

$$\mathbf{a} \cdot \mathbf{b} = a_1b_1 + a_2b_2 + a_3b_3 = 0.$$

The problem of expressing a vector $\mathbf{v}$ as a linear combination of $\mathbf{a}$, $\mathbf{b}$, and $\mathbf{c}$ (Example 1) becomes simpler if $\mathbf{a}$, $\mathbf{b}$, and $\mathbf{c}$ are orthogonal in pairs.

Example 2   Express $\mathbf{v} = 3\mathbf{i} - 5\mathbf{j} + 2\mathbf{k}$ as a linear combination of $\mathbf{a} = 2\mathbf{i} + 3\mathbf{j} + \mathbf{k}$, $\mathbf{b} = \mathbf{i} - \mathbf{j} + \mathbf{k}$, and $\mathbf{c} = 4\mathbf{i} - \mathbf{j} - 5\mathbf{k}$.

Solution   By taking dot products we can check that $\mathbf{a} \cdot \mathbf{b} = \mathbf{b} \cdot \mathbf{c} = \mathbf{c} \cdot \mathbf{a} = 0$. Thus $\mathbf{a}$, $\mathbf{b}$, and $\mathbf{c}$ are orthogonal.

To find r, s, and t in $\mathbf{v} = r\mathbf{a} + s\mathbf{b} + t\mathbf{c}$, dot multiply both sides of the equation in turn by $\mathbf{a}$, $\mathbf{b}$, and $\mathbf{c}$.

$$
\begin{aligned}
\mathbf{a} \cdot \mathbf{v} &= r\mathbf{a} \cdot \mathbf{a} + s\mathbf{a} \cdot \mathbf{b} + t\mathbf{a} \cdot \mathbf{c} \\
&= r(\mathbf{a} \cdot \mathbf{a}) + s(\mathbf{a} \cdot \mathbf{b}) + t(\mathbf{a} \cdot \mathbf{c}) \\
&= r(\mathbf{a} \cdot \mathbf{a}) + 0 + 0 \\
r &= \frac{\mathbf{a} \cdot \mathbf{v}}{\mathbf{a} \cdot \mathbf{a}} = \frac{-7}{14} = -\frac{1}{2}
\end{aligned}
$$

$$\mathbf{b} \cdot \mathbf{v} = r\mathbf{b} \cdot \mathbf{a} + s\mathbf{b} \cdot \mathbf{b} + t\mathbf{b} \cdot \mathbf{c}$$
$$= r(\mathbf{b} \cdot \mathbf{a}) + s(\mathbf{b} \cdot \mathbf{b}) + t(\mathbf{b} \cdot \mathbf{c})$$
$$= 0 + s(\mathbf{b} \cdot \mathbf{b}) + 0$$
$$s = \frac{\mathbf{b} \cdot \mathbf{v}}{\mathbf{b} \cdot \mathbf{b}} = \frac{10}{3}$$

$$\mathbf{c} \cdot \mathbf{v} = r\mathbf{c} \cdot \mathbf{a} + s\mathbf{c} \cdot \mathbf{b} + t\mathbf{c} \cdot \mathbf{c}$$
$$= r(\mathbf{c} \cdot \mathbf{a}) + s(\mathbf{c} \cdot \mathbf{b}) + t(\mathbf{c} \cdot \mathbf{c})$$
$$= 0 + 0 + t(\mathbf{c} \cdot \mathbf{c})$$
$$t = \frac{\mathbf{c} \cdot \mathbf{v}}{\mathbf{c} \cdot \mathbf{c}} = \frac{7}{42} = \frac{1}{6}$$

Therefore, $\mathbf{v} = -\dfrac{1}{2}\mathbf{a} + \dfrac{10}{3}\mathbf{b} + \dfrac{1}{6}\mathbf{c}$

Example 3 Find a nonzero vector that is orthogonal to both $\mathbf{a} = 2\mathbf{i} - \mathbf{j} - 3\mathbf{k}$ and $\mathbf{b} = \mathbf{i} + 2\mathbf{j} + 2\mathbf{k}.$

Solution We seek a vector $\mathbf{r} = x\mathbf{i} + y\mathbf{j} + z\mathbf{k}$ such that

$$\mathbf{a} \cdot \mathbf{r} = 2x + (-1)y + (-3)z = 0$$
$$\mathbf{b} \cdot \mathbf{r} = 1x + 2y + 2z = 0$$

We solve this system for x and y in terms of z.

$$2x - y = 3z$$
$$x + 2y = -2z$$
$$x = \frac{4}{5}z, \quad y = -\frac{7}{5}z$$

In order to avoid fractions, set $z = 5$ and obtain

$$x = 4, \quad y = -7, \quad z = 5.$$

Therefore, $\mathbf{r} = 4\mathbf{i} - 7\mathbf{j} + 5\mathbf{k}.$

The answer to Example 3 is not unique. Any nonzero scalar multiple of $\mathbf{r}$ has the required property.

Exercises 9-2

A 1. Let $\mathbf{a} = \mathbf{i} + 4\mathbf{j} - \mathbf{k}$ and $\mathbf{b} = \mathbf{i} - 2\mathbf{j} + 2\mathbf{k}$. (a) Find the linear combinations $\mathbf{a} + 2\mathbf{b}$ and $2\mathbf{b} - \mathbf{a}$. (b) Find $\|\mathbf{a}\|$, $\|\mathbf{b}\|$, $\|\mathbf{a} + \mathbf{b}\|$, and $\|\mathbf{a} - \mathbf{b}\|$.

2. Let $\mathbf{a} = \mathbf{i} + \mathbf{j} - 3\mathbf{k}$ and $\mathbf{b} = \mathbf{i} - 2\mathbf{j} + \mathbf{k}$. (a) Find the linear combinations $3\mathbf{a} - 2\mathbf{b}$ and $2\mathbf{a} - 3\mathbf{b}$. (b) Find $\|\mathbf{a}\|$, $\|\mathbf{b}\|$, $\|\mathbf{a} + \mathbf{b}\|$, and $\|\mathbf{a} - \mathbf{b}\|$.

In Exercises 3–6, (a) find the linear combination $3\mathbf{a} - 2\mathbf{b} + \mathbf{c}$. (b) Express **v** as a linear combination of **a**, **b**, and **c**.

3. $\mathbf{a} = \mathbf{i} + \mathbf{j} + \mathbf{k}, \mathbf{b} = \mathbf{j} + \mathbf{k}, \mathbf{c} = \mathbf{k}; \mathbf{v} = -3\mathbf{i} - 2\mathbf{j} + 4\mathbf{k}$

4. $\mathbf{a} = \mathbf{i} + \mathbf{j}, \mathbf{b} = \mathbf{j} + \mathbf{k}, \mathbf{c} = \mathbf{i} + \mathbf{k}; \mathbf{v} = 2\mathbf{i} + 3\mathbf{j}$

5. $\mathbf{a} = \mathbf{i} + \mathbf{j}, \mathbf{b} = \mathbf{i} - \mathbf{j}, \mathbf{c} = \mathbf{i} + \mathbf{j} + \mathbf{k}; \mathbf{v} = 2\mathbf{i} + 4\mathbf{j} + 5\mathbf{k}$

6. $\mathbf{a} = \mathbf{i} - 2\mathbf{j} + 3\mathbf{k}, \mathbf{b} = \mathbf{j} - \mathbf{k}, \mathbf{c} = \mathbf{k}; \mathbf{v} = \mathbf{i} - 2\mathbf{j} + 3\mathbf{k}$

7. Which pairs of the following vectors are (a) orthogonal? (b) parallel? (Recall that two vectors are parallel if and only if one is a scalar multiple of the other.)
   $\mathbf{a} = 4\mathbf{i} - 2\mathbf{j} + 2\mathbf{k}$ $\mathbf{b} = 2\mathbf{i} + 3\mathbf{j} + \mathbf{k}$
   $\mathbf{c} = 2\mathbf{i} + 3\mathbf{j} - \mathbf{k}$ $\mathbf{d} = -2\mathbf{i} + \mathbf{j} - \mathbf{k}$

8. A vector makes equal acute angles with **i**, **j**, and **k**. Find the measure of the angle to the nearest tenth of a degree.

In Exercises 9–12, find the angle between **a** and **b**.

Example $\mathbf{a} = 2\mathbf{i} + 3\mathbf{j} - \mathbf{k}, \mathbf{b} = \mathbf{i} - 2\mathbf{j} + 2\mathbf{k}$

Solution $\cos \theta = \dfrac{\mathbf{a} \cdot \mathbf{b}}{\|\mathbf{a}\| \|\mathbf{b}\|}$

$$= \frac{2 \cdot 1 + 3(-2) + (-1)2}{\sqrt{2^2 + 3^2 + (-1)^2}\sqrt{1^2 + (-2)^2 + 2^2}}$$

$$= \frac{-6}{3\sqrt{14}} = -\frac{2}{\sqrt{14}}$$

Therefore, $\theta = 122.3°$

9. $\mathbf{a} = \mathbf{i} + \mathbf{j} + \mathbf{k}, \mathbf{b} = 2\mathbf{i} + \mathbf{j} + 2\mathbf{k}$

10. $\mathbf{a} = \mathbf{i} - 2\mathbf{j} + 3\mathbf{k}, \mathbf{b} = 2\mathbf{i} + \mathbf{j}$

11. $\mathbf{a} = 3\mathbf{i} + 2\mathbf{j} + 2\mathbf{k}, \mathbf{b} = \mathbf{i} - 2\mathbf{j} - \mathbf{k}$

12. $\mathbf{a} = \mathbf{i} - \mathbf{j} + \mathbf{k}, \mathbf{b} = \mathbf{i} + \mathbf{j} - \mathbf{k}$

In Exercises 13 and 14 express **v** as a linear combination of **a**, **b**, and **c**. Use the fact that **a**, **b**, and **c** are orthogonal in pairs.

13. $\mathbf{a} = 3\mathbf{i} + 2\mathbf{j} + \mathbf{k}, \mathbf{b} = \mathbf{i} - \mathbf{j} - \mathbf{k}, \mathbf{c} = \mathbf{i} - 4\mathbf{j} + 5\mathbf{k}; \mathbf{v} = \mathbf{i} + \mathbf{j} + 2\mathbf{k}$

14. $\mathbf{a} = 2\mathbf{i} + 2\mathbf{j} + \mathbf{k}, \mathbf{b} = \mathbf{i} - 2\mathbf{j} + 2\mathbf{k}, \mathbf{c} = 2\mathbf{i} - \mathbf{j} - 2\mathbf{k}; \mathbf{v} = 3\mathbf{i} - \mathbf{j} + 4\mathbf{k}$

In Exercises 15 and 16, find a vector that is orthogonal to both **a** and **b**.

15. $\mathbf{a} = \mathbf{i} + \mathbf{j} + \mathbf{k}, \mathbf{b} = 2\mathbf{i} - 5\mathbf{j} + 3\mathbf{k}$

16. $\mathbf{a} = \mathbf{i} + \mathbf{j} - \mathbf{k}, \mathbf{b} = \mathbf{i} - 3\mathbf{j} - 3\mathbf{k}$

B 17. Sometimes the product $\mathbf{v} \cdot \mathbf{v}$ is denoted by $\mathbf{v}^2$. Show that in this notation we have the following familiar formulas.

$$(\mathbf{a} + \mathbf{b}) \cdot (\mathbf{a} - \mathbf{b}) = \mathbf{a}^2 - \mathbf{b}^2; \; (\mathbf{a} + \mathbf{b})^2 = \mathbf{a}^2 + 2\mathbf{a} \cdot \mathbf{b} + \mathbf{b}^2$$

18. Recalling the fact that $\|\mathbf{v}\|^2 = \mathbf{v} \cdot \mathbf{v}$, show that

$$\|\mathbf{a} + \mathbf{b}\|^2 + \|\mathbf{a} - \mathbf{b}\|^2 = 2\|\mathbf{a}\|^2 + 2\|\mathbf{b}\|^2.$$

(Since $\mathbf{a} + \mathbf{b}$ and $\mathbf{a} - \mathbf{b}$ are the diagonals of the parallelogram having sides $\mathbf{a}$ and $\mathbf{b}$, this equation shows that *the sum of the squares of the diagonals of any parallelogram equals the sum of the squares of the sides.*)

C 19. Is the following statement true or false? Give a reason for your answer. Given any three vectors $\mathbf{a}$, $\mathbf{b}$, and $\mathbf{c}$ in space, there is a nonzero vector $\mathbf{v}$ such that $\mathbf{a} \cdot \mathbf{v} = \mathbf{b} \cdot \mathbf{v} = \mathbf{c} \cdot \mathbf{v}$.

20. Let $\mathbf{a}$, $\mathbf{b}$, and $\mathbf{c}$ be *unit* vectors which are orthogonal in pairs and let $\mathbf{v}$ be an arbitrary vector. Show that

$$\mathbf{v} = (\mathbf{a} \cdot \mathbf{v})\mathbf{a} + (\mathbf{b} \cdot \mathbf{v})\mathbf{b} + (\mathbf{c} \cdot \mathbf{v})\mathbf{c}.$$

(Hint: Let $\mathbf{v} = r\mathbf{a} + s\mathbf{b} + t\mathbf{c}$ and determine r, s, and t as in Example 2.)

Lines and Planes

9-3 *Lines in Space*

In many applications it is convenient to denote the vector from the point $P_0(x_0, y_0, z_0)$ to $P_1(x_1, y_1, z_1)$ by $\overrightarrow{P_0P_1}$. As a linear combination of $\mathbf{i}$, $\mathbf{j}$, and $\mathbf{k}$,

$$\overrightarrow{P_0P_1} = (x_1 - x_0)\mathbf{i} + (y_1 - y_0)\mathbf{j} + (z_1 - z_0)\mathbf{k}.$$

For example, given the points $P(1, 2, 5)$ and $Q(4, -1, 3)$, we have

$$\overrightarrow{PQ} = (4 - 1)\mathbf{i} + (-1 - 2)\mathbf{j} + (3 - 5)\mathbf{k} = 3\mathbf{i} - 3\mathbf{j} - 2\mathbf{k}.$$

We can use vectors to describe the positions of points. We first choose a fixed reference point O, which we call the **origin** (whether or not a coordinate system is present). Then the location of any point P is described by the vector $\overrightarrow{OP}$, which is called the **position vector** of P. It is important to remember that *every position vector must have its initial point at O*. If a coordinate system is present, the position vector of the point $P(x, y, z)$ is given by

$$\overrightarrow{OP} = x\mathbf{i} + y\mathbf{j} + z\mathbf{k}.$$

Lines in space cannot be described by a single first-degree scalar equation as they are in the plane. However, they do have simple *vector* equations. For example, suppose we want to find an equation of the line l that passes through a point P_0 and is parallel to a given vector $\mathbf{m}$ (Figure 9-5).

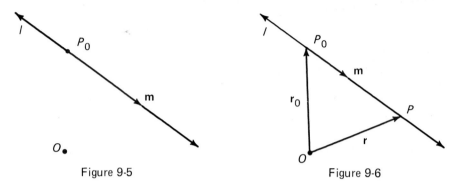

Figure 9-5   Figure 9-6

A point P is on l if and only if the vector $\overrightarrow{P_0P}$ is a scalar multiple of $\mathbf{m}$ as shown in Figure 9-6, that is, if and only if

$$\overrightarrow{P_0P} = t\mathbf{m}$$

for some real number t. Thus:

$$\overrightarrow{OP} = \overrightarrow{OP_0} + \overrightarrow{P_0P} = \overrightarrow{OP_0} + t\mathbf{m}$$

Denoting the position vectors of P and P_0 by $\mathbf{r}$ and $\mathbf{r}_0$, respectively, we obtain:

$$\mathbf{r} = \mathbf{r}_0 + t\mathbf{m} \qquad (\mathbf{r} = \overrightarrow{OP},\ \mathbf{r}_0 = \overrightarrow{OP_0})$$

This is a **vector equation of the line** l. In the equation, $\mathbf{r}_0$ and $\mathbf{m}$ are given constant vectors, t is a scalar variable, and $\mathbf{r}$ is a variable vector whose initial point is always at O. As t varies, so does $\mathbf{r}$ in such a way that its terminal point traces l.

Example 1 Find a vector equation of the line l that passes through the point $P_0(1, 4, 3)$ and is parallel to $3\mathbf{i} - 2\mathbf{j} + \mathbf{k}$.

Solution Let $\mathbf{r}_0 = \overrightarrow{OP_0} = \mathbf{i} + 4\mathbf{j} + 3\mathbf{k}$ and let $\mathbf{m} = 3\mathbf{i} - 2\mathbf{j} + \mathbf{k}$. Then $\mathbf{r} = \mathbf{r}_0 + t\mathbf{m}$ becomes

$$\mathbf{r} = (\mathbf{i} + 4\mathbf{j} + 3\mathbf{k}) + t(3\mathbf{i} - 2\mathbf{j} + \mathbf{k}),$$

or

$$\mathbf{r} = (1 + 3t)\mathbf{i} + (4 - 2t)\mathbf{j} + (3 + t)\mathbf{k}.$$

Next, we will find a vector equation of the line l determined by two points P_0 and P_1 with position vectors $\mathbf{r}_0$ and $\mathbf{r}_1$, respectively (Figure 9-7).

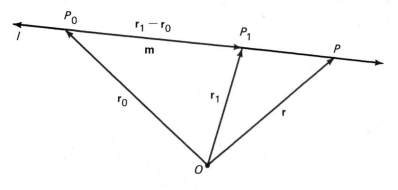

Figure 9-7

We note that

$$\mathbf{m} = \overrightarrow{P_0 P_1} = \mathbf{r}_1 - \mathbf{r}_0$$

is parallel to l and therefore $\mathbf{r} = \mathbf{r}_0 + t\mathbf{m}$ becomes

$$\mathbf{r} = \mathbf{r}_0 + t(\mathbf{r}_1 - \mathbf{r}_0),$$

or

$$\mathbf{r} = (1 - t)\mathbf{r}_0 + t\mathbf{r}_1$$

This is the desired vector equation of l. Figure 9-8 indicates how $\mathbf{r}$ varies as t increases. Setting $t = \dfrac{1}{2}$, we can see that the midpoint of $\overrightarrow{P_0 P_1}$ has position vector $\dfrac{1}{2}(\mathbf{r}_0 + \mathbf{r}_1)$.

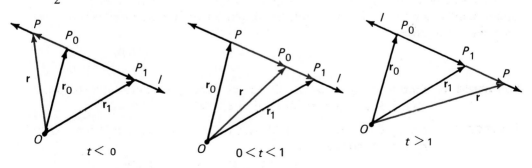

Figure 9-8

Example 2 (a) Find a vector equation of the line *l* determined by $P_0(1, 4, 0)$ and $P_1(2, 2, 3)$. Then (b) find the point where *l* intersects the *xz*-plane, and (c) draw the part of *l* lying in the first octant.

Solution (a) Let $\mathbf{r}_0 = \overrightarrow{OP_0} = \mathbf{i} + 4\mathbf{j}$ and $\mathbf{r}_1 = \overrightarrow{OP_1} = 2\mathbf{i} + 2\mathbf{j} + 3\mathbf{k}$. Then $\mathbf{r} = (1 - t)\mathbf{r}_0 + t\mathbf{r}_1$ becomes

$$\mathbf{r} = (1 - t)(\mathbf{i} + 4\mathbf{j}) + t(2\mathbf{i} + 2\mathbf{j} + 3\mathbf{k})$$

or

$$\mathbf{r} = (1 + t)\mathbf{i} + (4 - 2t)\mathbf{j} + 3t\mathbf{k}. \tag{$*$}$$

(b) The vector $\mathbf{r}$ positions a point in the *xz*-plane if and only if the coefficient of $\mathbf{j}$ is 0.

Therefore, $4 - 2t = 0$, or $t = 2$.

Substituting $t = 2$ into $(*)$, we have $\mathbf{r} = 3\mathbf{i} + 6\mathbf{k}$. Therefore, the desired point is $(3, 0, 6)$.

(c) $P_1(2, 2, 3)$ is in the first octant. Since *l* intersects the *xz*-plane at $(3, 0, 6)$ and the *xy*-plane at $P_0(1, 4, 0)$, these are the exit points in the first octant for *l*.

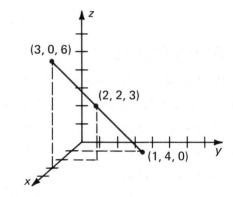

Example 3 Find the distance from the origin to the line *l* whose equation is given by

$$\mathbf{r} = (2 + 3t)\mathbf{i} + (1 - t)\mathbf{j} + (1 + 2t)\mathbf{k}.$$

Solution Put the given equation into the form $\mathbf{r} = \mathbf{r}_0 + t\mathbf{m}$.

$$\mathbf{r} = (2\mathbf{i} + \mathbf{j} + \mathbf{k}) + t(3\mathbf{i} - \mathbf{j} + 2\mathbf{k})$$

We see that $\mathbf{m} = 3\mathbf{i} - \mathbf{j} + 2\mathbf{k}$ is parallel to *l*. A point *P* on *l* is closest to *O* when

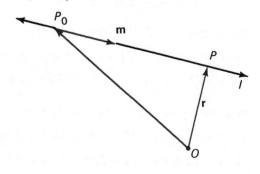

$r = \overrightarrow{OP}$ is orthogonal to l, that is, when $\mathbf{m} \cdot \mathbf{r} = 0$. Substituting for $\mathbf{m}$ and $\mathbf{r}$ gives

$$(3\mathbf{i} - \mathbf{j} + 2\mathbf{k}) \cdot [(2 + 3t)\mathbf{i} + (1 - t)\mathbf{j} + (1 + 2t)\mathbf{k}] = 0.$$

Solving this equation for t results in

$$3(2 + 3t) - 1(1 - t) + 2(1 + 2t) = 0$$
$$7 + 14t = 0$$
$$t = -\frac{1}{2}.$$

Substituting into the equation of l, we obtain

$$\mathbf{r} = \frac{1}{2}\mathbf{i} + \frac{3}{2}\mathbf{j}.$$

Therefore, $d = \|\mathbf{r}\| = \frac{1}{2}\sqrt{10}.$

Exercises 9-3

In Exercises 1–8, find a vector equation of the line having the stated properties.

A   1. Passes through $(1, 2, 0)$ and is parallel to $\mathbf{i} - \mathbf{j} + 3\mathbf{k}$

2. Passes through $(1, 1, 1)$ and is parallel to $-\mathbf{i} - \mathbf{j} - \mathbf{k}$

3. Passes through the origin and $(2, 3, -1)$

4. Passes through $(1, -2, 4)$ and $(2, 1, 3)$

5. Passes through $(1, 2, 0)$ and makes equal angles with $\mathbf{i}$, $\mathbf{j}$, and $\mathbf{k}$

6. Passes through the origin and makes equal angles with $\mathbf{i} + \mathbf{j}$, $\mathbf{j} + \mathbf{k}$, and $\mathbf{i} + \mathbf{k}$

7. Passes through $(0, -1, 3)$ and is parallel to the line through $(2, 0, 1)$ and $(3, 2, 0)$

8. Passes through $(1, 2, 3)$ and is parallel to the line $\mathbf{r} = (1 + t)\mathbf{i} + 2t\mathbf{j} + (1 - t)\mathbf{k}$

In Exercises 9 and 10, find the point where the line intersects the given plane.

9. $\mathbf{r} = (3 + t)\mathbf{i} + 2t\mathbf{j} + (2 - t)\mathbf{k}$; xy-plane

10. $\mathbf{r} = (3 + t)\mathbf{i} + (1 + t)\mathbf{j} + 2t\mathbf{k}$; xz-plane

In Exercises 11 and 12, find the distance from the origin to the line l.

11. $\mathbf{r} = (3 + t)\mathbf{i} + (4 - 2t)\mathbf{j} + (-2 + 2t)\mathbf{k}$

12. $\mathbf{r} = (2 + t)\mathbf{i} + (4 + 2t)\mathbf{j} + (-2 - t)\mathbf{k}$

In Exercises 13 and 14, find the distance from the point P_1 to the line l. (Hint: Replace the condition $\mathbf{m} \cdot \mathbf{r} = 0$ in Example 3 by $\mathbf{m} \cdot (\mathbf{r} - \mathbf{r}_1) = 0$, where $\mathbf{r}_1 = \overrightarrow{OP_1}$.)

B 13. $P_1(2, 0, 1)$, l: $\mathbf{r} = (3 + t)\mathbf{i} + (4 + t)\mathbf{j} + t\mathbf{k}$

14. $P_1(2, 2, 1)$, l: $\mathbf{r} = (3 + t)\mathbf{i} + (4 - 2t)\mathbf{j} + (1 - t)\mathbf{k}$

In Exercises 15 and 16, show that l_1 and l_2 are the same line. (Hint: Find two points on l_1 and show that they are on l_2.)

15. l_1: $\mathbf{r} = (2 + t)\mathbf{i} + (-1 - 2t)\mathbf{j} + (1 - t)\mathbf{k}$
l_2: $\mathbf{r} = (1 - s)\mathbf{i} + (1 + 2s)\mathbf{j} + (2 + s)\mathbf{k}$

16. l_1: $\mathbf{r} = (5 - 2t)\mathbf{i} + 3t\mathbf{j} + (2 - t)\mathbf{k}$
l_2: $\mathbf{r} = (1 + 2s)\mathbf{i} + (6 - 3s)\mathbf{j} + s\mathbf{k}$

In Exercises 17–20, find the points, if any, where the given line intersects the given sphere.

Example $\mathbf{r} = (1 + t)\mathbf{i} + 2\mathbf{j} - t\mathbf{k}$, $x^2 + y^2 + z^2 - 4y - 4z - 33 = 0$

Solution   Let $\mathbf{r} = x\mathbf{i} + y\mathbf{j} + z\mathbf{k}$. Then

$$x = 1 + t, \; y = 2, \; z = -t.$$

Substitute into the equation of the sphere:

$$(1 + t)^2 + 2^2 + (-t)^2 - 4 \cdot 2 - 4(-t) - 33 = 0$$

This equation reduces to

$$t^2 + 3t - 18 = 0, \text{ or } (t + 6)(t - 3) = 0.$$

Thus $t = -6$ or $t = 3$, and the points of intersection are $(-5, 2, 6)$ and $(4, 2, -3)$.

17. $\mathbf{r} = 2t\mathbf{i} + t\mathbf{j} - t\mathbf{k}$, $x^2 + y^2 + z^2 - 4x + 2y - 12 = 0$

18. $\mathbf{r} = (1 + t)\mathbf{i} + 2t\mathbf{k}$, $x^2 + y^2 + z^2 + 2x - 4y + 3z - 18 = 0$

19. $\mathbf{r} = \mathbf{i} + 2t\mathbf{j} + (2 - t)\mathbf{k}$, $x^2 + y^2 + z^2 - 6x - 2y + 2z + 2 = 0$

20. $\mathbf{r} = (2 + t)\mathbf{i} + t\mathbf{k}$, $x^2 + y^2 + z^2 - 2y = 0$

C 21. Let $\mathbf{r} = x\mathbf{i} + y\mathbf{j} + z\mathbf{k}$, $\mathbf{r}_0 = x_0\mathbf{i} + y_0\mathbf{j} + z_0\mathbf{k}$, and $\mathbf{m} = d\mathbf{i} + e\mathbf{j} + f\mathbf{k}$. Show that the vector equation $\mathbf{r} = \mathbf{r}_0 + t\mathbf{m}$ of the line l is equivalent to

$$x = x_0 + dt, \; y = y_0 + et, \; z = z_0 + ft.$$

These are the **scalar parametric equations** of l.

22. Show that if the line l described in Exercise 21 is not parallel to any coordinate axis, then it can be described by equations of the form

$$\frac{x - x_0}{d} = \frac{y - y_0}{e} = \frac{z - z_0}{f}.$$

These are the **symmetric equations** of l.

9-4 *Planes*

Let P_0 be a point and **n** be a nonzero vector. The set of all points P such that $\overrightarrow{P_0P}$ is perpendicular to **n**, that is, such that $\mathbf{n} \cdot \overrightarrow{P_0P} = 0$ (Figure 9-9), is called the *plane through P with **normal vector n.*** (The term normal vector is used to describe a vector that is perpendicular to a plane.) Letting $\mathbf{r}_0 = \overrightarrow{OP_0}$ and $\mathbf{r} = \overrightarrow{OP}$, we have the following as a vector equation of Q.

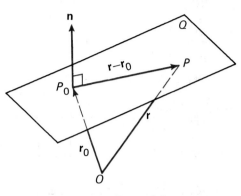

$$\mathbf{n} \cdot (\mathbf{r} - \mathbf{r}_0) = 0$$

Figure 9-9

Now introduce a coordinate system and let $\mathbf{n} = a\mathbf{i} + b\mathbf{j} + c\mathbf{k}$, $\mathbf{r}_0 = x_0\mathbf{i} + y_0\mathbf{j} + z_0\mathbf{k}$, and $\mathbf{r} = x\mathbf{i} + y\mathbf{j} + z\mathbf{k}$. Then $\mathbf{n} \cdot (\mathbf{r} - \mathbf{r}_0) = 0$ becomes

$$(a\mathbf{i} + b\mathbf{j} + c\mathbf{k}) \cdot [(x - x_0)\mathbf{i} + (y - y_0)\mathbf{j} + (z - z_0)\mathbf{k}] = 0.$$

Therefore, a scalar equation of Q is given by:

$$a(x - x_0) + b(y - y_0) + c(z - z_0) = 0$$

This equation also can be written in the form

$$ax + by + cz + d = 0,$$

where $d = -(ax_0 + by_0 + cz_0)$.

It can be shown (Exercise 24, page 251) that *the graph of every equation of this form is a plane having* $a\mathbf{i} + b\mathbf{j} + c\mathbf{k}$ *as a normal vector* provided a, b, and c are not all 0.

In drawing a plane it often is sufficient to show only the part of the plane that lies in the first octant. We do this by drawing its **traces.** These are the lines in which the given plane intersects the coordinate planes. It often is helpful to plot the **intercepts** of the given plane. These are the points where the coordinate axes intersect the given plane.

Example 1 Graph the planes whose equations are given.

(a) $4x + 3y + 2z = 8$ (b) $z = x + y - 1$ (c) $2x + \dfrac{3}{2}y = 6$

(Solution on next page.)

Solution

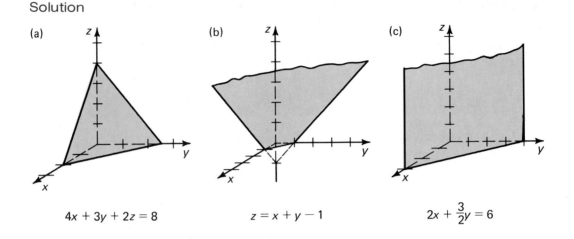

(a) $4x + 3y + 2z = 8$

(b) $z = x + y - 1$

(c) $2x + \dfrac{3}{2}y = 6$

Note that in space the graph of $2x + \dfrac{3}{2}y = 6$ in Example 1 is a *plane, not* a line. The absence of z simply means that the plane is parallel to the z-axis.

Example 2   Find a vector equation of the line l that passes through the point $(0, 0, 3)$ and is perpendicular to the plane $4x - 2y + z - 5 = 0$.

Solution   A vector perpendicular to the plane and therefore parallel to l is

$$\mathbf{m} = 4\mathbf{i} - 2\mathbf{j} + \mathbf{k}.$$

The position vector of $(0, 0, 3)$ is $\mathbf{r}_0 = 3\mathbf{k}$. Substitute these values into $\mathbf{r} = \mathbf{r}_0 + t\mathbf{m}$ to obtain

$$\mathbf{r} = 3\mathbf{k} + t(4\mathbf{i} - 2\mathbf{j} + \mathbf{k}),$$

or

$$\mathbf{r} = 4t\mathbf{i} - 2t\mathbf{j} + (t + 3)\mathbf{k}.$$

Example 3   Find an equation of the plane Q that contains the point $(2, -1, 1)$ and is perpendicular to the line $\mathbf{r} = (2 + t)\mathbf{i} - 3t\mathbf{j} + (1 + 2t)\mathbf{k}$.

Solution   Writing the equation of the line as

$$\mathbf{r} = (2\mathbf{i} + \mathbf{k}) + t(\mathbf{i} - 3\mathbf{j} + 2\mathbf{k}),$$

we see that $\mathbf{i} - 3\mathbf{j} + 2\mathbf{k}$ is parallel to the line and hence perpendicular to Q. Thus, $a(x - x_0) + b(y - y_0) + c(z - z_0) = 0$ becomes

$$1(x - 2) - 3(y + 1) + 2(z - 1) = 0,$$

or

$$x - 3y + 2z - 7 = 0.$$

Example 4   Find an equation of the plane that passes through $(1, -1, -3)$, $(0, 3, 2)$, and $(0, 0, -2)$.

Solution   Substitute the coordinates of the given points into $ax + by + cz + d = 0$ to obtain

$$a - b - 3c + d = 0$$
$$3b + 2c + d = 0$$
$$-2c + d = 0$$

Solve this system for c, b, and a in terms of d.

$$c = \frac{d}{2}$$

$$b = \frac{1}{3}(-2c - d) = \frac{1}{3}(-d - d) = -\frac{2d}{3}$$

$$a = b + 3c - d = -\frac{2d}{3} + \frac{3d}{2} - d = -\frac{d}{6}$$

To avoid fractions, set $d = -6$ and obtain $a = 1$, $b = 4$, $c = -3$. Therefore, an equation of the plane is

$$x + 4y - 3z - 6 = 0.$$

Next we will show that the distance from the point $P_1(x_1, y_1, z_1)$ to the plane $ax + by + cz + d = 0$ is given by:

$$D = \frac{|ax_1 + by_1 + cz_1 + d|}{\sqrt{a^2 + b^2 + c^2}}$$

First, choose any point $P_0(x_0, y_0, z_0)$ of the plane and let $\mathbf{n} = a\mathbf{i} + b\mathbf{j} + c\mathbf{k}$. We see from Figure 9-10 that

$$D = \|\overrightarrow{P_0P_1}\||\cos\theta| = \frac{|\mathbf{n} \cdot \overrightarrow{P_0P_1}|}{\|\mathbf{n}\|}.$$

Since
$$\overrightarrow{P_0P_1} = (x_1 - x_0)\mathbf{i} + (y_1 - y_0)\mathbf{j} + (z_1 - z_0)\mathbf{k},$$
this becomes

$$D = \frac{|a(x_1 - x_0) + b(y_1 - y_0) + c(z_1 - z_0)|}{\sqrt{a^2 + b^2 + c^2}}$$

$$= \frac{|ax_1 + by_1 + cz_1 - (ax_0 + by_0 + cz_0)|}{\sqrt{a^2 + b^2 + c^2}}.$$

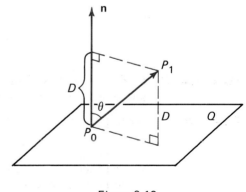

Figure 9-10

Since the quantity in parentheses is $-d$, the formula is established.

Example 5 Find the distance between the parallel planes Q_1: $3x - y + 2z = 8$ and Q_2: $3x - y + 2z = 13$.

Solution Choose $P_1(0, 0, 4)$ on Q_1. The distance we seek is the distance D from P_1 to Q_2.

$$\text{Therefore, } D = \frac{|3 \cdot 0 - 0 + 2 \cdot 4 - 13|}{\sqrt{3^2 + 1^2 + 2^2}} = \frac{5}{\sqrt{14}}.$$

Exercises 9-4

In Exercises 1–4, (a) find an equation of the plane through P_0 that is perpendicular to the given vector or line. (b) Then draw the first-octant part of the plane.

A 1. $P_0(0, 0, 2)$; $2\mathbf{i} + \mathbf{j} + 2\mathbf{k}$

2. $P_0(4, 3, 2)$; $2\mathbf{j} + 3\mathbf{k}$

3. $P_0(2, 3, 4)$; $\mathbf{r} = (1 + t)\mathbf{i} + 2t\mathbf{k}$

4. $P_0(1, 1, 1)$; $\mathbf{r} = (-1 + 2t)\mathbf{i} + t\mathbf{j} + 3t\mathbf{k}$

In Exercises 5–8, find a vector equation of the line through P_1 that is perpendicular to Q.

5. $P_1(3, 2, 1)$; Q: $2x + y - 2z + 6 = 0$

6. $P_1(2, 0, 1)$; Q: $x - 2y + 3z = 0$

7. $P_1(0, -2, 2)$; Q: $3x + 2y - 6z - 4 = 0$

8. $P_1(1, 2, 3)$; Q: $3x - 4z = 5$

9–12. Find the distance from P_1 to Q in Exercises 5–8.

In Exercises 13 and 14, find an equation of the plane that passes through the given points.

13. $(1, 3, 2)$, $(0, 1, -1)$, $(0, 1, 1)$

14. $(1, 2, 3)$, $(0, 1, 2)$, $(0, 0, -1)$

In Exercises 15 and 16, (a) verify that the given planes are parallel and then (b) find the distance between them.

15. $x - 2y + 2z + 6 = 0$; $2x - 4y + 4z - 5 = 0$

16. $2x + 4y - 5z = 0$; $2x + 4y - 5z + 10 = 0$

In Exercises 17 and 18, find the point where the line l intersects the plane Q. (Hint: See the example preceding Exercise 17, Section 9-3, page 246.)

B 17. l: $\mathbf{r} = (3 - t)\mathbf{i} - 2t\mathbf{j} + (1 + t)\mathbf{k}$
Q: $2x + y + 3z = 6$

18. l: $\mathbf{r} = 2t\mathbf{i} + (2 - t)\mathbf{j} + \mathbf{k}$
Q: $x + y - 3z = 1$

In Exercises 19 and 20, show that the line *l* is parallel to the plane *Q*. (Hint: Show that *l* and *Q* have no point in common.)

19. *l*: $\mathbf{r} = (2 + t)\mathbf{i} + 4t\mathbf{j} + (1 + 2t)\mathbf{k}$
    Q: $2x - y + z + 3 = 0$

20. *l*: $\mathbf{r} = (3 - 2t)\mathbf{i} + (1 + t)\mathbf{j} + 3t\mathbf{k}$
    Q: $x - y + z = 0$

In Exercises 21 and 22, find a vector equation of the line of intersection of the given planes. (Hint: First find two points that belong to both planes; that is, find two simultaneous solutions of the given equations.)

21. $x + y - z = 3;\ x - y + 3z = 1$ 22. $x + 2y - z = 1;\ x - 4y + z = 5$

23. Let *l* be a line, *P* be a point not on *l*, and *Q* be a plane that does not contain *P* and is neither parallel nor perpendicular to *l*. How many planes are there (none, one, or infinitely many) that
    (a) contain *P* and are perpendicular to *l*?
    (b) contain *P* and are parallel to *l*?
    (c) contain *P* and are perpendicular to *Q*?
    (d) contain *P* and are parallel to *Q*?
    (e) contain *l* and are perpendicular to *Q*?
    (f) contain *l* and are parallel to *Q*?

C 24. Show that the graph of any equation of the form $ax + by + cz + d = 0$ (*a*, *b*, and *c* not all 0) is a plane. (Hint: First show that there is a solution (x_0, y_0, z_0) of the given equation. Then show that the given equation is equivalent to $a(x - x_0) + b(y - y_0) + c(z - z_0) = 0$ and hence to one of the form $\mathbf{n} \cdot (\mathbf{r} - \mathbf{r}_0) = 0$.)

Challenge

The cross product of two vectors $\mathbf{t} = t_1\mathbf{i} + t_2\mathbf{j} + t_3\mathbf{k}$ and $\mathbf{v} = v_1\mathbf{i} + v_2\mathbf{j} + v_3\mathbf{k}$ is given by

$$\mathbf{t} \times \mathbf{v} = (t_2 v_3 - t_3 v_2)\mathbf{i} + (t_3 v_1 - t_1 v_3)\mathbf{j} + (t_1 v_2 - t_2 v_1)\mathbf{k}.$$

1. Show that the cross product of two parallel vectors is the zero vector.

2. Show that if **a** and **b** are vectors then $\mathbf{a} \times \mathbf{b}$ is perpendicular to each of the vectors **a** and **b**.

CHAPTER SUMMARY

1. A rectangular coordinate system for three-dimensional space is formed by adding a z-axis perpendicular to the origin of the xy-system. The plane determined by the x-axis and the z-axis is called the xz-plane. Similarly, the planes determined by the x- and y-axes and the y- and z-axes are called the xy- and yz-planes, respectively. The distance between $P_1(x_1, y_1, z_1)$ and $P_2(x_2, y_2, z_2)$ is given by

$$P_1P_2 = \sqrt{(x_2 - x_1)^2 + (y_2 - y_1)^2 + (z_2 - z_1)^2}.$$

The equation for a sphere with center (h, k, l) and radius r is

$$(x - h)^2 + (y - k)^2 + (z - l)^2 = r^2.$$

2. A vector $\mathbf{a}$ in space may be represented as a linear combination of the three unit vectors $\mathbf{i}$, $\mathbf{j}$, and $\mathbf{k}$; that is,

$$\mathbf{a} = a_1\mathbf{i} + a_2\mathbf{j} + a_3\mathbf{k}.$$

The norm of $\mathbf{a}$ is given by

$$\|\mathbf{a}\| = \sqrt{a_1{}^2 + a_2{}^2 + a_3{}^2}.$$

The dot product of $\mathbf{a}$ and $\mathbf{b}$ ($\mathbf{b} = b_1\mathbf{i} + b_2\mathbf{j} + b_3\mathbf{k}$) is defined by

$$\mathbf{a} \cdot \mathbf{b} = a_1b_1 + a_2b_2 + a_3b_3.$$

The formula $\cos\theta = \dfrac{\mathbf{a} \cdot \mathbf{b}}{\|\mathbf{a}\|\|\mathbf{b}\|}$ enables us to find the angle between two nonzero vectors. Two nonzero vectors $\mathbf{a}$ and $\mathbf{b}$ are orthogonal if and only if $\mathbf{a} \cdot \mathbf{b} = 0$.

3. Given a point P_0 and a vector $\mathbf{m}$, the vector equation for the line l through P_0 parallel to $\mathbf{m}$ is given by

$$\mathbf{r} = \mathbf{r}_0 + t\mathbf{m},$$

where $\mathbf{r} = \overrightarrow{OP}$ for O not on l, $\mathbf{r}_0 = \overrightarrow{OP_0}$, and t is a scalar variable. The vector equation for a line determined by two points P_0 and P_1 is

$$\mathbf{r} = (1 - t)\mathbf{r}_0 + t\mathbf{r}_1,$$

where $\mathbf{r}_0 = \overrightarrow{OP_0}$, $\mathbf{r}_1 = \overrightarrow{OP_1}$, and t is a scalar variable.

4. Through a given point P_0 there is only one plane Q perpendicular to a nonzero vector $\mathbf{n}$. The vector equation for the plane Q is $\mathbf{n} \cdot (\mathbf{r} - \mathbf{r}_0) = 0$, where $\mathbf{r} = \overrightarrow{OP}$ and $\mathbf{r}_0 = \overrightarrow{OP_0}$. If $\mathbf{n} = a\mathbf{i} + b\mathbf{j} + c\mathbf{k}$, $\mathbf{r}_0 = x_0\mathbf{i} + y_0\mathbf{j} + z_0\mathbf{k}$, and $\mathbf{r} = x\mathbf{i} + y\mathbf{j} + z\mathbf{k}$, the scalar equation for Q is

$$a(x - x_0) + b(y - y_0) + c(z - z_0) = 0.$$

This equation can be written as $ax + by + cz + d = 0$, where $d = -(ax_0 + by_0 + cz_0)$. The graph of every equation of this form is a plane having $a\mathbf{i} + b\mathbf{j} + c\mathbf{k}$ as a normal vector provided a, b, and c are not all zero. The distance D from the plane $ax + by + cz + d = 0$ to $P_1(x_1, y_1, z_1)$ is given by

$$D = \frac{|ax_1 + by_1 + cz_1 + d|}{\sqrt{a^2 + b^2 + c^2}}.$$

CHAPTER TEST

Exercises 1–3 refer to the sphere with the equation
$x^2 + y^2 + z^2 - 6x + 4y - 10z = 11$.

9-1 1. Find the center and radius of the given sphere.

2. Find the distance from the center of the given sphere to the origin.

3. Find the distance from the center of the given sphere to the xz-plane.

Exercises 4–6 refer to the vectors $\mathbf{c} = 5\mathbf{i} + 7\mathbf{j} + \mathbf{k}$ and $\mathbf{d} = 6\mathbf{j} + 8\mathbf{k}$.

9-2 4. Find the norm of $\mathbf{c} - \mathbf{d}$.

5. Find the angle between $\mathbf{c}$ and $\mathbf{d}$.

6. Find a vector orthogonal to both $\mathbf{c}$ and $\mathbf{d}$.

Exercises 7–9 refer to the line *l* that contains the point (2, 2, 3) and is parallel to $\mathbf{i} - \mathbf{j} - 2\mathbf{k}$.

9-3 7. Find a vector equation for *l*.

8. Find the distance from *l* to the origin.

9. Find the point where *l* intersects the xz-plane.

Exercises 10–12 refer to the plane that contains the point (6, 2, −2) and is perpendicular to $2\mathbf{i} - 3\mathbf{j} + 6\mathbf{k}$.

9-4 10. Give an equation for the plane in (a) vector form and (b) scalar form.

11. Draw the first-octant part of the plane.

12. Find the distance from the point (0, 0, 0) to the plane.

CUMULATIVE REVIEW *Chapters 8 and 9*

Chapter 8

Exercises 1 and 2 refer to the circle having the equation

$$x^2 + y^2 - 6x + 8y + 12 = 0.$$

1. Find (a) the equations of the translation that moves the origin to the point (1, 2) and (b) the new equation of the given circle.

2. Find a new equation of the given circle by rotating the axes 45°.

3. (*Optional*) Draw a diagram illustrating the alias and alibi interpretations of the transformation $x = x' + 2$, $y = y' - 1$ on the point (2, 2).

4. (*Optional*) Find the equations of a rotation that will transform the equation $x^2 + 2xy + y^2 + 2x - 4y + 5 = 0$ into an equation in x' and y' without an $x'y'$-term.

Chapter 9

1. Check whether the triangle having vertices $A(0, 6, 4)$, $B(4, 4, 2)$, and $C(2, 2, 0)$ is isosceles, right, both, or neither.

2. Find the center and radius of the sphere having the equation $x(x - 6) + y(y - 8) + z(z - 24) = 0$.

3. Let $\mathbf{a} = \mathbf{i} + 3\mathbf{j} - \mathbf{k}$ and $\mathbf{b} = \mathbf{i} + \mathbf{j} + 2\mathbf{k}$. Find (a) the linear combinations $2\mathbf{a} + \mathbf{b}$ and $2\mathbf{a} - \mathbf{b}$ and (b) $\|\mathbf{a} + \mathbf{b}\|$.

4. Find the angle between $\mathbf{a} = 3\mathbf{i} + 2\mathbf{j} - \mathbf{k}$ and $\mathbf{b} = \mathbf{i} - \mathbf{j} + 2\mathbf{k}$ to the nearest degree.

5. Find the vector equation for the line that passes through (1, 1, 1) and is parallel to $4\mathbf{i} + 3\mathbf{j} - 12\mathbf{k}$.

6. Find an equation of the plane that passes through the points (2, 1, 3), $(-1, -2, 4)$, and (4, 2, 1).

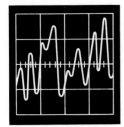

THE HYDROSTATIC PARADOX

Suppose you poured water into a U-shaped tube. Would the water settle at the same height on both sides of the tube or at different heights? Suppose one side of the tube was wider than the other. Would the water settle differently? The fact that the water levels on both sides of the tube would be the same regardless of the shapes of the sides, is known as the hydrostatic paradox.

Consider, for example, the U-shaped tube in the given diagram. Whether water moves from side A to side B or the other way depends on the pressure of the water in each side. Pressure P is defined as force F per unit area A; that is, $P = F/A$. The pressure at a certain depth in the tube is the weight of the water standing above the horizontal area at that depth, divided by the area. The weight w (force of gravitation) of the water above the horizontal area can be found by the formula $w = Ahdg$, where h is the height of the water, d is the density of the water, and g is the acceleration due to gravity at Earth's surface. Therefore, $P = hdg$. Since d and g are constants, the pressure at any point in one side of the U equals the pressure at a point at the same depth in the other side.

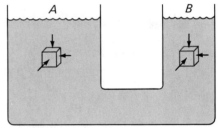

If we consider the force vectors acting on small cubes of water at the same depth in side A and side B of the tube, we can note that all horizontal components of the force vectors acting on these cubes add up to zero. The vertical components, however, are due entirely to the depth of the water above the cubes. This fact explains, for example, why the strength of a dam necessary to withhold a lake of a given depth is the same whether the lake is very large or is only a thin layer of water.

FINDING THE ANGLE BETWEEN TWO VECTORS

The program on the next page will find the angle between two vectors in 1, 2, or 3 dimensions.

Exercises

1–4. RUN the program for Exercises 9–12 on page 240.

RUN the program for the following pairs of vectors in Exercise 7 on page 240.

5. **a, b**   6. **a, c**   7. **a, d**

8. **b, c**   9. **b, d**   10. **c, d**

Find the angle between the pairs of vectors in two dimensions.

Example $a = i + j, b = -i + j$

Solution   RUN

```
TO FIND THE ANGLE BETWEEN TWO VECTORS:
INPUT NO. OF DIMENSIONS?2
INPUT COORDINATES OF FIRST VECTOR:
?1
?1
INPUT COORDINATES OF SECOND VECTOR:
?−1
?1
ANGLE = 90 DEG.
```

11. $a = i, b = i + j$ 12. $a = i + j, b = i - j$

13. $a = 2i - j, b = i - 2j$ 14. $a = -i - 2j, b = -2i + j$

```
10    PRINT "TO FIND THE ANGLE BETWEEN TWO VECTORS:"
20    PRINT "INPUT NO. OF DIMENSIONS";
30    INPUT D
40    PRINT "INPUT COORDINATES OF FIRST VECTOR:"
50    FOR I = 1 TO D
60    INPUT U[I].
70    NEXT I
80    PRINT "INPUT COORDINATES OF SECOND VECTOR:"
90    FOR I = 1 TO D
100   INPUT V[I].
110   NEXT I
115   REM*** P = DOT PRODUCT; N1, N2 ARE NORMS OF VECTORS
120   LET P = 0
130   FOR I = 1 TO D
140   LET P = P + U[I]*V[I]
150   NEXT I
160   IF P = 0 THEN 360
170   LET N1 = 0
180   LET N2 = 0
190   FOR I = 1 TO D
200   LET N1 = N1 + U[I]*U[I]
210   LET N2 = N2 + V[I]*V[I]
220   NEXT I
230   LET N1 = SQR(N1)
240   LET N2 = SQR(N2)
245   REM*** C = COSINE OF ANGLE
250   LET C = P/(N1*N2)
260   IF ABS(C − 1) < .00001 THEN 380
270   IF ABS(C + 1) < .00001 THEN 400
280   LET S = SQR(1 − C*C)
290   LET T = ATN(S/C)
300   LET P1 = 3.14159
310   IF T > = 0 THEN 330
320   LET T = P1 + T
330   LET T = T*180/P1
340   PRINT "ANGLE = ";INT (100*T + .5)/100; "DEG."
350   STOP
360   PRINT "ANGLE = 90 DEG."
370   STOP
380   PRINT "ANGLE = 0 DEG."
390   STOP
400   PRINT "ANGLE = 180 DEG."
410   END
```

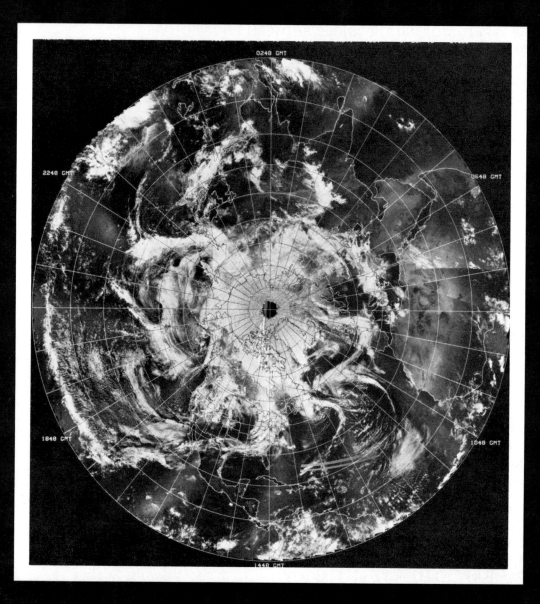

This satellite photo of the Northern Hemisphere shows
the latitude and longitude of clouds in a global weather
system.

SPHERICAL TRIGONOMETRY

OBJECTIVES

1. To define various figures on the surface of a sphere.

2. To apply trigonometry to these figures.

3. To apply spherical trigonometry to navigation.

Spherical Geometry

10-1 *Some Spherical Geometry*

In this chapter we fix our attention on a definite sphere S having center O and radius r. We shall often take S to be the surface of Earth, for even though Earth is not truly spherical, it is nearly so. In this case r would equal 6368 km.

A plane passing through O cuts the spherical surface S in a **great circle;** any other plane cuts S in a small circle or a point, if at all. On Earth a well-known great circle is the equator (Figure 10-1). Great circles containing the geographic North and South Poles are called **meridians.** Small circles cut by planes parallel to the equator are called **parallels.**

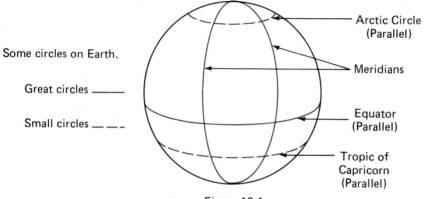

Some circles on Earth.

Great circles _____

Small circles _ _ _ _

Arctic Circle
(Parallel)

Meridians

Equator
(Parallel)

Tropic of
Capricorn
(Parallel)

Figure 10-1

Any two points *A* and *B* of a sphere that are not at the ends of a diameter determine a unique great circle because the noncollinear points *A*, *B*, and *O* determine a unique plane. The shorter arc of this circle, with endpoints *A* and *B*, is denoted by $\overset{\frown}{AB}$ (Figure 10-2). This arc could be measured in units of length, but it usually will be more convenient to use degrees and assign to the arc the measure of $\angle AOB$. Thus, the measure of a great-circle arc with one end at the North Pole and the other on the equator is 90°.

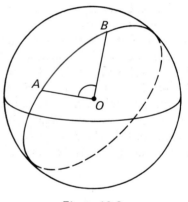

Figure 10-2

Example 1   Find the length in kilometers of a 40° great-circle arc on Earth.

Solution   We use the fact that an arc subtending an angle of θ radians on a circle of radius r has length $r\theta$. Since $r = 6368$ km, the required length is

$$6368 \cdot \frac{40\pi}{180} = 4446 \text{ km.}$$

The great-circle arcs, $\overset{\frown}{AB}$ and $\overset{\frown}{AC}$, having a common endpoint *A* form a **spherical angle.** We assign to this angle the measure of the plane angle formed by the rays tangent to the two arcs at A (Figure 10-3). On Earth, for example, any angle formed by an arc of the equator and an arc of a meridian is a right angle (Figure 10-4).

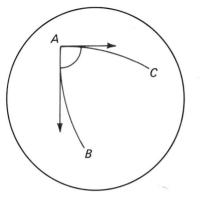

Figure 10-3

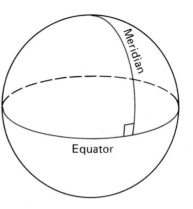

Figure 10-4

A **spherical triangle** ABC is formed by any three great-circle arcs $\widehat{AB}$, $\widehat{BC}$, and $\widehat{CA}$ (Figure 10-5). We shall consider only triangles with each angle less than 180°. We usually designate the angles of a spherical triangle by the same letters A, B, and C as the vertices. We designate the side opposite each angle by the corresponding lower-case letter a, b, or c. We shall see in subsequent sections that, given any three of these six parts, it is always possible to solve the triangle.

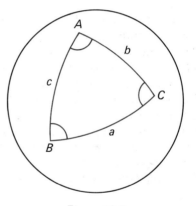

Figure 10-5

Example 2 Solve $\triangle ABC$, given $a = 100°$, $\angle B = 90°$, $\angle C = 90°$.

Solution We can think of the triangle as being on Earth with side a along the equator, as shown in the figure. Then, since angles B and C are right angles, sides b and c lie along meridians, and hence vertex A is at the North Pole. We can now see that $\angle A = 100°$, $b = 90°$, and $c = 90°$.

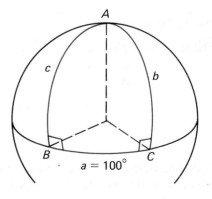

Notice that the sum of the angles of the triangle in Example 2 is greater than 180°. It can be shown that the sum of the angles of any spherical triangle is between 180° and 540°.

The two points that are 90° away from the great-circle arc $\overset{\frown}{BC}$ are called the **poles** of $\overset{\frown}{BC}$. (Think of an arc of the equator and the North and South Poles.) Given a triangle ABC, let A' be the pole of $\overset{\frown}{BC}$ on the same side of $\overset{\frown}{BC}$ as A (Figure 10-6). Also, let B' be the pole of $\overset{\frown}{AC}$ on the same side of $\overset{\frown}{AC}$ as B, and C' be the pole of $\overset{\frown}{AB}$ on the same side of $\overset{\frown}{AB}$ as C. Then $\triangle A'B'C'$ (shown in color in Figure 10-7) is the **polar triangle** of $\triangle ABC$.

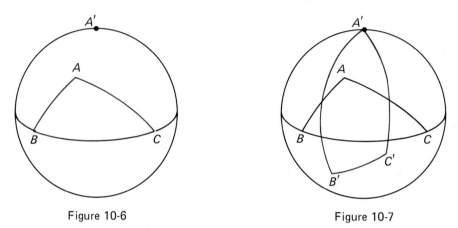

Figure 10-6   Figure 10-7

We shall use the following facts about polar triangles.

(1) If $\triangle A'B'C'$ is the polar triangle of $\triangle ABC$, then $\triangle ABC$ is the polar triangle of $\triangle A'B'C'$.

(2) Any angle of a spherical triangle is the supplement of the corresponding side of its polar triangle. That is, $\angle A + a' = 180°$, $\angle A' + a = 180°$, and so on.

Proof of (1): Since C' is 90° away from $\overset{\frown}{AB}$ and B' is 90° away from $\overset{\frown}{AC}$, both B' and C' are 90° away from A. Thus A is 90° away from $\overset{\frown}{B'C'}$ and therefore is the pole of $\overset{\frown}{B'C'}$ on the same side of $\overset{\frown}{B'C'}$ as A'.

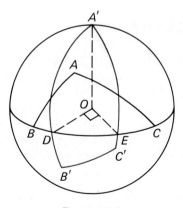

Proof of (2): Let $\overset{\frown}{A'B'}$ intersect $\overset{\frown}{BC}$ in the point D, and $\overset{\frown}{A'C'}$ intersect $\overset{\frown}{BC}$ in E (Figure 10-8). Then the arc length DE equals the measure of $\angle A'$. Since E is on $\overset{\frown}{A'C'}$, of which B is a pole, $\overset{\frown}{BE} = 90°$. Similarly $\overset{\frown}{DC} = 90°$. Therefore,

$$a + \angle A' = \overset{\frown}{BC} + \overset{\frown}{DE} = \overset{\frown}{BE} + \overset{\frown}{DC}$$
$$= 90° + 90° = 180°.$$

Figure 10-8

Exercises 10-1

Find the length in kilometers of a great-circle arc on Earth with the given angular measure.

A 1. 90°   2. 1°   3. 15°45′   4. 160°30′

Find the angular measure of a great-circle arc on Earth with the given length.

5. 1000 km   6. 10,002.8 km   7. 111.14 km   8. 3334.3 km

Solve the spherical triangle *ABC*, given the following.

9. $\angle A = 30°$, $\angle B = 90°$, $\angle C = 90°$ 10. $\angle A = 90°$, $\angle B = 40°$, $\angle C = 90°$

11. $a = 90°$, $b = 90°$, $\angle C = 70°$ 12. $a = 90°$, $\angle B = 90°$, $c = 90°$

13. $a = 150°$, $b = 90°$, $c = 90°$ 14. $a = 90°$, $b = 90°$, $c = 75°$

B 15. Describe a spherical triangle the sum of whose angles is nearly 540°.

16. Describe a spherical triangle the sum of whose angles is just over 180°.

17. What can be said about the poles of a meridian circle on Earth?

18. Two great circles intersect at a 30° angle. Find the shortest distance in degrees from a pole of one circle to the other circle.

C 19. Show that the sum of the angles of the spherical triangle *ABC* is less than 540°. (Hint: Let $\triangle A'B'C'$ be the polar triangle of $\triangle ABC$. Explain why $\angle A + a' + \angle B + b' + \angle C + c' = 540°$. Now solve this equation for $\angle A + \angle B + \angle C$.)

20. Assuming the fact that the sum of the sides of a spherical triangle is less than 360°, show that the sum of the angles is greater than 180°. (See the hint for Exercise 19.)

Challenge

Just as the location of a point *P* in space can be given by the coordinates (x, y, z), the location of *P* can also be given by the spherical coordinates (ρ, ϕ, θ), where $\rho \geq 0$ is the distance from the origin *O* to the point *P*, $0 \leq \phi \leq \pi$ is the angle measured down from the *z*-axis to the line *OP*, and $0 \leq \theta \leq 2\pi$ is the angle from the *xz*-plane to the plane through *P* and the *z*-axis. Verify each of the following relationships between the rectangular and spherical coordinate systems:

$$x = \rho \sin \phi \cos \theta \qquad y = \rho \sin \phi \sin \theta \qquad z = \rho \cos \phi$$

(Hint: Recall that $x = r \cos \theta$ and $y = r \sin \theta$, and find polar coordinates for *r* and *z* in terms of ϕ. A diagram may be helpful.)

Solving Right Spherical Triangles

10-2 *Right Spherical Triangles*

We shall now consider right spherical triangles ABC. In this section and the next, $\angle C$ will always be the right angle (Figure 10-9). The following ten formulas enable us to solve any right triangle.

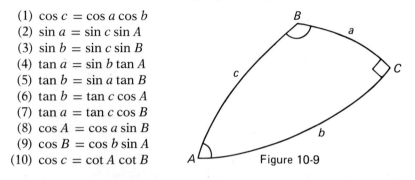

(1) $\cos c = \cos a \cos b$
(2) $\sin a = \sin c \sin A$
(3) $\sin b = \sin c \sin B$
(4) $\tan a = \sin b \tan A$
(5) $\tan b = \sin a \tan B$
(6) $\tan b = \tan c \cos A$
(7) $\tan a = \tan c \cos B$
(8) $\cos A = \cos a \sin B$
(9) $\cos B = \cos b \sin A$
(10) $\cos c = \cot A \cot B$

Figure 10-9

(These formulas need not be memorized. In the next section we give an easy way to reproduce them.) Before proving the formulas let us see how to use them. We shall use degrees and minutes because most navigational work is done in that system.

Example 1 Solve the right triangle ABC, given $a = 54°20'$ and $\angle B = 72°30'$. Give answers to the nearest ten minutes.

Solution To find the missing sides and angles, we look for formulas containing the given parts a and $\angle B$. We use the following formulas in solving the triangle.

$\tan b = \sin a \tan B$ (5) $\cos A = \cos a \sin B$ (8)

$\tan a = \tan c \cos B$ (7)   or $\tan c = \dfrac{\tan a}{\cos B}$

From these we have:

$\tan b = \sin 54°20' \cdot \tan 72°30' = 2.5769;$ $b = 68°50'$

$\cos A = \cos 54°20' \cdot \sin 72°30' = 0.5561;$ $\angle A = 56°10'$

$\tan c = \dfrac{\tan 54°20'}{\cos 72°30'} = 4.633;$ $c = 77°50'$

As a check, we can substitute the calculated values in the two sides of $\tan b = \tan c \cos A$ (6).

$\tan 68°50' = 2.583$

$\tan 77°50' \cdot \cos 56°10' = 2.582$

Rounding the answers to the nearest ten minutes caused the difference in the thousandths' place.

Example 2   Solve $\triangle ABC$ given $\angle A = 102°10'$, $\angle B = 155°50'$, and $\angle C = 90°$. Give answers to the nearest ten minutes.

Solution   We use these formulas:

$$\cos a = \frac{\cos A}{\sin B} \quad (8) \qquad \cos b = \frac{\cos B}{\sin A} \quad (9) \qquad \cos c = \cot A \cot B \quad (10)$$

$$a = \text{Cos}^{-1}\frac{\cos 102°10'}{\sin 155°50'} = \text{Cos}^{-1}(-0.5149) = 121°0'$$

$$b = \text{Cos}^{-1}\frac{\cos 155°50'}{\sin 102°10'} = \text{Cos}^{-1}(-0.9334) = 159°0'$$

$$c = \text{Cos}^{-1}(\cot 102°10' \cdot \cot 155°50') = \text{Cos}^{-1}(0.4806) = 61°20'$$

We can check these values by using $\cos c = \cos a \cos b$ (1).

$$\cos 61°20' = 0.4797$$

$$\cos 121°0' \cdot \cos 159°0' = 0.4808$$

We now prove several of the basic formulas (1)–(10).

Let $\triangle ABC$ be a right triangle with $a < 90°$ and $b < 90°$. Figure 10-10 shows the triangle and the three radii from O to A, B, and C. By the definition of the measure of a great-circle arc, the three plane angles at O are a, b, and c. The plane BDE is drawn through B perpendicular to $\overline{OA}$. Thus $\angle ODB$, $\angle ODE$, $\angle OEB$, and $\angle BED$ are right angles. Moreover, $\angle BDE = \angle A$ because the segments DB and DE are parallel, respectively, to the tangents to the arcs $\overset{\frown}{AB}$ and $\overset{\frown}{AC}$ at A.

From right triangles BDO, BEO, and EDO, we have

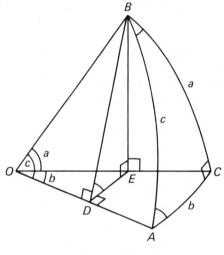

Figure 10-10

$$\cos c = \frac{OD}{OB} = \frac{OE}{OB} \cdot \frac{OD}{OE} = \cos a \cos b. \quad (1)$$

This formula is called the **Pythagorean theorem** of spherical trigonometry because it relates the hypotenuse and the legs of a right triangle.

From right triangles BEO, BDO, and BED, we have

$$\sin a = \frac{BE}{OB} = \frac{BD}{OB} \cdot \frac{BE}{BD} = \sin c \sin A. \quad (2)$$

From right triangles BEO, EDO, and BED, we have

$$\tan a = \frac{BE}{OE} = \frac{DE}{OE} \cdot \frac{BE}{DE} = \sin b \tan A. \quad (4)$$

Formula (6) is proved similarly (Exercise 13). Formulas (3), (5), and (7) follow from (2), (4), and (6), respectively, by interchanging the roles of a and b and of A and B.

To prove formula (8), first multiply formulas (3) and (6).

$$\sin b(\tan c \cos A) = \sin c \sin B(\tan b)$$

$$\frac{\sin b \sin c \cos A}{\cos c} = \frac{\sin c \sin b \sin B}{\cos b}$$

$$\cos A = \frac{\cos c \sin B}{\cos b}$$

Now use formula (1) to obtain

$$\cos A = \frac{\cos a \cos b \sin B}{\cos b} = \cos a \sin B. \qquad (8)$$

Proofs of the remaining formulas are left as exercises.

Exercises 10-2

Throughout this exercise set, $\angle C = 90°$.
In Exercises 1–10, solve the right spherical triangle *ABC*. In Exercises 1–6, give answers to the nearest ten minutes. In Exercises 7–10, give answers to the nearest tenth of a degree.

A 1. $a = 115°30'$, $b = 64°20'$ 2. $a = 97°0'$, $c = 92°10'$

3. $a = 35°40'$, $\angle B = 78°0'$ 4. $c = 44°10'$, $\angle A = 110°30'$

5. $\angle A = 30°0'$, $\angle B = 105°0'$ 6. $a = 90°0'$, $b = 50°20'$

7. $b = 62.4°$, $c = 112.5°$ 8. $\angle A = 38.8°$, $\angle B = 71.2°$

9. $c = 75.0°$, $\angle B = 40.0°$ 10. $b = 116.5°$, $\angle A = 162.5°$

B 11. Show that $\cos A = \dfrac{\cos a \sin b}{\sin c}$.

12. Show that $\tan A = \dfrac{\sin a}{\tan b \cos c}$.

13. Prove formula (6).   14. Prove formula (9).

15. Prove formula (10).

Exercises 16 and 17 indicate how we can remove the restrictions $a < 90°$, $b < 90°$ in the proof of formulas (1)–(10).

C 16. Suppose $a < 90°$, $b > 90°$. Extend the edges $\overparen{AB}$ and $\overparen{AC}$ to form right triangle $A'BC$ as shown in the figure, where $\angle A = \angle A'$, $b_s = 180° - b$, $c_s = 180° - c$, and $\angle B_s = 180° - \angle B$. In triangle $A'BC$, $a < 90°$ and $b_s < 90°$, so that formulas (1)–(10) hold. When we apply (6) to $\triangle A'BC$,

for example, we obtain:
$$\tan b_s = \tan c_s \cos A'$$
$$\tan (180° - b) = \tan (180° - c) \cos A$$

Therefore $-\tan b = -\tan c \cos A$, or $\tan b = \tan c \cos A$.
This proves (6) for the case where $a < 90°$, $b > 90°$. Prove the other formulas for this case.

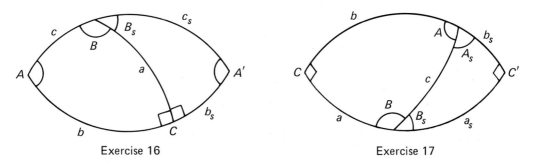

Exercise 16   Exercise 17

17. Prove formulas (1)–(10) for the case $a > 90°$, $b > 90°$ by using $\triangle ABC'$ shown in the figure.

10-3 *Napier's Rules*

The Scottish mathematician John Napier (1550–1617) discovered the following device for remembering formulas (1)–(10) of Section 10-2. Figure 10-11 shows a right spherical triangle in which the label C has been omitted, and A, c, and B have been replaced by co-$A = 90° - \angle A$, co-$c = 90° - c$, and co-$B = 90° - \angle B$, respectively. Figure 10-12 gives an alternative way of showing the relative positions of the five **circular parts** a, b, co-A, co-c, and co-B. Fixing our attention on any one of the parts, which we call the **middle part**, we see that there are two **adjacent parts** and two **opposite parts.** Thus if b is taken to be the middle part, then a and co-A are the adjacent parts and co-B and co-c are the opposite parts.

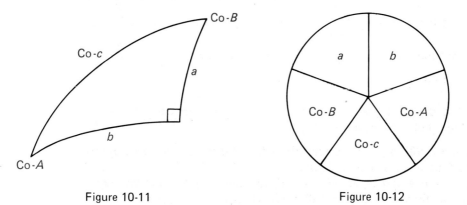

Figure 10-11   Figure 10-12

We can now state Napier's rules:

---

1. The sine of any middle part equals the product of the tangents of its adjacent parts.
2. The sine of any middle part equals the product of the cosines of its opposite parts.

For example, if we take a as the middle part and apply Rule 1, we obtain

$$\sin a = \tan b \tan (90° - \angle B) = \tan b \cot B,$$

which is equivalent to formula (5). Applying Rule 2, we have

$$\sin a = \cos (90° - c) \cos (90° - \angle A) = \sin c \sin A,$$

which is equivalent to formula (2).

Any three circular parts can be arranged so that one of them is the middle part and the other two are either adjacent to or opposite to this middle part. This arrangement enables us to use Napier's rules to solve right spherical triangles.

Example 1   Solve the right triangle ABC, given that $a = 52°$ and $\angle A = 75°$. Give answers to the nearest degree.

Solution   To find b, we let b be the middle part. Then a and co-A are the adjacent parts (refer to Figure 10-12), and by Rule 1,

$$\sin b = \tan a \tan (90° - \angle A) = \tan a \cot A$$

$$\sin b = \frac{\tan a}{\tan A} = \frac{\tan 52°}{\tan 75°} = 0.3430.$$

Therefore, $b = 20°$ or $b = 160°$.

To express $\angle B$ in terms of the given data, we choose co-A as the middle part and apply Rule 2.

$$\sin (90° - \angle A) = \cos a \cos (90° - \angle B), \text{ or } \cos A = \cos a \sin B$$

$$\sin B = \frac{\cos A}{\cos a} = \frac{\cos 75°}{\cos 52°} = 0.4204$$

Therefore, $\angle B = 25°$ or $\angle B = 155°$.

With a as the middle part, Rule 2 gives

$$\sin a = \cos (90° - c) \cos (90° - \angle A) = \sin c \sin A$$

$$\sin c = \frac{\sin a}{\sin A} = \frac{\sin 52°}{\sin 75°} = 0.8158.$$

Therefore, $c = 55°$ or $c = 125°$.

The quadrantal rules stated below (and to be proved in Exercises 18 and 19) enable us to decide how to group the values of b, $\angle B$, and c found above.

(i) Side a and $\angle A$ are in the same quadrant. Similarly for b and $\angle B$.
(ii) Sides a and b are in the same quadrant if and only if $c < 90°$.

In Example 1, Rule (i) tells us to pair $b_1 = 20°$ with $\angle B_1 = 25°$ and $b_2 = 160°$ with $\angle B_2 = 155°$. Rule (ii) tells us that since $a = 52°$ and $b_1 = 20°$ are in the same quadrant, $c_1 = 55°$. Also, since $a = 52°$ and $b_2 = 160°$ are in different quadrants, $c_2 = 125°$. Thus Example 1 has two solutions, namely

$$b_1 = 20°,\ \angle B_1 = 25°,\ c_1 = 55°;\ \text{and}\ b_2 = 160°,\ \angle B_2 = 155°,\ c_2 = 125°.$$

When the given parts of a right spherical triangle are an angle and the side opposite that angle, there may be two solutions, as in Example 1, or there may be one solution or none (Exercises 22 and 23). Figure 10-13 illustrates the two-solution case.

A **quadrantal triangle** is a triangle having a 90° side. It can be solved with the help of its polar triangle, which, by property (2), page 262, is a right triangle.

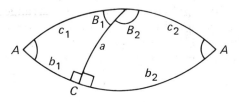

Figure 10-13

Example 2 Solve the quadrantal triangle ABC, given that $a = 80°$, $\angle B = 130°$, and $c = 90°$. Give answers to the nearest degree.

Solution Let $\triangle A'B'C'$ be the polar triangle of $\triangle ABC$. Then by property (2), page 262,

$$\angle A' = 180° - a = 180° - 80° = 100°$$
$$b' = 180° - \angle B = 180° - 130° = 50°$$
$$\angle C' = 180° - c = 180° - 90° = 90°.$$

Since $\triangle A'B'C'$ is a right triangle we can solve it:

$$\tan a' = \sin b' \tan A' = \sin 50° \tan 100° = -4.344; \qquad a' = 103°$$
$$\cos B' = \cos b' \sin A' = \cos 50° \sin 100° = 0.6330; \qquad \angle B' = 51°$$
$$\tan c' = \frac{\tan b'}{\cos A'} = \frac{\tan 50°}{\cos 100°} = -6.863; \qquad c' = 98°$$

(Solution continued on next page.)

By Property (1), page 262, $\triangle ABC$ is the polar triangle of $\triangle A'B'C'$. Hence, by property (2),

$$\angle A = 180° - a' = 180° - 103° = 77°$$
$$b = 180° - \angle B' = 180° - 51° = 129°$$
$$\angle C = 180° - c' = 180° - 98° = 82°$$

Exercises 10-3

Use Napier's rules to write formulas that can be used to solve right triangle *ABC* with $\angle C = 90°$ given the following.

A 1. a and b 2. b and c 3. a and $\angle A$

4. b and $\angle A$ 5. c and $\angle B$ 6. $\angle A$ and $\angle B$

Solve right triangle *ABC* with $\angle C = 90°$, giving answers to the nearest degree. Find all solutions. If there is no solution, so state.

7. $a = 37°$, $\angle A = 73°$ 8. $b = 165°$, $\angle B = 125°$

9. $a = 76°$, $\angle A = 106°$ 10. $b = 93°$, $\angle B = 93°$

Solve the following quadrantal triangles. Give answers to the nearest ten minutes. Note that $\angle C$ is not necessarily 90°.

11. $a = 78°20'$, $b = 122°30'$, $c = 90°$

12. $a = 35°40'$, $\angle B = 78°20'$, $c = 90°$

13. $b = 113°20'$, $\angle C = 62°0'$, $c = 90°$

14. $\angle A = 94°0'$, $\angle C = 72°30'$, $c = 90°$

15. $\angle A = 24°50'$, $\angle B = 77°10'$, $c = 90°$

16. $a = 70°0'$, $\angle A = 40°0'$, $c = 90°$ (two solutions)

Exercises 17–23 refer to a spherical triangle *ABC* in which $\angle C = 90°$. Use formulas (1)–(10) on page 264 to show the following.

B 17. Show that if either one of a or $\angle A$ is 90°, then so is the other. (Hint: Use formula (8).)

18. Show that if neither a nor $\angle A$ is 90°, then they are in the same quadrant.

19. Show that a and b are in the same quadrant if and only if $c < 90°$.

20. Explain why there are infinitely many solutions if $a = c = 90°$.

C 21. Show that if a and $\angle A$ are both acute, then $a \leq \angle A$. What is the situation if a and $\angle A$ are both in the second quadrant?

22. If a and $\angle A$ are given, under what conditions is there exactly one solution?

23. If a and $\angle A$ are given, under what conditions is there no solution?

Solving Oblique Spherical Triangles

10-4 *Oblique Spherical Triangles*

In solving general spherical triangles, there are six cases to consider.

    I. Given three sides
    II. Given three angles
    III. Given two sides and the included angle
    IV. Given two angles and the included side
    V. Given two sides and an angle opposite one of them
    VI. Given two angles and a side opposite one of them

The formulas given below are helpful in solving oblique triangles, that is, triangles having no right angle.

The Law of Sines

$$\frac{\sin a}{\sin A} = \frac{\sin b}{\sin B} = \frac{\sin c}{\sin C}$$

The Law of Cosines for Sides

$$\cos a = \cos b \cos c + \sin b \sin c \cos A$$

with analogous formulas for $\cos b$ and $\cos c$.

The Law of Cosines for Angles

$$\cos A = -\cos B \cos C + \sin B \sin C \cos a$$

with analogous formulas for $\cos B$ and $\cos C$.

To prove the law of sines, consider Figure 10-14, in which $\overset{\frown}{CD}$ is perpendicular to $\overset{\frown}{AB}$. Applying Napier's rules to right triangles ADC and BDC, we have

$$\sin h = \sin b \sin A, \text{ and } \sin h = \sin a \sin B.$$

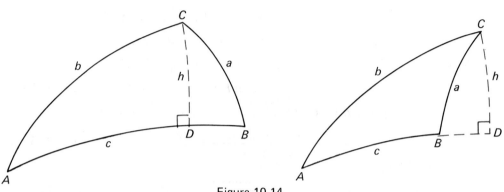

Figure 10-14

Thus, $\sin a \sin B = \sin b \sin A$, from which we obtain

$$\frac{\sin a}{\sin A} = \frac{\sin b}{\sin B}.$$

In a similar way we can show that $\dfrac{\sin b}{\sin B} = \dfrac{\sin c}{\sin C}$, and the law of sines follows.

We defer the proofs of the laws of cosines to the end of the section.

Example 1 Solve $\triangle ABC$, given

$$\angle A = 72°30', \ \angle B = 124°10', \ \angle C = 37°50'.$$

Give answers to the nearest ten minutes.

Solution We can solve this Case II problem by three applications of the law of cosines for angles. From $\cos A = -\cos B \cos C + \sin B \sin C \cos a$, we obtain

$$\cos a = \frac{\cos A + \cos B \cos C}{\sin B \sin C} = -0.2815; \qquad a = 106°20'$$

$$\cos b = \frac{\cos B + \cos A \cos C}{\sin A \sin C} = -0.5540; \qquad b = 123°40'$$

$$\cos c = \frac{\cos C + \cos A \cos B}{\sin A \sin B} = 0.7869; \qquad c = 38°10'$$

We can use the law of sines to check our results:

$$\frac{\sin a}{\sin A} = 1.0062, \quad \frac{\sin b}{\sin B} = 1.0059, \quad \frac{\sin c}{\sin C} = 1.0075.$$

In solving Case I problems, we use the method of Example 1, but with the law of cosines for sides.

Example 2 illustrates a Case III situation.

Example 2   Solve $\triangle ABC$, given

$$a = 68.4°, \ b = 131.3°, \ \angle C = 111.6°.$$

Give answers to the nearest tenth of a degree.

Solution   We find c first by using the law of cosines for sides.

$$\cos c = \cos a \cos b + \sin a \sin b \cos C = -0.5001; \qquad c = 120.0°.$$

From $\cos a = \cos b \cos c + \sin a \sin b \cos A$ we have

$$\cos A = \frac{\cos a - \cos b \cos c}{\sin b \sin c} = 0.0586; \qquad \angle A = 86.6°.$$

Similarly,

$$\cos B = \frac{\cos b - \cos a \cos c}{\sin a \sin c} = 0.5911; \qquad \angle B = 126.2°.$$

Note that if we had used the law of sines to find $\angle A$ and $\angle B$, we would have been faced with the ambiguity that $\angle A = 86.6°$, or $93.4°$, and $\angle B = 53.8°$, or $126.2°$.

We can solve Case IV problems using the method of Example 2, but with the law of cosines for angles. We can solve Case V and Case VI problems by dividing the given triangle into right triangles.

Example 3   Solve $\triangle ABC$ given

$$a = 65.0°, \ b = 110.0°, \ \angle A = 70.0°.$$

Give answers to the nearest tenth of a degree.

Solution   We find $\angle B$ using the law of sines:

$$\sin B = \frac{\sin b \sin A}{\sin a} = 0.9743;$$

$$\angle B_1 = 77.0° \ \text{or} \ \angle B_2 = 103.0°$$

In the figure $\overset{\frown}{CD}$ is drawn perpendicular to $\overset{\frown}{AB_1}$. From right triangle ADC we have:

$$\tan p = \tan b \cos A = 0.9397$$
$$p = 136.8°$$

From right triangle B_1DC:

$$\tan q = \tan a \cos B_1 = 0.4824$$
$$q = 25.8°$$

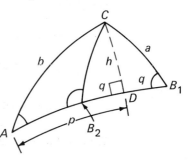

(Solution continued on next page.)

Thus we have:

$$\angle B_1 = 77.0°;\ c_1 = p + q = 162.6°$$
$$\angle B_2 = 103.0°;\ c_2 = p - q = 111.0°$$

To find the values of $\angle C$, we use the law of cosines for sides:

$$\cos C_1 = \frac{\cos c_1 - \cos a \cos b}{\sin a \sin b} = -0.9507$$

$$\angle C_1 = 161.9°$$

$$\cos C_2 = \frac{\cos c_2 - \cos a \cos b}{\sin a \sin b} = -0.2511$$

$$\angle C_2 = 104.5°$$

Thus there are two solutions of the triangle:

$$\angle B_1 = 77.0°,\ \ c_1 = 162.6°,\ \angle C_1 = 161.9°$$
$$\angle B_2 = 103.0°,\ c_2 = 111.0°,\ \angle C_2 = 104.5°$$

Cases V and VI are called the **ambiguous cases** because there may be two solutions, one solution, or no solution. For example, in the figure of Example 3, there will be no solution if $a < h$. Also if $p + q > 180°$ or $p - q < 0°$, there will be one solution or no solution.

It should be noted that there are many other formulas that can be used to solve spherical triangles. Some of these enable us to solve Case V and Case VI problems directly without having to use right triangles.

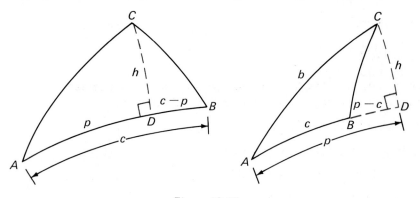

Figure 10-15

We shall now prove the laws of cosines for sides and angles.

In Figure 10-15, $\widehat{CD}$ is perpendicular to $\widehat{AB}$. Let the measures of $\widehat{CD}$ and $\widehat{AD}$ be h and p, respectively. From right triangle ADC, we have:

(i) $\sin p = \tan h \cot A$ (ii) $\cos p = \dfrac{\cos b}{\cos h}$ (iii) $\sin h = \sin b \sin A$

Now apply the Pythagorean theorem to right triangle BDC, noting that the measure of $\widehat{BD}$ is either $c - p$ or $p - c$. Since $\cos (c - p) = \cos (p - c)$, we obtain:

$$\begin{aligned}
\cos a &= \cos h \cos (c - p) \\
&= \cos h(\cos c \cos p + \sin c \sin p) \\
&= \cos h \left(\cos c \, \frac{\cos b}{\cos h} + \sin c \tan h \cot A \right) \\
&= \cos c \cos b + \sin c \sin h \cot A \\
&= \cos b \cos c + \sin c (\sin b \sin A) \frac{\cos A}{\sin A} \\
&= \cos b \cos c + \sin b \sin c \cos A
\end{aligned}$$

using (i) and (ii)

using (iii)

Applying the law of cosines for sides to the polar triangle $A'B'C'$ of $\triangle ABC$, we have

$$\cos a' = \cos b' \cos c' + \sin b' \sin c' \cos A'.$$

By property (2), page 262, $a' = 180° - \angle A$, $b' = 180° - b$, and so on. Thus

$$\cos a' = -\cos A, \quad \cos b' = -\cos B, \quad \sin b' = \sin b, \text{ and so on.}$$

Therefore we have

$$-\cos A = (-\cos B)(-\cos C) + \sin B \sin C(-\cos a),$$

or $$\cos A = -\cos B \cos C + \sin B \sin C \cos a.$$

Exercises 10-4

Solve the following spherical triangles. Give your answers to the nearest ten minutes or nearest tenth of a degree as indicated by the given data.

A   1. $a = 63°20'$, $b = 114°40'$, $c = 98°30'$

2. $\angle A = 123°10'$, $\angle B = 72°30'$, $\angle C = 63°20'$

3. $\angle A = 42°10'$, $\angle B = 142°20'$, $c = 116°20'$

4. $\angle A = 118°30'$, $b = 39°0'$, $c = 87°30'$

5. $a = 110.6°$, $\angle B = 63.3°$, $c = 37.5°$

6. $a = 30.0°$, $b = 40.0°$, $c = 50.0°$

7. $\angle A = 84.5°$, $\angle B = 84.5°$, $\angle C = 53.8°$

8. $a = 52.5°$, $\angle B = 102.9°$, $\angle C = 94.6°$

Exercises 9–14 are Case V problems. If there are two solutions, give both; if there is no solution, so state. A sketch similar to the figure of Example 3 will be helpful.

B   9. $a = 42°0'$, $b = 57°30'$, $\angle A = 43°30'$

10. $a = 68°0'$, $b = 57°30'$, $\angle A = 43°30'$

11. $a = 31°0'$, $b = 57°30'$, $\angle A = 43°30'$

12. $b = 71.2°$, $c = 65.8°$, $\angle C = 63.4°$

13. $b = 71.2°$, $c = 83.1°$, $\angle C = 63.4°$

14. $b = 71.2°$, $c = 52.7°$, $\angle C = 63.4°$

Exercises 15 and 16 are Case VI problems. Solve them by first solving the polar triangles. Give your answers to the nearest degree.

C 15. $a = 127°$, $\angle A = 75°$, $\angle B = 50°$

16. $a = 62°$, $\angle A = 48°$, $\angle B = 106°$

10-5   *Great-Circle Navigation*

The standard method of describing the position of a point P on Earth is to give the *latitude* and *longitude* of the point. The **latitude** of P is its distance, ϕ, from the equator, measured in degrees. Latitude lies in the interval $0° \leq \phi \leq 90°$ and is described as north latitude or south latitude. The **longitude**, λ, of P is the angle between the meridian of P and the prime meridian, which is the meridian of Greenwich, England. Longitude lies in the interval $0° \leq \lambda \leq 180°$ and is designated as east longitude or west longitude. For example, the position of Chicago is $41°50'$ N, $87°40'$ W and the position of Sydney, Australia is $34°0'$ S, $151°0'$ E.

An important problem in navigation is finding the distance between two points, A and B, on Earth and planning the great-circle course between them. To solve such a problem we use the **terrestrial triangle** that has A, B, and the North Pole N as vertices (Figure 10-16). Using the latitudes of A and B, we can find the sides b and a. The angle at N can be found from the longitudes of A and B. Thus, finding the great-circle distance d from A to B is just a matter of applying the law of cosines for sides. In navigation, distances are usually given in nautical miles. A **nautical mile** is the length of a one-minute great-circle arc. It equals about 1.15 land miles or 1.85 kilometers.

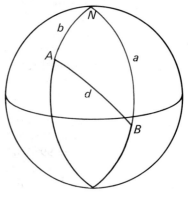

Figure 10-16

Example 1 (a) Find the great-circle distance from Chicago, 41°50′ N, 87°40′ W to Rome, 41°50′ N, 12°15′ E.

(b) Since the two cities have the same latitude, it is possible to fly from Chicago to Rome by flying due east along a parallel of latitude. What is the distance of this trip? The radius of the earth is 3438 nautical miles.

Solution (a) Letting C, R, and N denote Chicago, Rome, and the North Pole, respectively, we have the terrestrial triangle CRN shown in the adjoining figure. Since both C and R have latitude 41°50′,

$$c = r = 90° - 41°50'$$
$$= 48°10'.$$

Since C has west longitude 87°40′ and R has east longitude 12°15′,

$$N = 87°40' + 12°15' = 99°55'.$$

By the law of cosines for sides:

$$\cos n = \cos c \cos r + \sin c \sin r \cos N = 0.3492$$

Therefore, $n = 69°34'$.

To convert to nautical miles, we express n in minutes:
$$n = 69 \times 60 + 34 = 4174'$$
Therefore, the great-circle distance is 4174 nautical miles.

(b) We must first find the radius ρ of the parallel of latitude 41°50′ N. Since the radius of the earth is 3438 nautical miles,
$$\rho = 3438 \cos 41°50'$$
$$= 2562 \text{ nautical miles.}$$

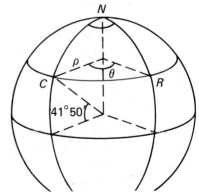

Next, we express
$$\theta = N = 99°55'$$
$$= 99.9167° \text{ in radians:}$$

$$\theta = \frac{99.9167\pi}{180} = 1.7439 \text{ radians.}$$

Therefore, the parallel-of-latitude distance is

$$\rho\theta = 2562 \times 1.7439 = 4468 \text{ nautical miles.}$$

Example 1 illustrates the advantage of using a great-circle track, especially on long trips. Many maps show parallels of latitude as horizontal lines. Figure 10-17 shows how this distortion disguises the true situation.

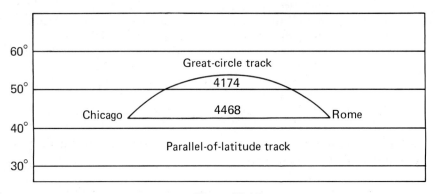

Figure 10-17

The **course** of a plane or ship is the angle θ, $0° \leq \theta < 360°$, that its path makes with the meridian, measured clockwise from north. Figure 10-18 shows the **initial course, θ_1,** and the **course on arrival, θ_2,** of the great-circle track from Los Angeles to Jakarta, Indonesia.

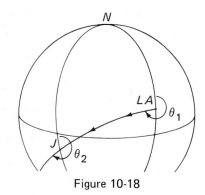

Figure 10-18

Example 2   A long-range jet is to fly the great-circle track from Los Angeles, 33°54′ N, 118°24′ W, to Jakarta, 6°18′ S, 104°48′ E. How far will the jet travel, and what will be its initial course and its course on arrival?

Solution   Let A, B, and C denote Los Angeles, Jakarta, and the North Pole, respectively. Then we have:

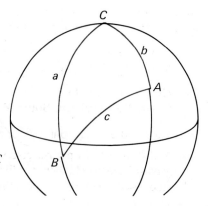

$$a = 90° + 6°18' = 96°18'$$
$$b = 90° - 33°54' = 56°6'$$
$$\angle C = 360° - (118°24' + 104°48')$$
$$= 136°48'$$
$$\cos c = \cos a \cos b + \sin a \sin b \cos C$$
$$= -0.6626$$
$$c = 131°30'$$

Therefore, the plane will travel $131 \times 60 + 30 = 7890$ nautical miles.

$$\cos A = \frac{\cos a - \cos b \cos c}{\sin b \sin c} = 0.4180$$

$$\angle A = 65°18'$$

Therefore, the initial course is $360° - 65°18' = 294°42'$.

$$\cos B = \frac{\cos b - \cos a \cos c}{\sin a \sin c} = 0.6516$$

$$\angle B = 49°20'$$

Therefore, the course at arrival is $180° + 49°20' = 229°20'$.

Note that although the plane is flying to the southern hemisphere, its initial course is north of west.

Exercises 10-5

	Latitude	Longitude
Boston	42°21′ N	71°0′ W
Chicago	41°50′ N	87°40′ W
Dar-es-Salaam	6°47′ S	39°15′ E
Houston	29°45′ N	95°22′ W
Mexico City	19°26′ N	99°7′ W
Paris	48°52′ N	2°20′ E
Rio de Janeiro	23°0′ S	43°20′ W
Rome	41°50′ N	12°15′ E
San Francisco	37°45′ N	122°27′ W
Sydney	34°0′ S	151°0′ E

Note: All tracks are great-circle arcs unless otherwise specified. Give your answers to the nearest nautical mile or to the nearest ten minutes.

A 1. How many nautical miles is Houston from (a) the equator? (b) the North Pole? (c) the South Pole?

2. How many nautical miles is Sydney from (a) the equator? (b) the North Pole? (c) the South Pole?

Exercises 3–8 can be done by solving *right* triangles.

3. A ship leaves San Francisco with initial heading 270° and follows a great-circle track. What will the ship's position be after it has traveled 2400 nautical miles? What will its course be then?

4. A plane leaves Houston with initial course 90° and flies a great-circle track. Find the longitude of the point where the plane crosses the equator. Find its course at that point.

5. A plane leaves Boston with initial course 63°30′. Find the latitude and longitude of the northernmost point of its track. (Hint: At this point the track is perpendicular to the meridian.)

6. What is the position of the northernmost point of the great-circle track from Chicago to Rome? (Hint: The triangle in Example 1(a) is isosceles.)

7. A ship crosses the equator with course 30°. What will be its greatest latitude?

8. A ship crosses the equator with course 45°. How far must the ship sail before its course is 60°? (Hint: The terrestrial triangle is quadrantal.)

In Exercises 9–14, a great-circle track is described. Find (a) its length, (b) the initial course, and (c) the course on arrival.

B 9. Boston to Rome

10. Paris to Mexico City

11. Boston to Dar-es-Salaam

12. San Francisco to Sydney

13. Houston to Rio de Janeiro

14. Rome to Rio de Janeiro

In Exercises 15 and 16, find the lengths of (a) the great-circle arc joining P and Q, and (b) the parallel-of-latitude arc joining P and Q.

C 15. P: 30° N, 100° W; Q: 30° N, 20° E   16. P: 40° N, 120° W; Q: 40° N, 120° E

CHAPTER SUMMARY

1. A plane passing through the center O of a sphere S cuts the spherical surface in a great circle. A spherical triangle ABC is formed by any three great-circle arcs $\widehat{AB}$, $\widehat{BC}$, and $\widehat{CA}$. The two points that are 90° away from the great-circle arc $\widehat{BC}$ are called the poles of B. If $\triangle A'B'C'$ is the polar triangle of $\triangle ABC$, then $\triangle ABC$ is the polar triangle of $\triangle A'B'C'$. Any angle of a spherical triangle is the supplement of the corresponding side of its polar triangle.

2. The following formulas enable us to solve any right triangle.

(1) $\cos c = \cos a \cos b$	(2) $\sin a = \sin c \sin A$
(3) $\sin b = \sin c \sin B$	(4) $\tan a = \sin b \tan A$
(5) $\tan b = \sin a \tan B$	(6) $\tan b = \tan c \cos A$
(7) $\tan a = \tan c \cos B$	(8) $\cos A = \cos a \sin B$
(9) $\cos B = \cos b \sin A$	(10) $\cos c = \cot A \cot B$

3. Napier's rules for solving right triangles are (1) the sine of any middle part equals the product of the tangents of its adjacent parts and (2) the sine of any middle part equals the product of the cosines of its opposite parts.

4. The following laws can be used to solve oblique spherical triangles.

$$\frac{\sin a}{\sin A} = \frac{\sin b}{\sin B} = \frac{\sin c}{\sin C} \quad \text{(law of sines)}$$

$\cos a = \cos b \cos c + \sin b \sin c \cos A$ (law of cosines for sides)
   with analogous formulas for $\cos b$ and $\cos c$

$\cos A = -\cos B \cos C + \sin B \sin C \cos a$ (law of cosines for angles)
   with analogous formulas for $\cos B$ and $\cos C$

5. The latitude of a point P on Earth is its distance from the equator, measured in degrees. The longitude of P is the angle between the meridian of P and the prime meridian. A nautical mile is the length of a one-minute great-circle arc. The course of a plane or ship is the angle that its path makes with the meridian, measured clockwise from north.

CHAPTER TEST

Solve the spherical triangle *ABC*, given the following.

10-1   1. $\angle A = 60°$, $\angle B = 90°$, $\angle C = 90°$ 2. $a = 150°$, $\angle B = 90°$, $c = 90°$

Solve the right spherical triangle *ABC* with $\angle C = 90°$. Give answers to the nearest ten minutes.

10-2   3. $a = 37°40'$, $\angle A = 37°40'$ 4. $a = 40°30'$, $b = 30°30'$

Use Napier's rules to write formulas that can be used to solve right triangle *ABC* with $\angle C = 90°$, given the following.

10-3   5. c and $\angle A$ 6. b and $\angle B$

Solve the following spherical triangles. Give your answers to the nearest tenth of a degree.

10-4   7. $a = 100.5°$, $\angle B = 63.0°$, $c = 35.5°$

        8. $a - 60.2°$, $b = 110.5°$, $c = 95.5°$

10-5   9. How many nautical miles is Chicago, 41°50′ N, 87°40′ W, from (a) the equator? (b) the North Pole? (c) the South Pole?

      10. A plane leaves San Francisco, 37°45′ N, 122°27′W, with initial course 90° and follows a great-circle track. Find the plane's position after it has traveled 2000 nautical miles. Give your answer to the nearest ten minutes.

SOLVING RIGHT SPHERICAL TRIANGLES

The solutions of right spherical triangles can be classified as shown below, where the given parts are in red. (*C* is always the right angle. If *b* and angle *B* are given, for instance, the triangle can be relettered to make them *a* and *A*.)

I		II		III		IV		V		VI	
a	*A*	*a*	*A*	*a*	*A*	*a*	*A*	*a*	*A*	*a*	*A*
b	*B*	*b*	*B*	*b*	*B*	*b*	*B*	*b*	*B*	*b*	*B*
c	*C*	*c*	*C*	*c*	*C*	*c*	*C*	*c*	*C*	*c*	*C*

To solve a Case II triangle, for example, we can use the following formulas in the given program. (Remember that BASIC has only the inverse tangent function.)

$$\tan b = \sin a \tan B \quad (5) \qquad \text{(See line 120.)}$$
$$\tan c = \tan a / \cos B \quad (7) \qquad \text{(See line 150.)}$$
$$\tan A = \tan a / \sin b \quad (4) \qquad \text{(See line 180.)}$$

Note that *a* is A1, *b* is B1, and *c* is C1 in the program.

```
10   PRINT "TO SOLVE A RIGHT SPHERICAL TRIANGLE"
20   PRINT "   WHEN C IS THE RIGHT ANGLE, GIVEN"
30   PRINT "   ANY OTHER TWO PARTS (<> 90 DEG.):"
35   REM * * * * * * INPUT SECTION
40   PRINT "INPUT GIVEN PARTS (NOTE THAT"
50   PRINT "   A1 MEANS 'SIDE OPPOSITE ANGLE A,'"
60   PRINT "   AND SO ON):"
70   PRINT "INPUT A1, B (DEGREES):";
80   INPUT A1,B
85   REM * * * * * * COMPUTATION SECTION
90   LET F = 180/3.14159
100    LET A1 = A1/F
110    LET B = B/F
115    REM * * * T = TAN B1
120    LET T = SIN(A1) * TAN(B)
130    GOSUB 290
140    LET B1 = T1
145    REM * * * T = TAN C1
150    LET T = TAN(A1)/COS(B)
160    GOSUB 290
```

```
170   LET C1 = T1
175   REM * * *T = TAN A
180   LET T = TAN(A1)/SIN(B1)
190   GOSUB 290
200   LET A = T1
205   REM * * * * * *OUTPUT SECTION
210   PRINT
220   PRINT "A = ";INT (100*A*F + .5)/100;"DEG."
230   PRINT TAB(20);"A1 = ";INT (100 * A1 * F + .5)/100;"DEG."
240   PRINT "B = ";INT (100 * B * F + .5)/100;"DEG."
250   PRINT TAB(20);"B1 = ";INT (100 * B1 * F + .5)/100;"DEG."
260   PRINT "C = 90 DEGREES"
270   PRINT TAB(20);"C1 = ";INT (100 * C1 * F + .5)/100;"DEG."
280   STOP
285   REM * * * * * *SUBROUTINE
290   LET T1 = ATN(T)
300   IF T1 >= 0 THEN 320
310   LET T1 = 3.14159 + T1
320   RETURN
330   END
```

Exercises

Use the preceding program to solve the following.

1. Exercise 3, page 266

2. Exercise 10, page 266

To solve a Case VI triangle, we can revise the program with the following lines.

```
70    PRINT "INPUT A, B (DEGREES):";
80    INPUT A,B
100   LET A = A/F
115   REM * * *C2 = COS B1, S = SIN B1, T = TAN B1
120   LET C2 = COS(B)/SIN(A)
122   LET S = SQR (1 − C2 * C2)
124   LET T = S/C2
126   LET X = T
150   LET T = X/COS(A)
175   REM * * *T = TAN A1
180   LET T = S * TAN(A)
200   LET A1 = T1
```

Use the revised program to solve the following.

3. Exercise 5, page 266

4. Exercise 8, page 266

The shape of the St. Louis Arch is a curve called a catenary, which can be described by a hyperbolic equation.

INFINITE SERIES

OBJECTIVES

1. To study Taylor and Fourier Series.
2. To see the relationship between power series and hyperbolic functions.

11-1 *Power Series*

Have you ever wondered how trigonometric tables such as Table 1 are compiled or how a calculator can quickly find the sine or the cosine of an angle? (These numbers are not stored in the calculator.) In order to evaluate any function, a calculator, a computer, or a human being can use only the four arithmetic operations of addition, subtraction, multiplication, and division. For example, in order to evaluate the sine of one (radian),

$$\sin 1$$

we must represent sin 1 as an expression that can be evaluated by arithmetic alone. No finite number of arithmetic operations will give the *exact* value of sin 1. However, an expression of the form

$$a_0 + a_1x + a_2x^2 + a_3x^3 + \cdots,$$

without end, where the a_i's are constants, will give an approximation of the function $\sin x$ for any value of x after a finite number of terms of the expression have been evaluated. Furthermore, it can be shown that if any degree of accuracy of $\sin x$ is specified (say to within ± 0.0001) for a particular value of x, this accuracy can be achieved, provided that enough terms are evaluated.

In this chapter we shall take an intuitive look at expressions like the one above. Such expressions are called **power series.** If a power series approximates a function to any desired degree of accuracy, we say that the series is a **power series expansion** of the function or, more simply, a **Taylor series** for the function. We also say that the power series **converges** to the function.

For the function $\sin x$, where x is any real number, the Taylor series is given by:

$$\sin x = x - \frac{x^3}{3!} + \frac{x^5}{5!} - \frac{x^7}{7!} + \cdots \qquad (1)$$

Recall that $n!$ means $n(n - 1)(n - 2) \times \cdots \times 4 \times 3 \times 2 \times 1$. Thus

$$\sin 1 = 1 - \frac{1^3}{3!} + \frac{1^5}{5!} - \frac{1^7}{7!} + \cdots$$

$$= 1 - \frac{1}{3!} + \frac{1}{5!} - \frac{1}{7!} + \cdots.$$

From a practical standpoint, if only the first four terms of the Taylor series for $\sin 1$ are used, the resulting value is accurate to about five decimal places.

The Taylor series for $\cos x$, where x is any real number, is given by:

$$\cos x = 1 - \frac{x^2}{2!} + \frac{x^4}{4!} - \frac{x^6}{6!} + \cdots \qquad (2)$$

Example 1   Evaluate $\cos 1.2$ to three decimal places using the first four terms of the Taylor series.

Solution $$\cos 1.2 = 1 - \frac{1.2^2}{2!} + \frac{1.2^4}{4!} - \frac{1.2^6}{6!} + \cdots$$

$$\approx 1 - \frac{1.44}{2} + \frac{2.0736}{24} - \frac{2.9860}{720} + \cdots$$

$$\approx 0.362$$

Among the most important functions in advanced mathematics is the **exponential function** e^x. (e is an irrational number which occurs naturally in the calculus; for example, see Exercises 7-9 on page 288.) It can be shown that, for all real x,

$$e^x = 1 + x + \frac{x^2}{2!} + \frac{x^3}{3!} + \frac{x^4}{4!} + \cdots \qquad (3)$$

Example 2 Find e to three decimal places using the first six terms of the Taylor series for e^x.

Solution Since $e = e^1$, we have:

$$e = 1 + \frac{1}{1!} + \frac{1^2}{2!} + \frac{1^3}{3!} + \frac{1^4}{4!} + \frac{1^5}{5!} + \frac{1^6}{6!} + \cdots$$

$$= 1 + 1 + \frac{1}{2} + \frac{1}{6} + \frac{1}{24} + \frac{1}{120} + \frac{1}{720} + \cdots$$

$$\approx 2.718$$

(An even more accurate approximation is $e \approx 2.718281828 \ldots$. This decimal expansion, however, does *not* continue to repeat.)

Note that the power series for e^x is made up of the terms of the series for $\sin x$ and $\cos x$, except for sign. This suggests that a relationship exists among these three functions. We shall discover what this relationship is in Section 11-2.

Exercises 11-1

A 1. Use the first three terms of the Taylor series for $\sin x$ to find the following:

(a) $\sin \frac{\pi}{2} \approx \sin 1.57$ (b) $\sin \frac{\pi}{6} \approx \sin 0.52$

(c) Compare the values you obtained in (a) and (b) with the known values of $\sin \frac{\pi}{2}$ and $\sin \frac{\pi}{6}$.

2. Use the first three terms of the Taylor series for $\cos x$ to find the following.

(a) $\cos \frac{\pi}{2} \approx \cos 1.57$ (b) $\cos \frac{\pi}{3} \approx \cos 1.05$

(c) Compare the values you obtained in (a) and (b) with the known values.

3. Use the first four terms of the Taylor series for $\sin x$ to evaluate each of the following to three decimal places.

(a) $\sin 2$ (b) $\sin 1.14$

(c) Compare the answers to (a) and (b). Give a reason for their relationship.

4. Use the first four terms of the Taylor series for $\cos x$ to evaluate each of the following to three decimal places.

    (a) $\cos 1.5$ (b) $\cos 1.64$

    (c) Compare the answers to (a) and (b). Give a reason for their relationship.

5. Use the first six terms of the Taylor series for e^x to evaluate $\dfrac{1}{e}$ to three decimal places.

6. Use the first six terms of the Taylor series for e^x to evaluate $\sqrt{e}$ to two decimal places.

B 7. Evaluate the expression $\left(1 + \dfrac{1}{n}\right)^n$ for each of the following.

    (a) $n = 10$ (b) $n = 50$ (c) $n = 100$ (d) $n = 1000$

    (You may wish to use the following to simplify the calculations.
    $\log 1.1 = 0.0414$ $\log 1.02 = 0.0086$ $\log 1.01 = 0.00432$ $\log 1.001 = 0.000434$)

    (e) Compare your answers with the value of e given on page 287.

8. Expand the expression $\left(1 + \dfrac{1}{n}\right)^n$, for n a positive integer, by the binomial theorem. Give the approximate value of the first few terms of this expansion when n is *very large*.

9. Based on the results of Exercise 8, justify the statement

$$\left(1 + \frac{1}{n}\right)^n \text{approaches } e,$$

    as n gets larger and larger. (This is essentially the definition of the number e given in more advanced math courses.)

C 10. Use the fact that $1° = \dfrac{\pi}{180}$ to find power series for (a) $\sin x°$ and (b) $\cos x°$. (The results suggest one reason radian measure is preferred to degree measure in higher mathematics.)

11. If you studied infinite geometric series in algebra, explain why the following power-series expansions are valid if $|x| < 1$.

    (a) $\dfrac{1}{1 - x} = 1 + x + x^2 + x^3 + \cdots$

    (b) $\dfrac{1}{1 + x} = 1 - x + x^2 - x^3 + \cdots$

    (c) $\dfrac{1}{1 - x^2} = 1 + x^2 + x^4 + x^6 + \cdots$

    (d) $\dfrac{x}{1 + x^2} = x - x^3 + x^5 - x^7 + \cdots$

11-2 *Hyperbolic Functions*

If we substitute $-x$ for x in Equation (3) of Section 11-1, we obtain:

$$e^{-x} = 1 + \frac{(-x)}{1!} + \frac{(-x)^2}{2!} + \frac{(-x)^3}{3!} + \frac{(-x)^4}{4!} + \frac{(-x)^5}{5!} + \cdots$$

$$= 1 - x + \frac{x^2}{2!} - \frac{x^3}{3!} + \frac{x^4}{4!} - \frac{x^5}{5!} + \cdots$$

Adding the sides of this last equation to the respective sides of Equation (3), we obtain

$$e^x + e^{-x} = 2\left(1 + \frac{x^2}{2!} + \frac{x^4}{4!} + \frac{x^6}{6!} + \cdots\right).$$

That is,

$$\frac{1}{2}(e^x + e^{-x}) = 1 + \frac{x^2}{2!} + \frac{x^4}{4!} + \frac{x^6}{6!} + \cdots.$$

Similarly,

$$\frac{1}{2}(e^x - e^{-x}) = x + \frac{x^3}{3!} + \frac{x^5}{5!} + \frac{x^7}{7!} + \cdots.$$

These last two Taylor series are the same as the series for cos x and sin x, respectively, except that the signs do not alternate. As you will show in the exercises of this section, the functions $\frac{1}{2}(e^x + e^{-x})$ and $\frac{1}{2}(e^x - e^{-x})$ also are similar to the sine and cosine functions in other ways. For instance, they are associated with the "unit hyperbola" $x^2 - y^2 = 1$ much as the circular functions sine and cosine are associated with the unit circle $x^2 + y^2 = 1$. (See Exercise 6 on page 291.) For this reason they are called **hyperbolic functions,** and we define the **hyperbolic cosine** (cosh) and **hyperbolic sine** (sinh) for all real x as follows:

$$\cosh x = \frac{1}{2}(e^x + e^{-x}) = 1 + \frac{x^2}{2!} + \frac{x^4}{4!} + \frac{x^6}{6!} + \cdots \qquad (4)$$

$$\sinh x = \frac{1}{2}(e^x - e^{-x}) = x + \frac{x^3}{3!} + \frac{x^5}{5!} + \frac{x^7}{7!} + \cdots \qquad (5)$$

The hyperbolic functions are used in physics and engineering. For example, a perfectly flexible, inextensible chain or cable suspended from two points hangs in a curve, called a **catenary,** that has an equation of the form

$$y = k \cosh\left(\frac{x}{k}\right),$$

for some positive real number k. (See pages 96 and 97.)

From the definitions of sinh x and cosh x, one can see that:

$$e^x = \cosh x + \sinh x \qquad (6)$$

During the eighteenth century, it was found that a similar relationship holds for the sine and cosine functions. If we formally replace x by ix ($i^2 = -1$) in the series for e^x, we have the following relationship:

$$e^{ix} = 1 + ix + \frac{(ix)^2}{2!} + \frac{(ix)^3}{3!} + \frac{(ix)^4}{4!} + \frac{(ix)^5}{5!} + \frac{(ix)^6}{6!} + \cdots$$

$$= 1 + ix - \frac{x^2}{2!} - i\frac{x^3}{3!} + \frac{x^4}{4!} + i\frac{x^5}{5!} - \frac{x^6}{6!} - i\frac{x^7}{7!} + \cdots$$

$$= \left(1 - \frac{x^2}{2!} + \frac{x^4}{4!} - \frac{x^6}{6!} + \cdots\right) + i\left(x - \frac{x^3}{3!} + \frac{x^5}{5!} - \frac{x^7}{7!} + \cdots\right)$$

The two Taylor series in parentheses in the last step are exactly those of $\cos x$ and $\sin x$, respectively. It therefore seems reasonable to define the use of an imaginary exponent by:

$$e^{ix} = \cos x + i \sin x \qquad (7)$$

As you will show in Exercises 7 and 8, the definition for raising a real number to an imaginary power obeys many of the same laws that hold for real exponents. For example,

$$e^{ai} \cdot e^{bi} = e^{(a+b)i}$$

$$(e^{ai})^k = e^{kai}, \text{ for } k \text{ a real number.}$$

It is *not* true, however, that $e^{ia} = e^{ib}$ implies $a = b$, as is true for real exponents.

Exercises 11-2

A 1. Use Equation (4) to show that $\cosh 0 = 1$.

2. Use Equation (5) to show that $\sinh 0 = 0$.

3. Use Equation (7) to show that $e^{i\pi} = -1$.

4. Find the value of the following.

   (a) $e^{2\pi i}$ (b) $e^{\frac{\pi i}{2}}$

5. Show that cosh x is an even function and that sinh x is an odd function.

B 6. We know that if $x = \cos t$ and $y = \sin t$, then $x^2 + y^2 = 1$. Show that if $x = \cosh t$ and $y = \sinh t$, then $x^2 - y^2 = 1$.

7. Use Equation (7) to show that for all real numbers a and b,
$$e^{ai} \cdot e^{bi} = e^{(a+b)i}.$$

8. Show that for any real number a and integer k,
$$(e^{ai})^k = e^{kai}.$$

9. Show that $\cosh 2x = \cosh^2 x + \sinh^2 x$.

10. Show that $\sinh 2x = 2 \sinh x \cosh x$.

C 11. Show that $e^{ai} = e^{bi}$ does not imply that $a = b$.

12. We can define e^z for any complex number $z = x + iy$ by
$$e^{x+iy} = e^x e^{iy} = e^x(\cos y + i \sin y).$$
Show that $e^z \cdot e^w = e^{z+w}$ for any two complex numbers z and w.

13. Show that

(a) $\cos x = \dfrac{e^{ix} + e^{-ix}}{2} = \cosh ix$

(b) $\sin x = \dfrac{e^{ix} - e^{-ix}}{2i} = -i \sinh ix$.

11-3 *Trigonometric Series*

Scientists often encounter wave phenomena such as light waves, sound waves, or water waves that, when analyzed mathematically, give rise to periodic functions. Since so much is known about the sine and cosine functions, it is natural to try to describe all such periodic functions in terms of sine and cosine. A method for doing this was first suggested by Jean Baptiste Joseph Fourier (1768–1830). Thus an infinite series of the form
$$a_0 + a_1 \cos x + b_1 \sin x + a_2 \cos 2x + b_2 \sin 2x$$
$$+ a_3 \cos 3x + b_3 \sin 3x + \cdots,$$
where the a's and b's are constants, is called a **Fourier series**. Like power series, Fourier series can be used to approximate a large class of functions to any desired degree of accuracy, if sufficiently many terms are evaluated. Although the above series has period 2π, it can be modified to have any positive period. (See Exercises 8 and 9.) For simplicity, however, we shall

assume that the function to be approximated has period 2π so that the series can be used as given. Thus, if such a function is defined in the interval $-\pi \le x \le \pi$, it will be defined for all real numbers by periodicity.

Finding the a's and b's in the Fourier series of a given function requires calculus. We shall, however, provide a few such series to work with in the example and exercises that follow.

Example Let $f(x)$ have period 2π, and be defined in the interval $-\pi \le x \le \pi$ by $f(x) = x$ for $-\pi < x < \pi$; $f(\pi) = f(-\pi) = 0$.

(a) Sketch the graph of $y = f(x)$.

(b) Using the Fourier series for $f(x)$ given below, represent $\dfrac{\pi}{4}$ as an infinite sum of rational numbers.

$$f(x) = 2\left(\frac{1}{1}\sin x - \frac{1}{2}\sin 2x + \frac{1}{3}\sin 3x - \frac{1}{4}\sin 4x + \cdots\right)$$

(The cosine terms all have coefficient 0; that is, the a's are all 0.)

Solution (a) For $-\pi < x < \pi$, the graph of $f(x)$ is represented by the line $y = x$. At both π and $-\pi$, $f(x) = 0$. Therefore, by periodicity, the entire graph of $f(x)$ is as indicated below.

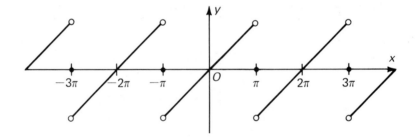

(b) We know that $f\left(\dfrac{\pi}{2}\right) = \dfrac{\pi}{2}$, since $f(x) = x$ for $-\dfrac{\pi}{2} < x < \dfrac{\pi}{2}$. Therefore, we have:

$$\frac{\pi}{2} = 2\left(\frac{1}{1}\sin\frac{\pi}{2} - \frac{1}{2}\sin 2\left(\frac{\pi}{2}\right) + \frac{1}{3}\sin 3\left(\frac{\pi}{2}\right) - \frac{1}{4}\sin 4\left(\frac{\pi}{2}\right) + \cdots\right)$$

$$= 2\left(1(1) - \frac{1}{2}(0) + \frac{1}{3}(-1) - \frac{1}{4}(0) + \frac{1}{5}(1) - \cdots\right)$$

$$= 2\left(1 - \frac{1}{3} + \frac{1}{5} - \frac{1}{7} + \cdots\right)$$

Hence,

$$\frac{\pi}{4} = 1 - \frac{1}{3} + \frac{1}{5} - \frac{1}{7} + \frac{1}{9} - \frac{1}{11} + \cdots.$$

Using the same function $f(x)$ defined in the example above, we can illustrate graphically how the partial Fourier sums give a better and better approximation of the function $f(x)$ as more terms are evaluated. The situation is illustrated in the figure.

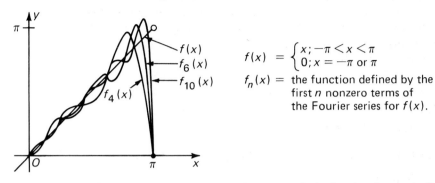

$$f(x) = \begin{cases} x; -\pi < x < \pi \\ 0; x = -\pi \text{ or } \pi \end{cases}$$

$f_n(x) = $ the function defined by the first n nonzero terms of the Fourier series for $f(x)$.

The Fourier series representations of some relatively simple periodic functions are given below. These will be used in the exercises.

(i) If $f(x) = |x|$ for $-\pi \le x < \pi$ and $f(x + 2\pi) = f(x)$, then

$$f(x) = \frac{\pi}{2} - \frac{4}{\pi}\left(\frac{1}{1^2}\cos x + \frac{1}{3^2}\cos 3x + \frac{1}{5^2}\cos 5x + \cdots\right).$$

(ii) If $f(x) = x^2$ for $-\pi \le x < \pi$ and $f(x + 2\pi) = f(x)$, then

$$f(x) = \frac{\pi^2}{3} + 4\left(-\frac{1}{1^2}\cos x + \frac{1}{2^2}\cos 2x - \frac{1}{3^2}\cos 3x + \cdots\right).$$

(iii) If

$$f(x) = \begin{cases} 0 \text{ for } -\pi < x < 0 \\ \pi \text{ for } \quad 0 < x < \pi \\ \frac{\pi}{2} \text{ for } x = -\pi, 0, \pi \end{cases}$$

and $f(x + 2\pi) = f(x)$, then

$$f(x) = \frac{\pi}{2} + 2\left(\frac{1}{1}\sin x + \frac{1}{3}\sin 3x + \frac{1}{5}\sin 5x + \cdots\right).$$

Exercises 11-3

A 1. Sketch the graph of the function f defined by $f(x) = |x|$, $-\pi \le x < \pi$, and $f(x + 2\pi) = f(x)$.

2. Use the Fourier series given in (i) to express $\frac{\pi^2}{8}$ as an infinite sum of rational numbers.

3. Sketch the graph of the function f defined by $f(x) = x^2$, $-\pi \le x < \pi$, and $f(x + 2\pi) = f(x)$.

4. Use the Fourier series given in (ii) to express $\dfrac{\pi^2}{12}$ as an infinite sum of rational numbers.

B   5. The number $\dfrac{6}{\pi^2}$ is exactly the probability that if two positive integers are chosen at random, they will be relatively prime, that is, they will have no common factor. Use the Fourier series given in (ii) to express the reciprocal of this probability, $\dfrac{\pi^2}{6}$, as an infinite sum of rational numbers.

6. Sketch the graph of the function defined in (iii).

7. Use the Fourier series given in (iii) to find the same expression of $\dfrac{\pi}{4}$ that was found in the example on page 292.

C   8. Show that if f is a periodic function with fundamental period p, then the function $f\!\left(\dfrac{px}{2\pi}\right)$ has period 2π. (This shows how a function can be modified to admit a Fourier series with arguments x, $2x$, $3x$, etc.)

9. Explain how the arguments of the terms of a Fourier series could be modified to represent a periodic function with fundamental period p where p is not necessarily equal to 2π.

10. If an *even* function has a Fourier series representation, the series contains only cosine terms. The Fourier series for an *odd* function contains only sine terms. Show that any function f that has a Fourier series representation can be written as the sum of an odd function and an even function.

CHAPTER SUMMARY

1. A power series may be used to approximate a function to any desired degree of accuracy. Such a power series is referred to as a Taylor series and is said to converge to the function. Some important Taylor series are:

$$\sin x = x - \frac{x^3}{3!} + \frac{x^5}{5!} - \frac{x^7}{7!} + \cdots$$

$$\cos x = 1 - \frac{x^2}{2!} + \frac{x^4}{4!} - \frac{x^6}{6!} + \cdots$$

$$e^x = 1 + x + \frac{x^2}{2!} + \frac{x^3}{3!} + \frac{x^4}{4!} + \cdots$$

2. The hyperbolic functions sinh and cosh are associated with the unit hyperbola much as sine and cosine are associated with the unit circle. Sinh and cosh are defined as follows:

$$\sinh x = \frac{1}{2}(e^x - e^{-x}) = x + \frac{x^3}{3!} + \frac{x^5}{5!} + \frac{x^7}{7!} + \cdots$$

$$\cosh x = \frac{1}{2}(e^x + e^{-x}) = 1 + \frac{x^2}{2!} + \frac{x^4}{4!} + \frac{x^6}{6!} + \cdots$$

The definitions lead to the facts that

$$e^x = \cosh x + \sinh x$$

and

$$e^{ix} = \cos x + i \sin x.$$

3. A Fourier series can also be used to approximate periodic functions. The general form of a Fourier series is

$$a_0 + a_1 \cos x + b_1 \sin x + a_2 \cos 2x + b_2 \sin 2x$$
$$+ a_3 \cos 3x + a_3 \sin 3x + \cdots,$$

where the *a*'s and *b*'s are constants.

CHAPTER TEST

11-1 1. Use the first four terms of the appropriate Taylor series to evaluate each of the following to three decimal places.

(a) $\sin 1.4$ (b) $\cos \frac{\pi}{4}$ (c) e^2

11-2 2. Use the first four terms of the Taylor series for $\cosh x$ or $\sinh x$ to evaluate the following to three decimal places.

(a) $\cosh 1$ (b) $\sinh 0.5$

3. Find the value of the following.

(a) $e^{\frac{\pi i}{4}}$ (b) $e^{\frac{\pi i}{3}}$

11-3 4. Use the Fourier series given in (i) on page 293 to express $\frac{\pi^2}{24}$ as an infinite sum of rational numbers.

EXPANSIONS
AND SERIES

Computers can be very useful in working with expansions and series. For example, the first four powers of $(1 + x)^n$, where n is a natural number, are:

$$1 + x$$
$$(1 + x)^2 = 1 + 2x + x^2$$
$$(1 + x)^3 = 1 + 3x + 3x^2 + x^3$$
$$(1 + x)^4 = 1 + 4x + 6x^2 + 4x^3 + x^4$$

In general,

$$(1 + x)^n = 1 + nx + \frac{n(n-1)}{1 \cdot 2}x^2 + \frac{n(n-1)(n-2)}{1 \cdot 2 \cdot 3}x^3 + \cdots + x^n.$$

We can write the coefficients of the powers of x in this formula as

$$\frac{n}{1}, \frac{n}{1} \cdot \left(\frac{n-1}{2}\right), \frac{n}{1} \cdot \frac{n-1}{2} \cdot \left(\frac{n-2}{3}\right), \ldots,$$

where each coefficient after the first is found by multiplying the preceding one by a factor involving n and the exponent of x in that term. The following program makes use of this fact in lines 80, 110, and 120.

```
10   PRINT "EXPANSION OF (1+X)↑N"
20   PRINT "INPUT N";
30   INPUT N
40   PRINT
50   PRINT "(1+X)↑"; N; "=1";
60   LET C=1
70   LET E=1
80   LET C=C*N/E
90   PRINT "+"; C; "X↑"; E;
100   IF N=1 THEN 140
110   LET E=E+1
120   LET N=N−1
130   GOTO 80
140   END
```

Exercises

1. RUN the preceding program for:
   (a) $n = 2$ (b) $n = 3$ (c) $n = 4$

Make the following changes in the computer program:

```
10   PRINT "EVALUATE (1+X)↑N"
34   PRINT "INPUT X";
36   INPUT X
50   PRINT "(1+"; X;")↑"; N;"="
60   LET T=1
74   LET S=1
76   PRINT "1"
80   LET T=T*N*X/E
85   LET S=S+T
90   PRINT "+"; T, "="; S
```

RUN the revised program using these values.

2. $N = 3$, $X = 1$. The final result is 2^3.

3. $N = 4$, $X = 1$. The final result is 2^4.

4. $N = 3$, $X = 2$. The final result is 3^3.

5. $N = 4$, $X = 2$. The final result is 3^4.

Change this line in the program

```
100   IF T<.000005 THEN 140
```

and RUN the program for these values.

6. $N = 10$, $X = 0.1$ 7. $N = 100$, $X = 0.01$

8. $N = 1000$, $X = 0.001$ 9. $N = 10,000$, $X = 0.0001$

The results of Exercises 6–9 show that:

(6) $\left(1 + \dfrac{1}{10}\right)^{10} \approx 2.59374$ (7) $\left(1 + \dfrac{1}{100}\right)^{100} \approx 2.70481$

(8) $\left(1 + \dfrac{1}{1000}\right)^{1000} \approx 2.71692$ (9) $\left(1 + \dfrac{1}{10000}\right)^{10000} \approx 2.71815$

The series that we have been using up to now have been finite. However, the computer can use the Taylor series for e^x, cos x, and sin x.

10. To compute an approximate value of e, make the following changes in the program:

```
10   PRINT "COMPUTE E"
50   PRINT "E="
80   LET T=T/E
```

Delete lines 20, 30, 34, 36, 84, and 120.

RUN the revised program.

11. Write a computer program using the series expression for cos x to compute the cosine values directly.

12. Repeat Exercise 11 for sin x.

CUMULATIVE REVIEW   *Chapters 10–11*

Chapter 10

1. Solve the spherical triangle ABC with $\angle A = 50°$, $b = 90°$, $c = 90°$.

2. Solve the right spherical triangle ABC with $\angle C = 90°$, $\angle A = 46.5°$, and $\angle B = 75.2°$.

3. Solve the quadrantal triangle ABC. Give answers to the nearest ten minutes. $\angle C$ is not necessarily $90°$. $a = 64°40'$, $b = 110°20'$, $c = 90°$

4. Solve the spherical triangle ABC with $a = 52.4°$, $b = 33.8°$, and $\angle B = 45.6°$.

5. Use the table on page 279 to answer the question. A plane leaves Chicago on a heading of $270°$ and follows a great-circle track. After the plane has flown 1500 nautical miles, what will be its position and course?

Chapter 11

Use the first four terms of the appropriate Taylor series to evaluate each of the following to three decimal places.

1. $\sin \dfrac{\pi}{3}$

2. $\cos 1.1$

3. $e^{0.3}$

4. $e^{1.5}$

5. $\cosh 1.8$

6. $\sinh 0.2$

7. Find the value of the following.

   (a) $e^{\frac{2\pi i}{3}}$ (b) $e^{\frac{\pi i}{6}}$

8. Use the Fourier series given in the Example on page 292 to represent $-\dfrac{\pi}{4}$ as an infinite sum of rational numbers.

COMPREHENSIVE TEST

Chapter 1

1. The terminal side of an angle in standard position passes through $(-5, 2\sqrt{6})$. Find the value of the six trigonometric functions of the angle.

2. Find $\csc 135° \cdot \cos(-45°) - \sec 240°$.

3. Give the trigonometric functions that are undefined for specific values of θ where $-90° \le \theta \le 0°$. Give both the function and the value of θ.

4. When the angle of elevation of the sun is 78°, a tree casts a 13 m shadow. How tall is the tree to the nearest meter?

5. Find θ if $\sin \theta = 0.5$, $\tan \theta < 0$, and $0° \le \theta < 360°$.

6. Using tables or a calculator, find θ and b if $\angle ABC = 90°$, $AB = 16.2$, and $BD = 5.1$.

7. Using tables or a calculator, find $\tan 16°15' + \cot 32.7°$ to the nearest hundredth.

8. In which quadrants will $\sec \theta \cdot \cot \theta$ be positive?

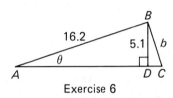

Exercise 6

Chapter 2

1. Convert $-30°$, $205°$, $18°$ to radian measure.

2. Convert $\dfrac{3\pi}{4}$, $-\dfrac{\pi}{3}$, 4 to degree measure.

3. Graph $y = 1 + \cos x$ for $-90° \le x \le 360°$.

4. Evaluate each of the following.

   (a) $\text{Tan}^{-1}(-1)$

   (b) $\text{Sec}\left(\sin^{-1}\dfrac{3}{5}\right)$

   (c) $\text{Cos}^{-1}\left(\cos\dfrac{5\pi}{6}\right)$

   (d) $\text{Cos}(\text{Tan}(\cos^{-1} 1))$

5. Graph $f(x) = -\cot x$ for $-\pi < x < \pi$. Then graph $g(x) = \tan x$ on the same axis. Identify all points of intersection.

6. Find the length of an arc of a circle of radius 15 cm that is intercepted by a central angle of 165°.

Chapter 3

1. If $\cos x = -0.6$ and $\sin x > 0$, find:

   (a) $\tan\left(x - \dfrac{\pi}{3}\right)$
   (b) $\cos\left(x + \dfrac{\pi}{6}\right)$
   (c) $\sin\left(x - \dfrac{\pi}{2}\right)$

2. Prove the identity $\sin 2x \cdot \tan x + \cos 2x = 1$.

3. Solve $2 \cos^4 x - 5 \cos^2 x + 2 = 0$ for $0 \le x \le \pi$.

4. Express $\dfrac{1 + \tan \theta (\csc \theta - \cot \theta \sec \theta)}{\cot \theta}$ as a single trigonometric function.

5. Prove the identity $\dfrac{\sec^2 x}{\sec x - 1} = \dfrac{\sec x + 1}{\sin^2 x}$.

6. Evaluate $\tan 15°$ using identities.

Chapter 4

1. In $\triangle TUB$, $TU = 13$, $UB = 8$, and $\angle U = 60°$. Find BT. Leave your answer in simplest radical form.

2. In $\triangle VAT$, $VA = 300$, $\angle V = 75°$, and $\angle A = 45°$. Find TV. Leave your answer in simplest radical form.

3. The measure of $\angle A$ in $\triangle JAR$ may be found by using either the law of sines or the law of cosines. Find $\angle A$ using both methods.

4. Find the area of $\triangle JAR$.

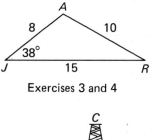

Exercises 3 and 4

5. Two road crews are clearing a trail through a forest from points A and B. The crews cannot see each other. However, they both can observe a tower at point C. A lookout at the tower reports that $\angle ACB$ is $115°$. If the distances to the tower are as marked, at what angles A and B should the crews begin their work?

Exercise 5

6. In $\triangle CAN$, $\angle C = 35°$ and $CN = 20$. For what values of NA is it true that there is a unique triangle for the given information?

Chapter 5

For each of the given functions (a) graph the function for $-2\pi \le x \le 2\pi$, and (b) give the period and amplitude.

1. $y = 5 \cos \dfrac{x}{2} + 2$

2. $y = \sin \left(2x + \dfrac{\pi}{3}\right)$

3. A flywheel is hung vertically 25 cm above the floor. At time $t = 0$, a point P is at the lowest point of the flywheel. If the wheel has a radius of 20 cm and is rotating at $\dfrac{3\pi}{2}$ radians/s,

   (a) write an equation for the height of P above the floor as a function of time.
   (b) graph the equation,
   (c) find the period, amplitude, and frequency of the point's motion.

4. Graph $y = \sin x + \cos 2x$ by plotting points for $0 \le x \le 2\pi$.

Chapter 6

1. Find a unit vector orthogonal to $-15\mathbf{i} - 20\mathbf{j}$.

2. Find the angle between $\mathbf{i} + 3\mathbf{j}$ and $\mathbf{i} - 3\mathbf{j}$.

3. A particle is acted on by the following forces:
   F_1: 10 N at 30° F_2: 15 N at 90° F_3: 4 N at 330°
   Find the bearing and magnitude of the additional force that is needed to keep the particle stationary.

4. A 100 kg box rests on a ramp with a 30° slope. What frictional force is needed to keep the box stationary on the ramp?

5. A ship sails due east for 300 km and then due south for 50 km. What are its distance and bearing from its starting point?

Chapter 7

1. Graph $r = 2 - \sin \theta$ using polar coordinates.

2. Find a polar equation for $x + \sqrt{2}y = 0$.

3. Simplify $\dfrac{i + (2 - i)(3 + 2i)}{(3 + i)}$.

4. Convert the complex number $(\sqrt{2} - i\sqrt{2})$ to polar form and find its reciprocal in polar form.

5. Find $(\sqrt{3} + i)^{-6}$. Express your answer in $x + yi$ form.

6. Express the three cube roots of -1 in both $x + yi$ form and polar form.

Chapter 8

1. Use a translation to simplify $y = 3 + \sin\left(x - \dfrac{\pi}{6}\right)$. Give the simplified equation and graph it, showing both sets of axes.

2. Simplify $x^2 + 2y^2 = 5 + \sqrt{3}xy$ by rotating the axes 30°. Then graph the equation, showing both sets of axes.

3. Simplify $8x^2 - 12xy - 8y^2 + 6\sqrt{10}x - 2\sqrt{10}y - 30 = 0$ by an appropriate rotation to eliminate the xy-term. Identify the resulting conic.

Chapter 9

1. Find a vector orthogonal to both $\mathbf{i} + \mathbf{j} - 2\mathbf{k}$ and $2\mathbf{i} - \mathbf{j}$.

2. How far is the center of the sphere $x^2 + y^2 + z^2 - 2y + 6z = 20$ from the point where the plane $x + 2y + z = 12$ intersects the y-axis?

3. Give the vector and scalar equations for the plane containing $(-1, 0, 3)$ and perpendicular to $2\mathbf{j} - \mathbf{k}$.

4. Give the vector equation for a line through $(1, 2, -4)$ perpendicular to the plane $y - z = 0$.

5. Find the distance from the point to the plane in Exercise 4.

Chapter 10

1. Solve the spherical triangle ABC with $c = 90°$, $b = 90°$, $\angle A = 65°$.

2. Solve the right spherical triangle ABC with $\angle A = 55°10'$, $\angle B = 102°30'$, and $\angle C = 90°$.

3. Solve the quadrantal triangle ABC with $\angle A = 75.2°$, $\angle B = 38.5°$, and $c = 90°$.

4. Solve the spherical triangle ABC with $a = 65°10'$, $b = 118°20'$, $c = 94°30'$.

5. A ship leaves Sydney, Australia with an initial heading of 90°. If the longitude of Sydney is 151°0' E and the latitude is 34°0' S, what will be the ship's position after it has traveled 1800 nautical miles?

Chapter 11

1. Use the first four terms of the appropriate Taylor series to find the following to three decimal places.

   (a) $\sin 2.6$ (b) $\cos \dfrac{2\pi}{3}$ (c) $e^{0.25}$ (d) $\cosh 1.8$ (e) $\sinh \dfrac{\pi}{4}$

2. Find the value of the following.

   (a) $e^{\frac{3\pi i}{4}}$ (b) $e^{-\frac{\pi i}{2}}$

3. Use the Fourier series given in (i) on page 293 to express $-\dfrac{\pi^2}{8}$ as an infinite sum of rational numbers.

COMPREHENSIVE TEST · MULTIPLE CHOICE

Chapter 1

1. Find sin 150°.

   (a) $-\dfrac{1}{2}$ (b) $\dfrac{1}{2}$ (c) $-\dfrac{\sqrt{3}}{2}$ (d) $\dfrac{\sqrt{3}}{2}$

2. The terminal side of an angle θ in standard position passes through $(8, -\sqrt{17})$. What is csc θ?

   (a) $-\dfrac{9\sqrt{17}}{17}$ (b) $-\dfrac{9\sqrt{17}}{8}$ (c) $-\dfrac{\sqrt{17}}{8}$ (d) $\dfrac{\sqrt{17}}{8}$

3. If $\tan \theta = \dfrac{3}{4}$, θ must be in which quadrant?

   (a) I or II   (b) I or III   (c) I or IV   (d) II or IV

4. Evaluate $\tan 120° \cdot \cos 210° - \csc 315° \cdot \sin 45°$.

   (a) 0   (b) $1\dfrac{1}{2}$ (c) 2   (d) $2\dfrac{1}{2}$

5. Find x to the nearest tenth.

   (a) 5.1   (b) 16.0

   (c) 40.1   (d) 43.9

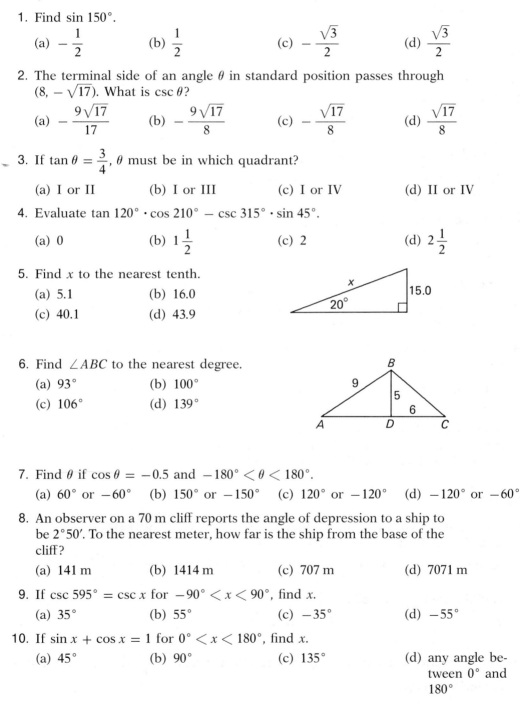

6. Find $\angle ABC$ to the nearest degree.

   (a) 93°   (b) 100°

   (c) 106°   (d) 139°

7. Find θ if $\cos \theta = -0.5$ and $-180° < \theta < 180°$.

   (a) 60° or −60°   (b) 150° or −150°   (c) 120° or −120°   (d) −120° or −60°

8. An observer on a 70 m cliff reports the angle of depression to a ship to be 2°50′. To the nearest meter, how far is the ship from the base of the cliff?

   (a) 141 m   (b) 1414 m   (c) 707 m   (d) 7071 m

9. If $\csc 595° = \csc x$ for $-90° < x < 90°$, find x.

   (a) 35°   (b) 55°   (c) −35°   (d) −55°

10. If $\sin x + \cos x = 1$ for $0° < x < 180°$, find x.

    (a) 45°   (b) 90°   (c) 135°   (d) any angle between 0° and 180°

Chapter 2

1. Convert 210° to radian measure.

   (a) $\dfrac{\pi}{6}$ (b) $\dfrac{5\pi}{6}$ (c) $\dfrac{3\pi}{4}$ (d) $\dfrac{7\pi}{6}$

2. Convert $\dfrac{7\pi}{4}$ to degree measure.

   (a) 315°   (b) 300°   (c) 270°   (d) 135°

3. Find $\tan \pi$.

   (a) 0   (b) 0.0549   (c) 1   (d) undefined

4. Which of the following is *not* an even function?

   (a) $f(x) = \cos x$ (b) $f(x) = |\sin x|$

   (c) $f(x) = \sin |x|$ (d) $f(x) = \tan x$

5. Evaluate $\mathrm{Tan}^{-1}\left(\sin \dfrac{\pi}{2}\right)$.

   (a) undefined   (b) $\dfrac{\pi}{2}$ (c) $\dfrac{\pi}{4}$ (d) $-\dfrac{\pi}{4}$

6. Evaluate $\mathrm{Sin}^{-1}\left(\sin \dfrac{\pi}{3}\right)$.

   (a) $\sin \dfrac{\pi}{3}$ (b) $\dfrac{\pi}{3}$ (c) $\sin\left(\mathrm{Sin}^{-1}\dfrac{\pi}{3}\right)$ (d) $\dfrac{2\pi}{3}$

7. Which graph never intersects the x-axis?

   (a) $y = |\cos x|$ (b) $y = \sec x$ (c) $y = \tan x$ (d) $y = \cot x$

8. Find the range of $y = \mathrm{Tan}^{-1}(x)$.

   (a) $0 < y < \dfrac{\pi}{2}$ (b) $-\dfrac{\pi}{2} < y < \dfrac{\pi}{2}$

   (c) $0 < y < \pi$ (d) $-\pi < y < \pi$

9. All of the following are undefined except:

   (a) $\csc\left(-\dfrac{\pi}{2}\right)$ (b) $\sec\left(\dfrac{\pi}{2}\right)$ (c) $\tan\left(\dfrac{\pi}{2}\right)$ (d) $\mathrm{Sin}^{-1}\left(\dfrac{\pi}{2}\right)$

10. A point on a circle of radius 10 m has a linear speed of 15 m/s. Find its angular speed.

    (a) 1.5 radians/s   (b) $0.\overline{6}$ radians/s

    (c) $\dfrac{15}{2\pi}$ radians/s   (d) $\dfrac{4\pi}{3}$ radians/s

Chapter 3

1. For all angles for which it is defined, $\dfrac{\cos \theta}{1 + \sin \theta} =$
   - (a) $\sec \theta + \tan \theta$
   - (b) $\sec \theta - \tan \theta$
   - (c) $\tan \theta - \sec \theta$
   - (d) $\sec \theta \tan \theta$

2. If $\tan x = \dfrac{1}{2}$, find $\tan 2x$.
   - (a) $\dfrac{1}{4}$
   - (b) $\dfrac{3}{4}$
   - (c) 1
   - (d) $\dfrac{4}{3}$

3. If $\cos x = \dfrac{3}{5}$ and $\sin x < 0$, evaluate $\cos \left(x + \dfrac{\pi}{3}\right)$.
   - (a) $\dfrac{3}{5} - \dfrac{2\sqrt{3}}{5}$
   - (b) $\dfrac{3}{10} - \dfrac{2\sqrt{3}}{5}$
   - (c) $\dfrac{3\sqrt{3}}{5} - \dfrac{2}{5}$
   - (d) $\dfrac{3}{10} + \dfrac{2\sqrt{3}}{5}$

4. For all angles for which it is defined, $(\csc x \tan x - \cos x) \cot x =$
   - (a) $\csc x - \sin x$
   - (b) $1 - \cot x$
   - (c) $\cos x$
   - (d) $\sin x$

5. If $-180° \leq \theta \leq 180°$, for how many values of θ will $\dfrac{1 - \sin \theta}{2} = \cos^2 \theta$?
   - (a) 0
   - (b) 1
   - (c) 2
   - (d) 3

6. If $\sin \left(x + \dfrac{\pi}{4}\right) = \dfrac{\sqrt{2}}{3}$, evaluate $\sin x + \cos x$.
   - (a) $\dfrac{1}{3}$
   - (b) $\dfrac{1}{2}$
   - (c) $\dfrac{2}{3}$
   - (d) 1

7. The sum of the solutions to $2 \sin^3 \theta - \sin \theta = 0$, $\dfrac{\pi}{2} < \theta < \dfrac{3\pi}{2}$, is:
   - (a) 3π
   - (b) 2π
   - (c) $\dfrac{7\pi}{4}$
   - (d) $\sqrt{2}$

8. $\cos \theta - \sqrt{3} \sin \theta =$
   - (a) $2 \cos (\theta + 60°)$
   - (b) $2 \sin (\theta + 60°)$
   - (c) $2 \cos (\theta - 60°)$
   - (d) $2 \sin (\theta - 60°)$

Chapter 4

1. According to the law of sines, in any $\triangle ABC$, $AB \cdot \sin A =$
   - (a) $AC \cdot \sin B$
   - (b) $BC \cdot \sin B$
   - (c) $AC \cdot \sin C$
   - (d) $BC \cdot \sin C$

2. State the law of cosines for $\triangle FGH$.
   (a) $(FG)^2 = (FH)^2 + (GH)^2 + 2(FH)(GH) \cos H$
   (b) $(FG)^2 = (FH)^2 + (GH)^2 - 2(FH)(GH) \cos H$
   (c) $(FG)^2 = (FH)^2 - (GH)^2 + 2(FH)(FG) \cos F$
   (d) $(FG)^2 = (FH)^2 - (GH)^2 - 2(FH)(GH) \cos H$

3. Two observation posts 3.74 km apart report the angle of elevation of an aircraft to be 37° and 23°. If the aircraft is between the two observation posts, how far is it from the nearer one?
   (a) 0.60 km   (b) 1.69 km   (c) 1.74 km   (d) 2.60 km

4. In $\triangle DEF$, if $DE = 6.2$, $EF = 7.1$, and $\angle E = 28°$, how many possible measures are there for $\angle F$?
   (a) none   (b) one   (c) two   (d) three

5. Two sides of a triangle have lengths 10 and 15. They meet at an angle of 75°. Find the length of the third side.
   (a) 12   (b) 14   (c) 16   (d) 18

6. Find $\cos \theta$.

   (a) $\dfrac{5}{8}$ (b) $\dfrac{13}{20}$

   (c) $-\dfrac{5}{16}$ (d) $\dfrac{5}{16}$

7. Find the area of the triangle shown.
   (a) 105   (b) 120
   (c) 135   (d) 160

8. In $\triangle RAT$, $\angle R = 35°$, $RA = 25$ m, and $AT = 15$ m. RT has been computed to the nearest meter. The following answers are suggested.
   (I) 34 m   (II) 25 m   (III) 16 m
   Which is (are) correct?
   (a) I and III   (b) II and III
   (c) II only   (d) I only

Chapter 5

1. The graph shows a portion of:

    (a) $y = \dfrac{1}{3} \sin 3x$

    (b) $y = 3 \sin 3x$

    (c) $y = \dfrac{1}{3} \sin \dfrac{1}{3} x$

    (d) $y = 3 \sin \dfrac{1}{3} x$

2. Which of the following does *not* have an amplitude of 4?

    (a) $y = 2 \sin x + 2$ (b) $y = 4 \cos x$

    (c) $y = |8 \cos x|$ (d) $y = 4 \sin \left(\dfrac{1}{2} x \right)$

3. Which of the following coincides with the graph of $y = \cos 2x$?

    (a) $y = \sin \left(2x - \dfrac{\pi}{2} \right)$ (b) $y = \sin 2 \left(x + \dfrac{\pi}{2} \right)$

    (c) $y = 2 \cos x$ (d) $y = \sin \left(2x + \dfrac{\pi}{2} \right)$

4. An object on the end of a spring is pulled 8 cm below its equilibrium point at time $t = 0$ and then released. If the object makes a complete up-and-down movement every 6 s, which of the following describes its motion?

    (a) $y = 8 \sin \left(3\pi t - \dfrac{\pi}{2} \right)$ (b) $y = 8 \sin \left(6t - \dfrac{\pi}{2} \right)$

    (c) $y = 8 \cos \left(\dfrac{\pi t}{3} + \pi \right)$ (d) $y = 6 \cos \left(\dfrac{\pi}{4} t - \pi \right)$

5. From an inspection of the graph of $y = \sin \dfrac{2x}{3} - \cos \left(2x - \dfrac{\pi}{2} \right)$, its amplitude would appear to be:

    (a) 1   (b) 2   (c) 4   (d) $\dfrac{1}{2}$

6. Which of the following has a period of 8?

    (a) $y = 8 \sin x$ (b) $y = \sin 8x$

    (c) $y = \sin 4\pi x$ (d) $y = \sin \dfrac{\pi x}{4}$

Chapter 6

1. If $(s\mathbf{i} + t\mathbf{j}) + (t\mathbf{i} + 2s\mathbf{j}) = 6\mathbf{i} - 4\mathbf{j}$, find $s - t$.

   (a) 6   (b) -6 (c) 26   (d) -26

2. A boat has a heading of $18°$ when crossing a stream that flows due south. If the boat reaches shore directly opposite its launching point and has a velocity of 14.0 km/h, what is the velocity of the current?

   (a) 4.3 km/h   (b) 4.5 km/h   (c) 13.3 km/h   (d) 14.7 km/h

3. Two forces act on the same point: F_1 of 12 N at a bearing of $90°$, and F_2 of 18 N at a bearing of $205°$. A third force is needed to keep the point stationary. What will be the magnitude of that force?

   (a) 13 N   (b) 17 N   (c) 21 N   (d) 285 N

4. Find the angle between $\mathbf{i} + 3\mathbf{j}$ and $4\mathbf{i} - 3\mathbf{j}$.

   (a) $72°$ (b) $98°$ (c) $102°$ (d) $108°$

5. $(2\mathbf{i} + 3\mathbf{j}) \cdot (3\mathbf{i} - 4\mathbf{j}) =$

   (a) 1   (b) -6 (c) $6\mathbf{i} - 12\mathbf{j}$ (d) $18\mathbf{i} + \mathbf{j}$

6. Find the work done by a force $F = 3\mathbf{i} - 4\mathbf{j}$ in moving a point from $(5, 2)$ to $(10, -10)$. Assume force is in newtons, distance is in meters.

   (a) 33 J   (b) 47 J   (c) 60 J   (d) 63 J

7. Which vector is not orthogonal to $2\mathbf{i} - 4\mathbf{j}$?

   (a) $4\mathbf{i} + 2\mathbf{j}$ (b) $-2\mathbf{i} - \mathbf{j}$ (c) $-4\mathbf{i} + 2\mathbf{j}$ (d) $-4\mathbf{i} - 2\mathbf{j}$

Chapter 7

1. Which of the following is the graph of $r = 2 \cos 2\theta$?

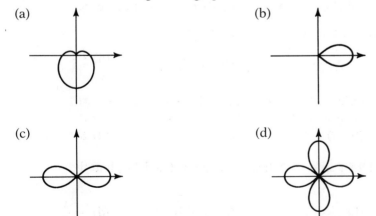

   (a)   (b)

   (c)   (d)

2. Simplify $\dfrac{(3 + i)(1 - 2i)}{2 + i}$.

(a) $1 - 3i$ (b) $\dfrac{7}{5} - \dfrac{11}{5}i$ (c) $\dfrac{5}{3} - 5i$ (d) $\dfrac{7}{3} - \dfrac{11}{3}i$

3. Find the reciprocal of $i\sqrt{3} - 1$ in polar form.

(a) $2(\cos 120° + i \sin 120°)$

(b) $-2(\cos 120° + i \sin 120°)$

(c) $\dfrac{1}{2}(\cos 240° + i \sin 240°)$

(d) $-\dfrac{1}{2}(\cos 120° + i \sin 120°)$

4. Evaluate $(\sqrt{3} - i)^{-12}$.

(a) $\dfrac{1}{4096}$

(b) $\dfrac{1}{4096}(\cos 30° + i \sin 30°)$

(c) $24(\cos (-30°) + i \sin (-30°))$

(d) $729 - 12i$

5. Which of the following is *not* a cube root of $4\sqrt{2} + 4i\sqrt{2}$?

(a) $-\sqrt{2} + i\sqrt{2}$

(b) $2(\cos 315° + i \sin 315°)$

(c) $2(\cos 255° + i \sin 255°)$

(d) $2(\cos 15° + i \sin 15°)$

Chapter 8

1. Suppose the axes of $y = 2 \sin 3x$ are translated so that the new origin is the point with old coordinates $(2, -5)$. The new equation is:

(a) $y' = 2 \sin (3x' - 2) + 5$

(b) $y' = 2 \sin (3x' + 6) - 5$

(c) $y' = 2 \sin (3x' + 2) + 5$

(d) $y' = 2 \sin (3x' + 6) + 5$

2. To simplify $y = 6 + \sin \left(2x + \dfrac{\pi}{3} \right)$, the axes should be shifted so that the origin is located at the point with old coordinates:

(a) $\left(\dfrac{\pi}{3}, -6 \right)$ (b) $\left(-\dfrac{\pi}{3}, 6 \right)$ (c) $\left(\dfrac{\pi}{6}, 6 \right)$ (d) $\left(-\dfrac{\pi}{6}, 6 \right)$

3. To identify the conic $6x^2 + y^2 + 5\sqrt{3}xy - 2x + 10 = 0$ the axes should be rotated by how many degrees?

(a) $15°$ (b) $30°$ (c) $45°$ (d) $60°$

4. Consider the graph of $y = 3x - 2$. If the axes are rotated $45°$, find the new y-intercept.

(a) 2 (b) $\dfrac{\sqrt{2}}{2}$ (c) $-\dfrac{\sqrt{2}}{2}$ (d) $-\dfrac{1}{2}$

Chapter 9

1. Find the distance between $(2, 5, -6)$ and $(6, 2, 6)$.

   (a) 13   (b) $\sqrt{19}$ (c) $\sqrt{37}$ (d) $\sqrt[3]{169}$

2. Which vector is *not* orthogonal to $\mathbf{i} + 2\mathbf{j} - \mathbf{k}$?

   (a) $5\mathbf{i} + 2\mathbf{j} + 9\mathbf{k}$ (b) $\mathbf{i} - 2\mathbf{j} + \mathbf{k}$

   (c) $7\mathbf{i} + 7\mathbf{k}$ (d) $\mathbf{i} - 2\mathbf{j} - 3\mathbf{k}$

3. Give the vector equation for a line through $(2, 0, 5)$ and $(3, -1, 2)$.

   (a) $2\mathbf{i} + 5\mathbf{k} + t(\mathbf{i} - \mathbf{j} + 3\mathbf{k})$ (b) $\mathbf{i} + \mathbf{j} + 3t\mathbf{k}$

   (c) $2\mathbf{i} + 5\mathbf{k} + t(\mathbf{i} - \mathbf{j} - 3\mathbf{k})$ (d) $t(-\mathbf{i} + \mathbf{j} + 3\mathbf{k})$

4. Find the distance from the origin to the line whose equation is $r = (1 + 4t)\mathbf{i} + t\mathbf{j} + (2 - t)\mathbf{k}$.

   (a) $2\dfrac{1}{3}$ (b) $\dfrac{\sqrt{398}}{9}$ (c) $\dfrac{\sqrt{43}}{3}$ (d) $\dfrac{1}{3}$

5. Give the scalar equation of the plane through the origin perpendicular to $3\mathbf{i} + 4\mathbf{j} - 12\mathbf{k}$.

   (a) $-3x + 4y - 12z = 0$ (b) $3x - 4y + 12z = 0$

   (c) $3x + 4y + 12z = 0$ (d) $3x + 4y - 12z = 0$

6. Give the scalar equation of the plane through $(1, 2, 1)$, $(0, 0, 4)$, and $(1, 5, -4)$.

   (a) $4x + 2y - 3z = 5$ (b) $11x + 2y + 5z = 20$

   (c) $-x + 5y + 3z = 12$ (d) $12x + 3y + 6z = 24$

Chapter 10

1. Solve spherical triangle ABC with $\angle A = 50°$, $\angle B = 90°$, $\angle C = 90°$.

   (a) $a = 130°, b = 90°, c = 90°$ (b) $a = 40°, b = 90°, c = 90°$

   (c) $a = 50°, b = 90°, c = 90°$ (d) $a = 50°, b = 10°, c = 10°$

2. Solve the right spherical triangle ABC with $\angle C = 90°$, $a = 37°40'$, and $\angle A = 37°40'$.

   (a) $\angle B = 37°40', b = 37°40', c = 90°$ (b) $\angle B = 90°, b = 90°, c = 90°$

   (c) $\angle B = 52°20', b = 52°20', c = 90°$ (d) $\angle B = 142°20', b = 142°20', c = 90°$

3. Which formula does not result from Napier's rules?

   (a) $\tan C = \sin a \sin b$ (b) $\cos B = \cos b \sin A$

   (c) $\sin a = \sin c \sin A$ (d) $\cos c = \cot A \cot B$

4. A spherical triangle ABC has the following parts: $a = 150°$, $\angle A = 120°$, $\angle C = 90°$. How many solutions are possible for this triangle?

   (a) 0   (b) 1   (c) 2   (d) 3

5. A plane left Houston, Texas, 29°45′ N, 95°22′ W, and flew to Boston, Massachusetts, 42°21′ N, 71°0′ W on a great-circle route. How many nautical miles did the plane travel?

    (a) 104.6   (b) 1395   (c) 2520   (d) 10,436

Chapter 11

1. $\sin 3 \approx$

    (a) 0.434   (b) 0.910   (c) 0.091   (d) 0.043

2. $\cos 1.7 \approx$

    (a) 0.131   (b) 0.013   (c) −0.013   (d) −0.131

3. $e^{1.2} \approx$

    (a) 3.294   (b) 3.182   (c) 2.718   (d) 3.718

4. $\cosh 2.1 \approx$

    (a) 0.4134   (b) −4.134   (c) −0.4134   (d) 4.134

5. $\sinh 1.5 \approx$

    (a) 0.2192   (b) 2.129   (c) −2.192   (d) −0.2192

6. Identify the graph of $f(x) = x^2$, $-\pi \le x < \pi$, and $f(x + 2\pi) = f(x)$.

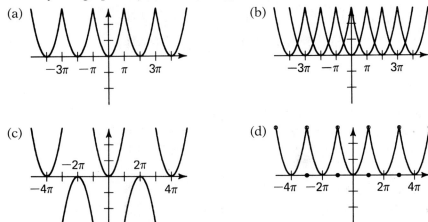

Extra Practice

For use after Chapter 1

Draw an angle in standard position having the given measure.

1. $60°$

2. $-60°$

3. $-390°$

4. $765°$

5. $-135°$

6. $1110°$

7. $-270°$

8. $200°$

Use Table 1-1 on page 8 and the square root table on page 369 (Table 4) as needed. Let θ be an angle in standard position whose terminal side passes through the given point. Find (a) $\sin \theta$ and $\cos \theta$ to four decimal places and (b) θ to the nearest 5°.

9. $(9, 12)$

10. $(2, 2\sqrt{3})$

11. $(3, 6\sqrt{2})$

12. $(5, 2\sqrt{6})$

13. $(4, 2\sqrt{5})$

14. $(2\sqrt{2}, 2\sqrt{2})$

15. $(2\sqrt{5}, 4\sqrt{5})$

16. $(\sqrt{13}, 2\sqrt{3})$

For the given angle θ, find $\sin \theta$ and $\cos \theta$ to four significant digits.

17. $23°40'$

18. $40°50'$

19. $65°10'$

20. $78°40'$

Find a value of θ that satisfies the given equation. Round your answer to the nearest minute or the nearest hundredth of a degree.

21. $\cos \theta = 0.2391$

22. $\sin \theta = 0.4950$

23. $\cos \theta = 0.5422$

Find the values of the six trigonometric functions for an angle θ in standard position whose terminal side passes through the given point. Simplify radicals and leave your answers in ratio form.

24. $(8, 6)$

25. $(12, 5)$

26. $(2, \sqrt{5})$

27. $(20, 21)$

28. $(2, 2\sqrt{3})$

29. $(\sqrt{2}, \sqrt{2})$

30. $(4, 5)$

31. $(3, 6)$

Solve $\triangle ABC$ in which $\angle C = 90°$.

32. $c = 65.0, a = 36.0$

33. $c = 120, b = 65$

34. $c = 42, \angle A = 50°$

35. $a = 35, b = 21$

36. $a = 110, \angle B = 40°$

37. $b = 72.0, \angle B = 36°$

38. $c = 140, \angle A = 29°$

39. $a = 18.0, b = 24.0$

Find exact values of the six trigonometric functions for each angle. Leave your answers in simplest form.

40. $120°$

41. $225°$

42. $315°$

43. $270°$

Use a reference triangle to find each of the following.

44. $\cos -42°$

45. $\tan 200°$

46. $\sec 115°30'$

47. $\sin 280.5°$

48. $\csc -50°$

49. $\cot 185.5°$

For use after Chapter 2

Convert the following to radian measure. Leave your answers in terms of π.

1. $225°$
2. $240°$
3. $-100°$
4. $20°$
5. $-60°$

Convert the following to degree measure.

6. $\dfrac{5\pi^R}{12}$
7. $\dfrac{3\pi^R}{4}$
8. $-\dfrac{\pi^R}{12}$
9. $\dfrac{7\pi^R}{6}$
10. $2\pi^R$

Find the exact values of sin t, cos t, tan t, sec t, csc t, and cot t for the given values of t.

11. $\dfrac{4\pi}{3}$
12. $\dfrac{\pi}{2}$
13. $-\dfrac{11\pi}{6}$
14. $\dfrac{7\pi}{4}$

15. $-\dfrac{7\pi}{3}$
16. $-\dfrac{7\pi}{6}$
17. $-\dfrac{\pi}{3}$
18. $\dfrac{13\pi}{6}$

Suppose that a point P travels uniformly at an angular speed of ω in a counterclockwise direction around a circular path of radius a with center at the origin. Find the coordinates of the final position of P if P starts at the point $(a, 0)$ and moves for t units of time.

19. $a = 3$ cm; $\omega = \dfrac{5\pi}{6}$ radians/s; $t = 12$ s

20. $a = 4$ cm; $\omega = \dfrac{3\pi}{4}$ radians/s; $t = 4$ s

21. $a = 2$ cm; $\omega = \dfrac{5\pi}{3}$ radians/s; $t = 9$ s

22. $a = 3$ cm; $\omega = 6$ rpm; $t = 24$ min

23. $a = 4$ cm; $\omega = 4$ rpm; $t = 2$ min

24. $a = 12$ cm; $\omega = \dfrac{4\pi}{3}$ radians/s; $t = 18$ s

25. $a = 4$ cm; $\omega = 8$ rpm; $t = 15$ s

26. $a = 6$ cm; $\omega = 3.5$ rpm; $t = 2$ min

By plotting multiples of $\dfrac{\pi}{4}$ and $\dfrac{\pi}{6}$, draw a detailed graph of the given function on the given domain.

27. $y = \cos x;\ -2\pi \le x \le 0$
28. $y = \sin x;\ 0 \le x \le 2\pi$

29. $y = -\cos x;\ -2\pi \le x \le 0$
30. $y = -\sin x;\ -2\pi \le x \le 0$

31. $y = \tan x;\ -2\pi \le x \le 0$
32. $y = 1 + \csc x;\ 0 \le x \le 2\pi$

33. $y = |\sec x|;\ 0 \le x \le 2\pi$
34. $y = |\cot x|;\ 0 \le x \le 2\pi$

Evaluate.

35. $\text{Cos}^{-1} \dfrac{\sqrt{2}}{2}$

36. $\text{Sin}^{-1} 0$

37. $\text{Tan}^{-1} \dfrac{\sqrt{3}}{3}$

38. $\sin \left(\text{Cos}^{-1} \left(-\dfrac{\sqrt{3}}{2} \right) \right)$

39. $\cos \left(\text{Sin}^{-1} 1 \right)$

40. $\text{Sin}^{-1} \left(\tan \dfrac{\pi}{4} \right)$

41. $\text{Cos}^{-1} \left(\tan \dfrac{\pi}{4} \right)$

42. $\tan \left(\text{Sin}^{-1} \dfrac{\sqrt{3}}{2} \right)$

43. $\text{Cos}^{-1} \left(\cos \left(-\dfrac{\pi}{2} \right) \right)$

For use after Chapter 3

Express the following in terms of the given trigonometric function. Give your answers in simplest form.

1. $\cot^2 x - 1$; $\sin x$

2. $\cos^2 x \csc x$; $\sin x$

3. $\csc x(1 - \sin x)(1 + \sin x)$; $\cos x$

4. $\csc^2 x + \cot^2 x$; $\cos x$

Express each of the following in terms of a single trigonometric function.

5. $\cot x(\sin x - \csc x)$

6. $\dfrac{1 + \tan^2 x}{\csc x \sec x}$

7. $\dfrac{\cos x \, (\sin x + \cos x \tan x)}{1 - \sin^2 x}$

8. $\dfrac{(\csc x - \cot x)(\csc x + \cot x)}{\sin x}$

Prove each identity.

9. $\csc^2 x(1 - \sin^2 x) = \cot^2 x$

10. $(\csc x + 1)(\sec x - \tan x) = \cot x$

11. $(\tan x - \sin x)(1 + \sec x) = \sin x \tan^2 x$

12. $(1 - \sin x)(\sec x + \tan x) = \cos x$

13. $\tan^2 x + \csc^2 x = \csc^2 x \sec^2 x - 1$

14. $\dfrac{1 - \cos^2 x}{\csc^2 x - 1} = \sin^4 x \sec^2 x$

15. $\dfrac{\sin x}{1 - \cos x} - \dfrac{\sin x}{1 + \cos x} = 2 \cot x$

16. $(\cot x + \cos x)(1 - \sin x) = \cot x(1 - \sin^2 x)$

Find each of the following if $\sin x_1 = -\dfrac{4}{5}$, $\sin x_2 = \dfrac{5}{13}$, $\pi < x_1 < \dfrac{3\pi}{2}$, and $0 < x_2 < \dfrac{\pi}{2}$.

17. $\sin (x_1 + x_2)$
    18. $\sin (x_1 - x_2)$
    19. $\cos (x_1 - x_2)$
    20. $\cos (x_1 + x_2)$

Find each of the following if $\cos x_1 = -\dfrac{3}{5}$, $\cos x_2 = \dfrac{5}{13}$, $\pi < x_1 < \dfrac{3\pi}{2}$, and $\dfrac{3\pi}{2} < x_2 < 2\pi$.

21. $\sin (x_1 + x_2)$ 22. $\cos (x_1 + x_2)$ 23. $\sin (x_1 - x_2)$ 24. $\cos (x_1 - x_2)$

For each angle θ satisfying the given condition in the given quadrant, find (a) $\sin 2\theta$ and (b) $\cos 2\theta$.

25. $\cos \theta = \dfrac{5}{13}$; I   26. $\sin \theta = \dfrac{12}{13}$; II   27. $\cos \theta = \dfrac{3}{5}$; IV

Use a half-angle formula to evaluate each expression.

28. $\sin \dfrac{x}{2}$, if $\cos x = -\dfrac{17}{25}$ and $0 < x < \pi$

29. $\cos \dfrac{x}{2}$, if $\cos x = -\dfrac{19}{36}$ and $\pi < x < 2\pi$

30. $\tan \dfrac{x}{2}$, if $\sin x = \dfrac{7}{25}$ and $0 < x < 2\pi$

31. $\tan \dfrac{x}{2}$, if $\cos x = \dfrac{24}{25}$ and $0 < x < \pi$

32. $\sin \dfrac{x}{2}$, if $\cos x = -\dfrac{24}{25}$ and $\pi < x < 2\pi$

Solve each equation for $0 \le x \le 2\pi$.

33. $\tan 2x = 2 \cos x$ 34. $2 \sin x \sec x = \sec x$

35. $\tan^2 x = \sin 2x$ 36. $3 \csc x - \sin x = 2$

Give the general solution for each equation.

37. $\sin x \cos x = 0$ 38. $\cos 2x + \sin x = 1$

39. $-\sin^2 x + 2 = \sin^4 x$ 40. $\tan 2x + \sec^2 2x = 1$

Rewrite each expression in the form $C \cos (\theta - \phi)$ for $-180° \le \phi < 180°$. Find ϕ to the nearest degree.

41. $3 \cos \theta - 4 \sin \theta$ 42. $\cos \theta + \sin \theta$

43. $5 \cos \theta - \sqrt{2} \sin \theta$ 44. $\sqrt{7} \cos \theta - 3 \sin \theta$

45. $3 \cos \theta - 3 \sin \theta$ 46. $-3 \cos \theta + 4 \sin \theta$

47. $\sqrt{2} \cos \theta + \sqrt{2} \sin \theta$ 48. $2 \cos \theta + \dfrac{3}{2} \sin \theta$

For use after Chapter 4

Complete Exercises 1–6 for $\triangle ABC$ with the given parts.

1. $a = 4$, $b = 7$, $c = 9$, $\angle B = ?$ 2. $a = 12$, $b = 9$, $\angle C = 45°$, $c = ?$

3. $b = 4$, $c = 5$, $\angle A = 60°$, $a = ?$ 4. $a = 21$, $b = 13$, $c = 24$, $\angle A = ?$

5. $a = 13$, $c = 17$, $\angle B = 60°$, $b = ?$ 6. $a = 5$, $b = 12$, $c = 16$, $\angle C = ?$

7. Find the lengths of the diagonals of a parallelogram with sides of lengths 15 and 40 and one angle of measure $120°30'$.

8. Find the lengths of the diagonals of a parallelogram with sides of lengths 5 and 11 and one angle of measure $80°$.

Solve each $\triangle ABC$ having the given angles and the given side. Give lengths to the nearest tenth.

9. $\angle A = 40°$, $\angle C = 62°$, $b = 7$

10. $\angle B = 112°$, $\angle C = 40°20'$, $a = 20$

11. $\angle A = 110°20'$, $\angle B = 38°$, $c = 13$

12. $\angle A = 47.5°$, $\angle C = 100°$, $b = 32$

13. $\angle B = 71°10'$, $\angle C = 54°$, $a = 22$

14. $\angle B = 65°50'$, $\angle C = 60°10'$, $a = 18$

15. $\angle A = 42°40'$, $\angle B = 63°$, $c = 17$

16. $\angle A = 82°10'$, $\angle C = 43°50'$, $b = 38$

For each $\triangle ABC$, state all possible values for $\angle B$ to the nearest 10' or tenth of a degree. Sketch a diagram to illustrate each solution. If no triangle can be formed, so state.

17. $\angle A = 48°$, $a = 12$, $b = 17$ 18. $\angle A = 30°$, $a = 21$, $b = 42$

19. $\angle A = 22°$, $a = 20$, $b = 21$ 20. $\angle A = 71°10'$, $a = 12$, $b = 11$

21. $\angle A = 30°$, $a = 67.5$, $b = 135$ 22. $\angle A = 64.5°$, $a = 28$, $b = 30.6$

23. $\angle A = 55°40'$, $a = 6$, $b = 9$ 24. $\angle A = 38°30'$, $a = 57$, $b = 43$

Find the area of $\triangle ABC$ with the given sides and angles. Estimate your answer to the nearest tenth.

25. $a = 9$, $b = 14$, $c = 21$ 26. $\angle A = 51°20'$, $\angle B = 43°$, $c = 18$

27. $c = 17$, $b = 28$, $\angle A = 84.5°$ 28. $a = 13$, $c = 15$, $\angle C = 72°$

29. $\angle B = 18°20'$, $\angle C = 56°$, $a = 5$ 30. $a = 22$, $b = 17.5$, $c = 34$

31. $a = 7$, $b = 8$, $\angle C = 101°$ 32. $b = 108$, $c = 74$, $\angle B = 82°$

For use after Chapter 5

Graph each function for $0 \leq x \leq 2\pi$.

1. $y = -3 \sin x$

2. $y = 4 \sin x$

3. $y = 2 \cos x$

4. $y = 2 \sin 3x$

5. $y = \cos \frac{1}{2}x$

6. $y = -3 \sin 2x$

7. $y = |\cos 2x|$

8. $y = \frac{1}{2} \cos \frac{3}{2}x$

9. $y = \sin \pi x$

Graph all three functions in each exercise on one set of axes. Label the graph of each function.

10. (a) $y = \sin x$ (b) $y = \sin \left(x - \frac{\pi}{2}\right)$ (c) $y = \sin \left(x + \frac{\pi}{2}\right)$

11. (a) $y = \cos x$ (b) $y = \cos (x + \pi)$ (c) $y = \cos x + 2$

12. (a) $y = 2 \sin x$ (b) $y = 2 \sin x - 2$ (c) $y = 2 \sin x + 2$

13. (a) $y = \cos x$ (b) $y = \sin \left(x - \frac{\pi}{2}\right)$ (c) $y = -\cos (x - \pi)$

Graph each of the following.

14. $y = 2 \sin (x + \pi) + 1$

15. $y = 3 \cos \left(2x + \frac{\pi}{2}\right) - 1$

16. $y = \frac{1}{2} \sin \left(x - \frac{\pi}{2}\right) + 2$

17. $y = 4 \cos (2x - \pi) + \frac{1}{2}$

For each of the following cases of simple harmonic motion, (a) write an equation for the position y of the object as a function of time t in the form $y = a \sin (\omega t + \beta)$, (b) graph the function, and (c) find the frequency of the oscillation. Neglect the effects of friction and air resistance.

18. An object attached to a spring is pulled downward 3 cm from its equilibrium position and released. It makes one complete oscillation every 3 seconds.

19. An object attached to a spring is pulled downward 7 cm from its equilibrium position and released. It makes one complete oscillation every 2 seconds.

20. The water line of a buoy starts from a position 20 cm below the surface of the water and bobs up to a maximum position 20 cm above the surface in 1 second.

21. The buoy in Exercise 20 covers the same vertical range in the same amount of time, but starts from a position 10 cm below the surface of the water and bobs up first.

22.–25. Express the functions in Exercises 18–21 in the form $y = a \cos(\omega t + \beta)$.

Graph each function for $0 \le x \le 2\pi$.

26. $y = \sin x + \cos 2x$ 27. $y = \sin 2x - \cos x$ 28. $y = -\sin x - \cos x$

29. $y = 3 \sin x - \sin 3x$ 30. $y = \cos x + \cos \dfrac{x}{2}$ 31. $y = \sin x - \sin \dfrac{x}{2}$

32. $y = \cos x + \sin \dfrac{x}{2}$ 33. $y = 2 \cos x - \cos 2x$ 34. $y = 2 \cos x + \cos 2x$

For use after Chapter 6

Draw the vectors **a**, **b**, **a** + **b**, and **a** − **b** with their initial points at the origin. Then find the norm of each of the four vectors.

1. **a** $= -$**i**, **b** $=$ **j**   2. **a** $= 6$**i** $+ 2$**j**, **b** $= 3$**j**

3. **a** $= 3$**i** $- 2$**j**, **b** $= -2$**i** $+$ **j**   4. **a** $= -2$**i** $+ 5$**j**, **b** $=$ **i** $- 4$**j**

For Exercises 5–8, find the specified linear combinations of **u** $= 2$**i** $-$ **j** and **v** $= -$**i** $+ 5$**j**.

5. 2**u** $+ 3$**v**   6. $\dfrac{1}{2}$**u** $+$ **v**

7. $\dfrac{2}{3}$**u** $+ \dfrac{1}{3}$**v**   8. -3**u** $+ \dfrac{1}{3}$**v**

9. A ship sails due south for 200 km, then due east for 50 km. What are the distance and bearing of the starting point from the ship?

10. A plane flies due north for 620 km and then on a heading of 80° for 225 km. What are its distance and bearing from its starting point?

11. A rescue ship has left port and travels 15 km on a heading of 40°. It hears a distress call from a ship 60 km from port, bearing 240°. What heading should the rescue ship take to reach the ship in distress and how far must it travel?

12. A plane heads due west with an air speed of 380 km/h. A 60 km/h wind is blowing from 210°. Find the plane's true course and ground speed.

13. If the plane in Exercise 12 was heading due south while everything else remained the same, what would be its true course and ground speed?

14. The speed in still water of a boat is 30 km/h. The boat heads directly east across a 75 m wide river that is flowing due south at 4.5 km/h.
(a) How far downstream will the boat land?
(b) What heading should the operator have used in order to land directly opposite the starting point?

15. Two planes take off in still air from the same airport at the same time. The speed of one plane is 480 km/h and its heading is 160°. The speed of the other plane is 430 km/h and its heading is 200°. How far apart are they at the end of one hour and what is the bearing of each from the other at that time?

16. A plane is flying on a heading of 300° at an air speed of 500 km/h. Its true course is 330° and the wind is blowing from 245°. What is the speed of the wind?

17. A 200 kg crate is on a plane inclined at 20° with the horizontal. Find the components of the gravitational force on the crate parallel and perpendicular to the plane.

18. A 180 kg motor slides with constant velocity down a ramp inclined at 23° with the horizontal. Find the frictional force acting on the motor.

19. A 275 kg object is suspended by two cables each of which makes an angle of 30° with the horizontal. Find the tension in each cable.

20. A 55 kg box is resting on a ramp that makes an angle of 20° with the horizontal. What is the frictional force that will keep the box from sliding down the ramp?

21. A 300 kg object is suspended by two cables making angles of 30° and 50°, respectively, with the horizontal. Find the tension in each cable.

22. A crate is resting on a ramp that makes an angle of 22° with the horizontal. A frictional force of 480 N is required to keep the crate from sliding down the ramp. Find the mass of the crate.

The given forces act on a particle P. Find the magnitude and bearing of the additional force that will keep P stationary.

23. F_1: 12 N, bearing 85°
    F_2: 18 N, bearing 220°
    F_3: 16 N, bearing 300°

24. F_1: 30 N, bearing 70°
    F_2: 20 N, bearing 200°
    F_3: 24 N, bearing 320°

Find the dot product of the given vectors. Are the vectors orthogonal? Are they parallel?

25. $a = -i - j$, $b = 4i$

26. $a = 2i - 3j$; $b = 3i + 2j$

27. $a = -i + 3j$; $b = 2i - 6j$

28. $a = 3i - j$; $b = 2i + 2j$

Resolve v into components parallel to a and b.

29. $v = i + 3j$; $a = 4i - 3j$; $b = 3i + 4j$

30. $v = 2i - 4j$; $a = -3i + 2j$; $b = 2i + 3j$

31. $v = -i + 2j$; $a = -3i + 4j$; $b = 4i + 3j$

32. $v = 2i + 5j$; $a = 5i - j$; $b = i + 5j$

In Exercises 33–40, forces are measured in newtons and distances in meters. Give answers in joules.

33. Find the work done by the force $\mathbf{F} = 3\mathbf{i} + 2\mathbf{j}$ in moving a particle from $A(3, 0)$ to $B(0, 5)$.

34. Find the work done by the force $\mathbf{F} = 3\mathbf{j}$ in moving an object from $A(2, 6)$ to $B(3, -3)$.

35. How much work is done by force $\mathbf{F} = -4\mathbf{i} + 6\mathbf{j}$ in moving an object (a) from $A(4, 0)$ to $B(2, 9)$ to $C(8, 0)$?   (b) from $A(4, 0)$ to $C(8, 0)$?

36. What is the combined work done by forces $\mathbf{F} = 3\mathbf{i} - 5\mathbf{j}$ and $\mathbf{G} = -2\mathbf{i} + 6\mathbf{j}$ in moving an object from $A(-3, 4)$ to $B(3, -6)$? What is the work done by $\mathbf{F} + \mathbf{G}$ in moving an object from A to B?

37. The mass of a loaded helicopter is 1500 kg. How much energy does its engine expend in ascending vertically for 75 m?

38. How much energy does a 60 kg climber expend in climbing a rock formation that is 25 m high and nearly vertical?

39. A crane loads 7500 kg containers onto the deck of a ship. How much energy does the crane expend in lifting a container 35 m off the ground?

40. How much energy does a 75 kg person expend in climbing a 20% grade for a horizontal distance of 6 km?

For use after Chapter 7

Find a polar equation for each of the following.

1. $y = 3$ 2. $x = -3$

3. $x^2 + y^2 = 9$ 4. $x - y = 1$

Graph the following polar equations.

5. $r \sin \theta = 2$ 6. $r = 4 \sin \theta$

7. $\theta = \dfrac{3\pi}{2}$ 8. $\theta + \dfrac{\pi}{3} = 0$

Graph the following polar equations.

9. $r = 1 + \sin 3\theta$ (three-leaved rose)

10. $r = 4 \cos \theta$ (circle)

11. $r = 2 \sin \theta$ (circle)

12. $r = 1 + \sin 2\theta$ (lemniscate)

13. $r = 2 \sin 2\theta$ (four-leaved rose)

14. $r = 3 \cos 3\theta$ (three-leaved rose)

15. $r = 1 - 2 \cos \theta$ (limaçon with small loop)

16. $r = 2\theta,\ \theta \geq 0$ (spiral)

Determine the type of each conic.

17. $r = \dfrac{4}{1 + \cos \theta}$

18. $r = \dfrac{4}{2 - \cos \theta}$

19. $r = \dfrac{6}{1 + 3 \sin \theta}$

20. $r = \dfrac{9}{3 - 2 \sin \theta}$

21.–24. Sketch the graphs of the conics in Exercises 17–20.

In Exercises 25–32, put all complex-number answers into the form $x + yi$.

Find (a) $w + z$, (b) wz, and (c) $\dfrac{w}{z}$.

25. $w = 2 + i,\ z = 3 - i$

26. $w = i,\ z = 5 - i$

27. $w = 2 - 2i,\ z = 3 + 3i$

28. $w = 5 - i,\ z = 5 + i$

Find (a) the conjugate, (b) the modulus, and (c) the reciprocal of z, and (d) z^2.

29. $z = 3 - i$

30. $z = \sqrt{2} + \sqrt{2}i$

31. $z = \dfrac{4}{5} + \dfrac{3}{5}i$

32. $z = \dfrac{\sqrt{3}}{3} + \dfrac{1}{3}i$

Graph the given complex number.

33. $\cos 45° + i \sin 45°$

34. $4 (\cos 60° + i \sin 60°)$

35. $2(\cos 120° + i \sin 120°)$

36. $3(\cos 30° + i \sin 30°)$

Write each complex number in polar form. If necessary, express angles to the nearest tenth of a degree.

37. $-1 + i$

38. $2i$

39. $3 - i\sqrt{3}$

40. $1 + 4i$

41. $-2 - 2i\sqrt{3}$

42. $5 - 5i$

For each of the following, find $z_1 z_2$.

43. $z_1 = \sqrt{2}(\cos 315° + i \sin 315°),\ z_2 = 2(\cos 30° + i \sin 30°)$

44. $z_1 = 2(\cos 270° + i \sin 270°),\ z_2 = 3(\cos 30° + i \sin 30°)$

45. $z_1 = \sqrt{2}(\cos 60° + i \sin 60°),\ z_2 = 3(\cos 30° + i \sin 30°)$

46. $z_1 = \sqrt{3}(\cos 90° + i \sin 90°),\ z_2 = \sqrt{2}(\cos 45° + i \sin 45°)$

47.–50. In Exercises 43–46, find $\dfrac{z_1}{z_2}$.

Find the reciprocal of z in (a) polar form and (b) the form $x + iy$.

51. $z = \cos 300° + i \sin 300°$

52. $z = \cos 60° + i \sin 60°$

53. $z = 1 - i$

54. $z = \sqrt{2} + i$

Use De Moivre's theorem to express each of the following in the form $x + iy$.

55. $[3(\cos 45° + i \sin 45°)]^3$

56. $[2(\cos 60° + i \sin 60°)]^5$

57. $\left[\dfrac{1}{2}(\cos 72° + i \sin 72°)\right]^5$

58. $(3 + 3i)^4$

59. $[2(\cos 15° + i \sin 15°)]^4\left[\dfrac{1}{2}(\cos 20° + i \sin 20°)\right]^6$

60. $\dfrac{(\cos 27° + i \sin 27°)^6}{(\cos 18° + i \sin 18°)^4}$

61. $\dfrac{(\sqrt{3} + i)^3}{(1 - i)^2}$

62. $\dfrac{(1 + i)^4}{(2 - 2i)^3}$

Find the indicated roots in the form specified, and sketch their graphs along with that of the given number.

63. The cube roots of $1 + i$ in the form $x + iy$

64. The sixth roots of i in the form $x + iy$

65. The fifth roots of $1 - i$ in the form $x + iy$

66. The seventh roots of $\sqrt{3} - i$ in polar form

67. The fourth roots of $-i$ in polar form

68. The cube roots of $\sqrt{3} + i$ in polar form

69. The fourth roots of $1 - i$ in polar form

70. The fifth roots of $\sqrt{2} + i$ in polar form

For use after Chapter 8

Find the translation that moves the origin to the center of the given circle. Then draw both sets of axes and the circle.

1. $x^2 + y^2 - 6x - 8y = 0$

2. $x^2 + y^2 + 4x - 10y - 20 = 0$

3. $x^2 + y^2 - 10x - 10y + 34 = 0$

By means of a translation, write the equation in the form $y' = a \sin \omega x'$ or $y' = a \cos \omega x'$. Then draw the curve and both sets of axes.

4. $y = \sin\left(2x + \dfrac{\pi}{3}\right)$

5. $y = 2 \cos \left(2x - \dfrac{3\pi}{4} \right)$

6. $y = \sin \left(3x + \dfrac{3\pi}{4} \right) - 1$

Choose a translation that changes the given equation into a simpler form. Then graph the equation showing both sets of axes.

7. $y - 2 = (x - 3)^2$ (parabola)

8. $x - 3 = (y - 1)^2$ (parabola)

9. $\dfrac{(x - 1)^2}{16} + \dfrac{(y - 1)^2}{4} = 1$ (ellipse)

Simplify the equation by rotating the axes through the specified angle. Then draw its graph.

10. $5x^2 - 6xy + 5y^2 - 32 = 0,\ \phi = 45°$

11. $8x^2 + 4xy + 5y^2 - 36 = 0,\ \phi = \operatorname{Sin}^{-1} \left(\dfrac{1}{\sqrt{5}} \right)$

12. $5x^2 + 8xy + 5y^2 - 81 = 0,\ \phi = \operatorname{Cos}^{-1} \left(\dfrac{1}{\sqrt{2}} \right)$

13. $x^2 - \sqrt{3}xy + 2y^2 - 10 = 0,\ \phi = 30°$

14. $x^2 - \sqrt{3}xy + 12 = 0,\ \phi = 60°$

15. $4x^2 + 3\sqrt{3}xy + y^2 = 22,\ \phi = 120°$

16. $52x^2 - 72xy + 73y^2 - 100 = 0,\ \phi = \operatorname{Sin}^{-1} \left(\dfrac{3}{5} \right)$

17. $5x^2 + 7y^2 - 2\sqrt{3}xy - 32 = 0,\ \phi = 30°$

Draw figures illustrating the alias and the alibi interpretations of the given transformation's action on the given point or curve.

18. $x = x' - 1,\ y = y' + 1$; the point $(1, 1)$

19. $x = x' - 2,\ y = y' + 3$; the point $(2, 1)$

20. $x = x' + 3,\ y = y' + 2$; the point $(4, 5)$

21. $x = \dfrac{x' - y'\sqrt{3}}{2},\ y = \dfrac{x'\sqrt{3} + y'}{2}$; the line $2x + y = 6$

22. $x = \dfrac{x' - y'\sqrt{3}}{2},\ y = \dfrac{x'\sqrt{3} + y'}{2}$; the ellipse $\dfrac{x^2}{4} + \dfrac{y^2}{9} = 1$

23. $x = \dfrac{x' - y'}{\sqrt{2}},\ y = \dfrac{x' + y'}{\sqrt{2}}$; the parabola $y = x^2$

24. $x = \dfrac{x' - y'\sqrt{3}}{2}$, $y = \dfrac{x'\sqrt{3} + y'}{2}$; the parabola $y = x^2$

25. $x = \dfrac{x' - y'}{\sqrt{2}}$, $y = \dfrac{x' + y'}{\sqrt{2}}$; the hyperbola $xy = 4$

Find the equations of a rotation that will transform the equation into an equation in x' and y' without an $x'y'$ term.

26. $\sqrt{3}x^2 - 3xy - 6 = 0$

27. $4x^2 + 5xy - 8y^2 + 8 = 0$

28. $x^2 - 3xy + y^2 + 8 = 0$

29. $4x^2 - 4xy + y^2 - 5 = 0$

30. $3x^2 - 4xy + 8x - 1 = 0$

31. $2x^2 + 5xy + 2y^2 = 8$

32. $5x^2 + 4xy = 16$

33. $x^2 - 4xy - 2y^2 = 24$

For use after Chapter 9

In Exercises 1 and 2, (a) plot the given points and draw the triangle that they determine. (b) Use the distance formula to check whether the triangle is isosceles, right, both or neither.

1. $(2, 4, 2)$, $(4, 5, 4)$, $(4, 6, 1)$

2. $(4, 5, -6)$, $(3, 6, -2)$, $(2, 4, -4)$

Find an equation of the sphere having the given center and radius. Give your answer in the form $x^2 + y^2 + z^2 + ax + by + cz + d = 0$.

3. center $(0, 0, 0)$, radius 3

4. center $(1, 0, 1)$, radius 2

5. center $(2, 0, 2)$, radius 5

6. center $(1, 2, 3)$, radius 4

Find the center and radius of the given sphere.

7. $x^2 + y^2 + z^2 = 16$

8. $x^2 + y^2 + z^2 - 2x - 2y = 7$

9. $x^2 + y^2 + z^2 + 2y - 4z + 1 = 0$

10. $x^2 + y^2 + z^2 + 2z = 24$

In Exercises 11 and 12, $\mathbf{a} = \mathbf{i} - 2\mathbf{j} + \mathbf{k}$ and $\mathbf{b} = 3\mathbf{j} - 2\mathbf{k}$. (a) Find the given linear combinations. (b) Find $\|\mathbf{a}\|$, $\|\mathbf{b}\|$, $\|\mathbf{a} + \mathbf{b}\|$, and $\|\mathbf{a} - \mathbf{b}\|$.

11. $4\mathbf{a} - 2\mathbf{b}$, $2\mathbf{a} + 4\mathbf{b}$

12. $3\mathbf{a} - 5\mathbf{b}$, $5\mathbf{a} + 3\mathbf{b}$

In Exercises 13–16, find the angle between **a** and **b**.

13. $\mathbf{a} = \mathbf{i} + \mathbf{j} - \mathbf{k}, \mathbf{b} = 2\mathbf{i} + \mathbf{j} + \mathbf{k}$ 14. $\mathbf{a} = 2\mathbf{i} + 2\mathbf{j} - \mathbf{k}, \mathbf{b} = 3\mathbf{i} + \mathbf{j} - \mathbf{k}$

15. $\mathbf{a} = \mathbf{i} - \mathbf{j} - \mathbf{k}, \mathbf{b} = \mathbf{i} + 2\mathbf{j} + \mathbf{k}$ 16. $\mathbf{a} = 2\mathbf{i} - \mathbf{j} + \mathbf{k}, \mathbf{b} = 3\mathbf{i} - 3\mathbf{j} + \mathbf{k}$

In Exercises 17 and 18, express **v** as a linear combination of **a**, **b**, and **c**. Use the fact that **a**, **b**, and **c** are orthogonal in pairs.

17. $\mathbf{a} = 2\mathbf{i} - 2\mathbf{j} + \mathbf{k}, \mathbf{b} = \mathbf{i} + 2\mathbf{j} + 2\mathbf{k}, \mathbf{c} = 2\mathbf{i} + \mathbf{j} - 2\mathbf{k}; \mathbf{v} = \mathbf{i} + \mathbf{j} + 2\mathbf{k}$

18. $\mathbf{a} = 4\mathbf{i} - 2\mathbf{j} + 2\mathbf{k}, \mathbf{b} = 2\mathbf{i} + 3\mathbf{j} - \mathbf{k}, \mathbf{c} = \mathbf{i} - 2\mathbf{j} - 4\mathbf{k}; \mathbf{v} = 3\mathbf{i} + \mathbf{j} + \mathbf{k}$

Find a vector equation of the line having the stated properties.

19. Passes through $(1, 0, 1)$ and is parallel to $\mathbf{i} + \mathbf{j} - 2\mathbf{k}$.

20. Passes through the origin and $(2, 2, 4)$.

21. Passes through $(1, 0, 3)$ and makes equal angles with $\mathbf{i}$, $\mathbf{j}$, and $\mathbf{k}$.

22. Passes through $(1, 2, 3)$ and is parallel to the line through $(1, 1, 3)$ and $(2, 0, 1)$.

Find the point where the line *l* intersects the given plane.

23. $\mathbf{r} = (1 + t)\mathbf{i} + 3t\mathbf{j} + (1 - t)\mathbf{k}$; xz-plane

24. $\mathbf{r} = (1 - t)\mathbf{i} + (2 + t)\mathbf{j} + 2t\mathbf{k}$; xy-plane

Find the distance from the origin to the line *l*.

25. $\mathbf{r} = (4 - t)\mathbf{i} + (3 + t)\mathbf{j} + (1 - t)\mathbf{k}$

26. $\mathbf{r} = (3 + 2t)\mathbf{i} + (-3 + t)\mathbf{j} + (2 + 2t)\mathbf{k}$

In Exercises 27 and 28, (a) find an equation of the plane through P_0 that is perpendicular to the given vector or line. (b) Then draw the first-octant part of the plane.

27. $P_0(1, 2, 1)$; $3\mathbf{i} + \mathbf{j} + 2\mathbf{k}$

28. $P_0(2, 0, 1)$; $\mathbf{r} = (2 + t)\mathbf{i} + (-1 + t)\mathbf{j} + t\mathbf{k}$

Find a vector equation of the line through P_1 that is perpendicular to Q.

29. $P_1(2, 2, 0)$; $Q: 3x - y + 2z = 2$

30. $P_1(1, 3, 1)$; $Q: 2x + 3y - z - 2 = 0$

31–32. Find the distance from P_1 to Q in Exercises 29 and 30.

Find an equation of the plane that passes through the given points.

33. $(4, 1, 2), (5, 2, 3), (-3, 3, 1)$ 34. $(0, 0, 1), (3, 1, 2), (2, 0, 2)$

In Exercises 35 and 36, (a) verify that the given planes are parallel and (b) find the distance between them.

35. $2x + y - 3z = 4; \; -2x - y + 3z = 6$

36. $x - y + 2z - 3 = 0; \; 2x - 2y + 4z - 5 = 0$

For use after Chapter 10

Find the length in kilometers of a great-circle arc on Earth with the given angular measure.

1. $20°$ 2. $50°30'$ 3. $120°45'$

Find the angular measure of a great-circle arc on Earth with the given length.

4. $180.5 \, \text{km}$ 5. $1774.2 \, \text{km}$ 6. $9840 \, \text{km}$

Solve the spherical triangle *ABC*, given the following.

7. $a = 120°, \; b = 90°, \; c = 90°$ 8. $\angle A = 90°, \; \angle B = 70°, \; \angle C = 90°$

9. $a = 60°, \; b = 90°, \; \angle C = 90°$ 10. $\angle A = 60°, \; \angle B = 120°, \; \angle C = 90°$

Solve the given right spherical triangle with $\angle C = 90°$. Give answers to the nearest ten minutes.

11. $c = 102°20', \; \angle A = 60°30'$

12. $a = 42°50', \; b = 68°30'$

13. $\angle A = 48°10', \; \angle B = 102°50'$

14. $a = 90°20', \; c = 82°30'$

15. $\angle A = 43°10', \; \angle B = 65°40'$

Solve the given right spherical triangle with $\angle C = 90°$. Give answers to the nearest tenth of a degree.

16. $b = 107.3°, \; \angle A = 120.7°$

17. $a = 74.5°, \; b = 109.2°$

18. $b = 131.5°, \; c = 98.4°$

19. $a = 35.4°, \; \angle B = 86.4°$

20. $b = 55.8°, \; \angle A = 103.6°$

Use Napier's rules to write formulas that can be used to solve right triangle *ABC* with $\angle C = 90°$ given the following.

21. a and c 22. a and $\angle B$ 23. c and $\angle A$

Solve right triangle *ABC* with $\angle C = 90°$ giving answers to the nearest degree. There may be two solutions or none. If there is no solution, so state.

24. $a = 42°, \; \angle A = 68°$ 25. $a = 65°, \; \angle A = 120°$

Solve the following quadrantal triangles. Give answers to the nearest ten minutes. Note that $\angle C$ is not necessarily 90°.

26. $a = 52°$, $b = 115°20'$, $c = 90°$

27. $a = 40°20'$, $\angle B = 84°10'$, $c = 90°$

28. $\angle A = 33°40'$, $\angle B = 62°10'$, $c = 90°$

29. $a = 97°20'$, $\angle A = 109°10'$, $c = 90°$

Solve the following spherical triangles. Give your answers to the nearest ten minutes.

30. $\angle A = 46°30'$, $\angle B = 135°10'$, $c = 112°10'$

31. $a = 71°40'$, $b = 108°30'$, $c = 87°10'$

32. $\angle A = 124°50'$, $\angle B = 75°20'$, $\angle C = 60°40'$

33. $\angle A = 116°20'$, $b = 37°20'$, $\angle C = 91°30'$

Solve the following spherical triangles. Give your answers to the nearest tenth of a degree.

34. $\angle A = 81.2°$, $\angle B = 82.6°$, $\angle C = 50.2°$

35. $a = 112.3°$, $\angle B = 59.9°$, $\angle C = 39.1°$

36. $a = 54.1°$, $\angle B = 105.3°$, $\angle C = 96.4°$

37. $a = 32.5°$, $b = 42.2°$, $\angle C = 55.0°$

In Exercises 38–43, all tracks are great-circle arcs unless otherwise specified.

38. How many nautical miles is San Juan, Puerto Rico, 18°27′ N, 66°4′ W, from (a) the equator? (b) the North Pole? (c) the South Pole?

39. How many nautical miles is Paris, 48°52′ N, 2°20′ E, from (a) the equator? (b) the North Pole? (c) the South Pole?

40. Find the great-circle distance from Montreal, 45°31′ N, 73°33′ W, to Portland, Oregon, 45°31′ N, 122°41′ W.

41. What is the position of the northernmost point of the great-circle track in Exercise 40 above? (Hint: The triangle is isosceles.)

42. A plane leaves Atlanta, Georgia, 33°45′ N, 84°24′ W with initial course 90° and flies a great-circle track. Find the longitude of the point where the plane crosses the equator. Find its course at that point.

43. A plane leaves New York, New York, 40°45′ N, 74° W with initial course 54°30′. Find the latitude and longitude of the northernmost point of its track. (Hint: At this point the track is perpendicular to the meridian.)

For use after Chapter 11

1. Use the first three terms of the Taylor series for sin x to find the following.

   (a) $\sin \dfrac{\pi}{3} \approx \sin 1.05$

   (b) $\sin \dfrac{\pi}{4} \approx \sin 0.79$

   (c) Compare the values you obtained in (a) and (b) with the known values of $\sin \dfrac{\pi}{3}$ and $\sin \dfrac{\pi}{4}$.

2. Use the first three terms of the Taylor series for cos x to find the following.

   (a) $\cos \dfrac{\pi}{6} \approx \cos 0.52$

   (b) $\cos \dfrac{2\pi}{3} \approx \cos 2.09$

   (c) Compare the values you obtained in (a) and (b) with the known values of $\cos \dfrac{\pi}{6}$ and $\cos \dfrac{2\pi}{3}$.

3. Use the first four terms of the Taylor series for sin x to evaluate each of the following to three decimal places.

   (a) $\sin 2.2$

   (b) $\sin (-0.94)$

   (c) Compare the answers to (a) and (b). Give a reason for their relationship.

4. Use the first four terms of the Taylor series for cos x to evaluate each of the following to three decimal places.

   (a) $\cos 2.5$

   (b) $\cos 2.3$

   (c) Compare the answers to (a) and (b). Give a reason for their relationship.

Find the value of the expression using the given equation.

5. $\cosh 1$; $\cosh x = 1 + \dfrac{x^2}{2!} + \dfrac{x^4}{4!} + \dfrac{x^6}{6!} + \cdots$

6. $\sinh 1$; $\sinh x = x + \dfrac{x^3}{3!} + \dfrac{x^5}{5!} + \dfrac{x^7}{7!} + \cdots$

7. $e^{-\pi i}$; $e^{ix} = \cos x + i \sin x$

Find the value of the following.

8. $e^{\frac{3\pi}{2}i}$

9. $e^{i\frac{\pi}{4}}$

10. Sketch the graph of the function f defined by $f(x) = |x|$, $-\pi \le x \le \pi$ and $f(x + 2\pi) = f(x)$.

Algebra and Geometry Review Exercises

The following exercises review some of the algebra and geometry you will use in trigonometry. The page references indicate the first time each review topic is needed in the text.

Similar Triangles (page 6)

Find the value of x to the nearest tenth of a unit.

Example

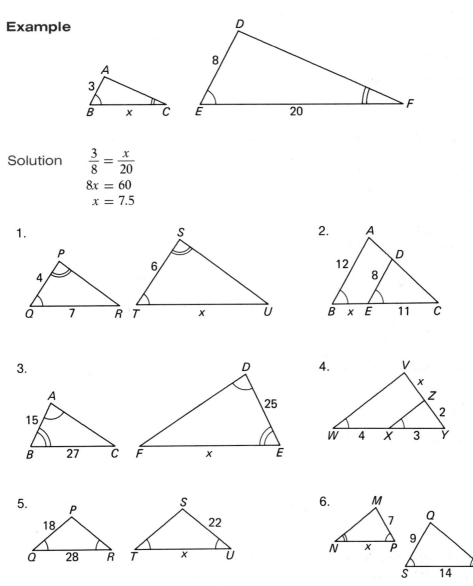

Solution $\dfrac{3}{8} = \dfrac{x}{20}$

$8x = 60$

$x = 7.5$

1.

2.

3.

4.

5.

6.

7.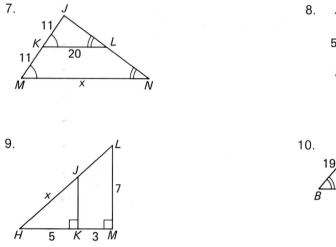

8.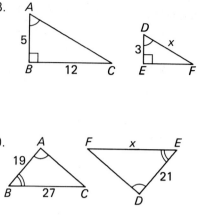

9.

10.

The Pythagorean Theorem (page 7)

ABC is a right triangle with $\angle C = 90°$, hypotenuse *c*, and legs *a* and *b*. Complete. Give answers to the nearest tenth of a unit.

Example $a = 7, c = 9, b = ?$

Solution $a^2 + b^2 = c^2$
$$49 + b^2 = 81$$
$$b^2 = 32$$
$$b = 5.7$$

1. $a = 13, b = 11, c = ?$ 2. $a = 42, b = 25, c = ?$

3. $a = 28, c = 46, b = ?$ 4. $c = 108, b = 39, a = ?$

5. $a = 52, b = 64, c = ?$ 6. $a = 9, b = 16, c = ?$

7. $c = 91, a = 47, b = ?$ 8. $a = 63, b = 41, c = ?$

9. $c = 21, a = 14, b = ?$ 10. $c = 13, b = 5, a = ?$

Simplifying Radical Expressions (page 12)

Simplify. Assume all variables represent nonnegative integers.

Example $\dfrac{\sqrt{x^3}}{5\sqrt{18}}$

Solution $\dfrac{\sqrt{x^3}}{5\sqrt{18}} = \dfrac{\sqrt{x^2 \cdot x}}{5\sqrt{9 \cdot 2}} = \dfrac{x\sqrt{x}}{15\sqrt{2}} = \dfrac{x\sqrt{x}\sqrt{2}}{15\sqrt{2}\sqrt{2}} = \dfrac{x\sqrt{2x}}{30}$

1. $\dfrac{3\sqrt{6}}{5\sqrt{3}}$

2. $\dfrac{1}{\sqrt{18}}$

3. $\dfrac{\sqrt{12}}{\sqrt{3}}$

4. $\dfrac{1}{\sqrt{2}-1}$

5. $\sqrt{2} + \sqrt{8} + \sqrt{6}$

6. $\dfrac{\sqrt{2}}{\sqrt{3}+1}$

7. $\dfrac{\sqrt{32x^4}}{\sqrt{6}}$

8. $\dfrac{3x}{\sqrt{9x^3}}$

9. $\dfrac{\sqrt{x^2y}}{y\sqrt{27}}$

Special Right Triangles (*page 12*)

Find the value of x. Leave answers in simplest radical form.

Example 1

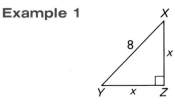

Solution $\triangle XYZ$ is isosceles and is a right triangle. Then $\triangle XYZ$ is a 45°-45°-90° triangle and $XY = XZ\sqrt{2}$.

$$x\sqrt{2} = 8$$
$$x = \frac{8}{\sqrt{2}} = \frac{8\sqrt{2}}{\sqrt{2}\cdot\sqrt{2}} = 4\sqrt{2}$$

Example 2

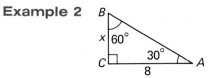

Solution $\triangle ABC$ is a 30°-60°-90° triangle so $AB = 2BC$ and $AC = BC\sqrt{3}$.

$$8 = x\sqrt{3}$$
$$x = \frac{8}{\sqrt{3}} = \frac{8\sqrt{3}}{\sqrt{3}\cdot\sqrt{3}} = \frac{8\sqrt{3}}{3}.$$

Find the value of *x*. Leave answers in simplest radical form.

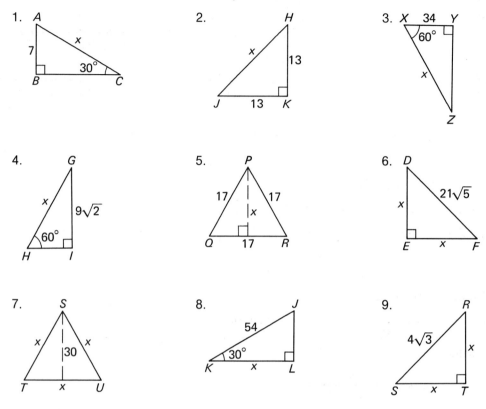

1. A
7
x
30°
B   C

2. H
x
13
J   13   K

3. X   34   Y
60°
x
Z

4. G
x
$9\sqrt{2}$
60°
H   I

5. P
17   x 17
Q   17   R

6. D
x $21\sqrt{5}$
E x F

7. S
x 30 x
T x U

8. J
54
30°
K x L

9. R
$4\sqrt{3}$ x
S x T

Slope of a Line (page 16)

Find the slope, if it is defined, of the given line.

Example

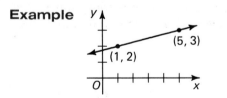

y
(5, 3)
(1, 2)
O   x

Solution   The slope *m* of a line through (x_1, y_1), (x_2, y_2) is given by $m = \dfrac{y_2 - y_1}{x_2 - x_1}$.

$$m = \frac{3 - 2}{5 - 1} = \frac{1}{4}$$

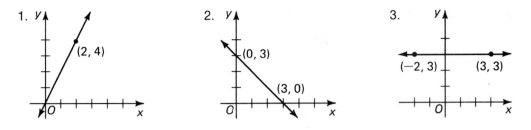

4. the line through the points (4, 3) and (4, −2)

5. the line through the points (2, −7) and (3, −2)

6. the line through (5, 8) and (9, 10)

7. the line through the origin and (0, 7)

8. the line through (−4, −11) and (2, −13)

Parallel Lines (page 22)

The lines *a* and *b* are parallel. Find the measure of θ.

Example

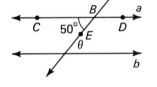

Solution 1   If two parallel lines are cut by a transversal, alternate interior angles are equal. ∠ *CBE* and θ are alternate interior angles. Then θ = 50°.

Solution 2   If two parallel lines are cut by a transversal, corresponding angles are equal. ∠ *ABD* and θ are corresponding angles. ∠ *ABD* and ∠ *CBE* are vertical angles so ∠ *CBE* = ∠ *ABD* = 50° = θ.

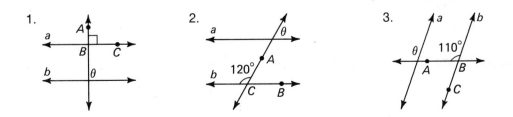

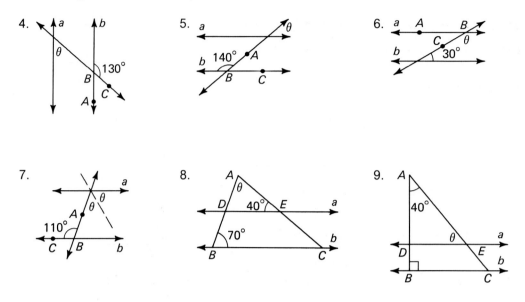

Graphing (page 49)

Graph the function.

Example $y = x^2 - 1$

Solution   Choose values of x and find the corresponding values of y. Graph the points (x, y) and connect them.

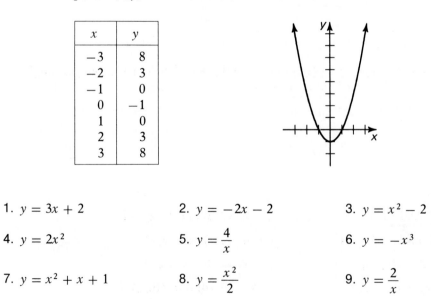

x	y
-3	8
-2	3
-1	0
0	-1
1	0
2	3
3	8

1. $y = 3x + 2$

2. $y = -2x - 2$

3. $y = x^2 - 2$

4. $y = 2x^2$

5. $y = \dfrac{4}{x}$

6. $y = -x^3$

7. $y = x^2 + x + 1$

8. $y = \dfrac{x^2}{2}$

9. $y = \dfrac{2}{x}$

Simplifying Polynomial Expressions (page 69)

Simplify. Identify any restrictions on the variables.

Example $\dfrac{a^2}{a-2} - \dfrac{a+2}{a-2}$

Solution $\dfrac{a^2}{a-2} - \dfrac{a+2}{a-2} = \dfrac{a^2-a-2}{a-2}$

$$= \dfrac{(a-2)(a+1)}{a-2}$$

$$= a+1 \qquad (a \neq 2)$$

1. $\dfrac{x^2+2x}{x^2+7x+10}$

2. $\dfrac{y-3}{y^2-6y+9}$

3. $\dfrac{2z^2-z-3}{3z+3}$

4. $\dfrac{u^2-5u}{2-u} + \dfrac{12-3u}{2-u}$

5. $(4x-4) \div \dfrac{x^2-1}{x}$

6. $\dfrac{1-t}{1+t} + \dfrac{1+t}{1-t}$

7. $\dfrac{x^2+x-2}{x^2+2x-3}$

8. $\dfrac{x^2-1}{x^2+3x+2}$

9. $\dfrac{a^4-5a^2+4}{a^2-a-2}$

The Distance Formula (page 76)

Find the distance between the given points. Give answers in simplest radical form.

Example (2, 3) and (4, 7)

Solution The distance d between two points (x_1, y_1) and (x_2, y_2) is
$d = \sqrt{(x_2-x_1)^2 + (y_2-y_1)^2}$.
$d = \sqrt{(4-2)^2 + (7-3)^2}$
$ = \sqrt{4+4}$
$ = \sqrt{8}$
$ = 2\sqrt{2}$

1. (4, 6) and (2, −3)

2. (−5, −1) and (3, 8)

3. (−1, −6) and (5, 5)

4. (3, 2) and (−4, −4)

5. (−3, −5) and (−1, −1)

6. (2, 7) and (6, 8)

7. (7, 12) and (−4, 13)

8. (1, 8) and (8, 1)

9. (−11, −8) and (9, 9)

10. (5, 10) and (10, 5)

Solving Equations (page 86)

Solve each equation.

Example $\dfrac{x}{x-2} + \dfrac{x+3}{x+1} = \dfrac{2}{x-2}$

Solution $(x-2)(x+1)\left[\dfrac{x}{x-2} + \dfrac{x+3}{x+1}\right] = (x-2)(x+1)\left(\dfrac{2}{x-2}\right)$

$(x+1)x + (x-2)(x+3) = (x+1)(2)$

$x^2 + x + x^2 + x - 6 = 2x + 2$

$2x^2 - 8 = 0$

$x^2 - 4 = 0$

$x = 2 \text{ or } x = -2$

Check: $\dfrac{-2}{-2-2} + \dfrac{-2+3}{-2+1} = \dfrac{1}{2} + \dfrac{1}{-1} = -\dfrac{1}{2}$

$\dfrac{2}{-2-2} = -\dfrac{1}{2}$

Note that 2 is not a solution since $\dfrac{2}{2-2}$ is not defined. The solution is $x = -2$.

1. $x^2 + 5x + 4 = 0$

2. $3x^2 - 8x - 3 = 0$

3. $6x^2 - 13x^2 + 6 = 0$

4. $x - 6 = \sqrt{x}$

5. $\dfrac{x}{x-2} - 7 = \dfrac{2}{x-2}$

6. $\dfrac{1}{x^2 - x} - \dfrac{3}{x} = -1$

7. $x^2 - 10x + 25 = 81$

8. $x^2 + 8x + 4 = -12$

9. $x - \sqrt{x} = 2$

10. $3 + \dfrac{10}{x^2 - 1} = \dfrac{5}{x-1}$

Area of a Polygon (page 112)

Find the area of the polygon.

Example

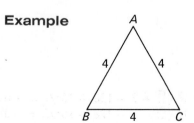

Solution　ABC is an equilateral triangle. Draw $\overline{AD}$, the altitude to $\overline{BC}$. ADB is a $30°$-$60°$-$90°$ triangle and $AD = 2\sqrt{3}$.

Area of $\triangle ADB = \dfrac{1}{2}bh = 2\sqrt{3}$

Area of $\triangle ABC = 2(2\sqrt{3}) = 4\sqrt{3}$

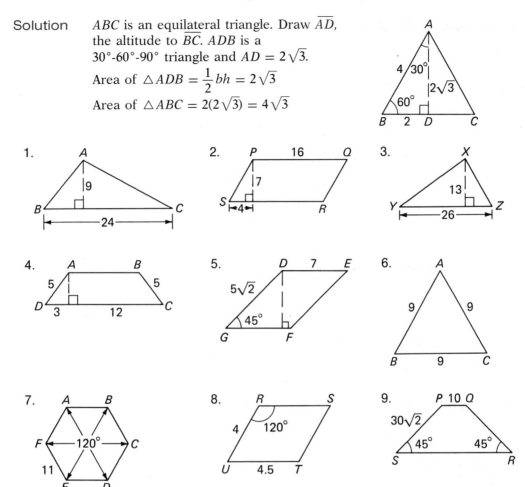

Completing the Square (page 214)

Solve by completing the square.

Example　$y^2 - 16y + 62 = 0$

Solution
$$y^2 - 16y + 62 = 0$$
$$y^2 - 16y = -62$$
$$y^2 - 16y + 8^2 = -62 + 8^2$$
$$(y - 8)^2 = -62 + 64$$
$$(y - 8)^2 = 2$$
$$y - 8 = \pm\sqrt{2}$$
$$y = 8 \pm\sqrt{2}$$

Check: $(8 + \sqrt{2})^2 - 16(8 + \sqrt{2}) = 64 + 16\sqrt{2} + 2 - 128 - 16\sqrt{2} = -62$
$(8 - \sqrt{2})^2 - 16(8 - \sqrt{2}) = 64 - 16\sqrt{2} + 2 - 128 + 16\sqrt{2} = -62$

Solve by completing the square.

1. $y^2 + 10y - 5 = 0$ 2. $x^2 - 6x - 2 = 0$ 3. $y^2 + 8y - 6 = 0$

4. $x^2 - 8x - 20 = 0$ 5. $x^2 - 6x + 4 = 0$ 6. $x^2 + 8x - 1 = 0$

7. $x^2 + 10x + 21 = 0$ 8. $x^2 + 6x = 31$ 9. $2x^2 + 5x = 3$

Conic Sections (page 217)

Classify each equation as the equation of a circle, an ellipse, a parabola, or a hyperbola.

Example $y = x^2 + 2x - 3$

Solution $y + 3 = x^2 + 2x$
$y + 4 = x^2 + 2x + 1$
$y + 4 = (x + 1)^2$
$\quad y = (x + 1)^2 - 4$

Form	Curve
$y = a(x - h)^2 + k$	parabola
$x = a(y - h)^2 + k$	parabola
$(x - h)^2 + (y - k)^2 = r^2$	circle
$\dfrac{x^2}{a^2} + \dfrac{y^2}{b^2} = 1$	ellipse
$\dfrac{x^2}{a^2} - \dfrac{y^2}{b^2} = 1$	hyperbola
$xy = k$	hyperbola

As shown in the table, a curve whose equation can be written in the form $y = a(x - h)^2 + k$ is a parabola. The given equation is the equation of a parabola.

1. $x^2 - 2x + y^2 - 4y = 4$ 2. $9x^2 + 16y^2 = 144$

3. $y = \dfrac{3}{x}$ 4. $x^2 + 2x + y^2 + 2y = 5$

5. $x^2 + 4y^2 + 2 = 6$ 6. $x^2 + y = 2x + 7$

7. $y - x^2 = 12x + 4$ 8. $25x^2 - 4y^2 - 25 = 75$

Systems of Equations (page 237)

Solve the system of equations.

Example 1 $-5x + 3y = 25$
$\quad\quad\quad\quad 4x + 2y = 2$

Solution   Multiply the first equation by -2, and the second by 3 to eliminate y and solve for x.

$$
\begin{aligned}
10x - 6y &= -50 \\
12x + 6y &= 6 \\
\hline
22x &= -44 \\
x &= -2
\end{aligned}
$$

Now substitute: $-5(-2) + 3y = 25$
$$3y = 15$$
$$y = 5$$

The solution is $(-2, 5)$.

1. $3x - 2y = 8$
   $3x - 6y = -16$

2. $5x + 6y = 16$
   $6x - 5y = 7$

3. $7x + 10y = 17$
   $8x + 15y = 23$

4. $5x + 2y = -1$
   $4x + 3y = 9$

5. $3x - 7y = 9$
   $2x - 5y = 7$

6. $10x + 3y = 2$
   $6x - 7y = -12$

Example 2 $x + y + z = 3$
$$x + 2y - z = -1$$
$$-2x - y + 3z = 11$$

Solution   Use the method of Example 1 to eliminate z from equations 1 and 2, then from equations 2 and 3.

$$
\begin{array}{ll}
x + y + z = 3 & 3x + 6y - 3z = -3 \\
\underline{x + 2y - z = -1} & \underline{-2x - y + 3z = 11} \\
2x + 3y = 2 & x + 5y = 8
\end{array}
$$

$$
\begin{array}{ll}
2x + 3y = 2 & 2x + 3y = 2 \\
x + 5y = 8 & \underline{-2x - 10y = -16} \\
 & 7y = -14 \\
 & y = -2
\end{array}
$$

Substitute to find x and z. $2x + 3(-2) = 2$
$$2x = 8$$
$$x = 4$$
$$4 + (-2) + z = 3$$
$$z = 1$$

The solution is $(4, -2, 1)$.

7. $x + y - z = 2$
   $4x - y + z = 0$
   $6x + y + z = 4$

8. $x + y - z = 4$
   $4x - y + z = 0$
   $6x + y + z = 8$

9. $x + y + z = 1$
   $x + 2y - z = -1$
   $2x - y - 3z = 8$

10. $-x - y + z = 7$
    $x + 2y + 3z = 7$
    $4x - 3y - 2z = -6$

11. $x + 2y + z = 11$
    $x + 3y + 3z = 19$
    $2x + y + z = 13$

12. $2x - y + 3z = 3$
    $x + y + 2z = 1$
    $x - 2y - z = -2$

APPENDIX

Computation with Logarithms

A-1 *Logarithms to the Base 10*

The properties of the function

$$y = \log_{10} x$$

may be useful in making numerical computations. This function is called the **logarithmic function to the base 10** and the y-values in the range are called **logarithms** to the base 10.

The logarithmic function to the base 10 can be defined as follows.

If x is a positive number, then
$\log_{10} x = y$ if and only if $10^y = x$.

Thus, $\log_{10} x$ (or y) is the exponent needed so that 10^y equals x. This definition leads immediately to the fundamental identity

$$10^{\log_{10} x} = x \qquad (*)$$

for all positive real numbers x.

The following laws of exponents should be familiar to you from your study of algebra.

If a, b, x, y are real numbers and a, $b > 0$, then

1. $a^x a^y = a^{x+y}$

2. $\dfrac{a^x}{a^y} = a^{x-y}$

3. $(ab)^x = a^x b^x$

4. $\left(\dfrac{a}{b}\right)^x = \dfrac{a^x}{b^x}$

5. $a^x = a^y$ if and only if $x = y$.

Using the laws of exponents, we can prove the following properties for $\log_{10} x$.

For k, x_1, x_2 real, and, $x_1 > 0$ and $x_2 > 0$,

1. $\log_{10} x_1 x_2 = \log_{10} x_1 + \log_{10} x_2$,

2. $\log_{10} \dfrac{x_1}{x_2} = \log_{10} x_1 - \log_{10} x_2$,

3. $\log_{10} x_1^k = k \log_{10} x_1$,

4. $\log_{10} x_1 = \log_{10} x_2$ if and only if $x_1 = x_2$.

To show that Property 1 is true, from identity ($*$) we have

$$x_1 x_2 = 10^{\log_{10} x_1 x_2}, \; x_1 = 10^{\log_{10} x_1}, \; x_2 = 10^{\log_{10} x_2},$$

so that

$$10^{\log_{10} x_1 x_2} = x_1 x_2 = 10^{\log_{10} x_1} 10^{\log_{10} x_2}.$$

By the first law of exponents, $10^{\log_{10} x_1} 10^{\log_{10} x_2} = 10^{\log_{10} x_1 + \log_{10} x_2}$. Thus, we have

$$10^{\log_{10} x_1 x_2} = 10^{\log_{10} x_1 + \log_{10} x_2}.$$

But, by the fifth law of exponents, this implies that

$$\log_{10} x_1 x_2 = \log_{10} x_1 + \log_{10} x_2,$$

as was to be shown.

In order to make computations with logarithms, it is necessary to apply the four properties listed above to express the logarithm of a product, quotient, power, or root, or a combination of these, in expanded form.

Example 1 Expand $\log_{10} \dfrac{(115)(37)}{218}$.

Solution By Property 2,

$$\log_{10} \frac{(115)(37)}{218} = \log_{10} (115)(37) - \log_{10} 218$$

By Property 1,

$$\log_{10} (115)(37) = \log_{10} 115 + \log_{10} 37$$

Therefore,

$$\log_{10} \frac{(115)(37)}{218} = \log_{10} 115 + \log_{10} 37 - \log_{10} 218$$

Example 2 Expand $\log_{10} \dfrac{(17)^3 \sqrt{29}}{\sqrt[4]{9}}$.

Solution   Rewriting the expression with fractional exponents gives
$$\frac{(17)^3(29)^{\frac{1}{2}}}{9^{\frac{1}{4}}}.$$

$$\text{Thus } \log_{10} \frac{(17)^3 \sqrt{29}}{\sqrt[4]{9}} = \log_{10} \frac{(17)^3(29)^{\frac{1}{2}}}{9^{\frac{1}{4}}}$$

$$= \log_{10}(17)^3(29)^{\frac{1}{2}} - \log_{10} 9^{\frac{1}{4}}$$

$$= \log_{10}(17)^3 + \log_{10}(29)^{\frac{1}{2}} - \log_{10} 9^{\frac{1}{4}}$$

$$= 3\log_{10} 17 + \frac{1}{2}\log_{10} 29 - \frac{1}{4}\log_{10} 9$$

Exercises A-1

Find the value of each of the following.

Example $\log_{10} 1000$

Solution   Let $y = \log_{10} 1000$. Then $10^y = 1000$, $10^y = 10^3$, and $y = 3$.

A 1. $\log_{10} 10$

2. $\log_{10} 100$

3. $\log_{10} 1$

4. $\log_{10} \dfrac{1}{10}$

5. $\log_{10} \dfrac{1}{100}$

6. $\log_{10} \sqrt{10}$

Expand the given logarithm.

7. $\log_{10}(127)(42)$

8. $\log_{10}(5.8)(11.4)$

9. $\log_{10}(17)^{15}$

10. $\log_{10} \sqrt[5]{47}$

11. $\log_{10} \dfrac{59.7}{23.2}$

12. $\log_{10} \dfrac{10.01}{2.032}$

13. $\log_{10} \dfrac{(14.1)(2.5)^2}{3.2}$

14. $\log_{10} \dfrac{(11.2)^2 \sqrt{71}}{\sqrt[3]{85}}$

15. $\log_{10} \dfrac{[(10.2)^2(5.1)]^{\frac{1}{3}}}{(4.1)(6.7)^3}$

16. $\log_{10} \dfrac{\sqrt{(48.1)(52)}}{(\sqrt[3]{17})(\sqrt[5]{92})}$

Express the logarithm of each of the following numbers as the sum of an integer and the logarithm of a number between 1 and 10.

Example 0.0549

Solution Since $0.0549 = 5.49 \times \dfrac{1}{100}$,

$$\log_{10} 0.0549 = \log_{10}\left(5.49 \times \dfrac{1}{100}\right)$$

$$= \log_{10} 5.49 + \log_{10} \dfrac{1}{100}$$

$$= \log_{10} 5.49 + \log_{10} 10^{-2}$$

$$= \log_{10} 5.49 + (-2)$$

B 17. 47   18. 1582   19. 0.736

20. 0.0301   21. 859.2   22. 726.8

A-2 *Finding Values for* $\log_{10}$

Values for $\log_{10}$ may be found either by using a calculator or a table. Table 5 on page 370 gives values to four significant digits of logarithms for certain numbers from 1 to 10. In order to extend the use of this table to numbers between 0 and 1 or to numbers greater than 10, it is helpful first to express such numbers in scientific notation. You should recall that scientific notation involves the product of a number between 1 and 10 and a power of 10. For example, in scientific notation,

$$452 = 4.52 \times 10^2, \qquad 0.452 = 4.52 \times 10^{-1},$$
$$45.2 = 4.52 \times 10^1, \qquad 0.00452 = 4.52 \times 10^{-3}.$$

To save space in the entries in Table 5, all decimal points have been omitted. A decimal point is understood to belong between each pair of digits in the left-hand column of the table, and before each four-digit entry in the body of the table.

Example 1 Find $\log_{10} 45.2$

Solution We write 45.2 as 4.52×10^1. To find $\log_{10} 4.52$ we locate 45 in the left-hand column of Table 5 and 2 in the top row. The intersection of the row containing 45 and the column containing 2 contains the number 6551. Therefore, $\log_{10} 4.52 = 0.6551$. Now

$$\log_{10} 45.2 = \log_{10}(4.52 \times 10^1) = \log_{10} 4.52 + \log_{10} 10^1,$$

and since $\log_{10} 10^1 = 1$, we have $\log_{10} 45.2 = (0.6551) + 1 = 1.6551$.

Notice in Example 1 that finding $\log_{10} 45.2$ requires knowing two numbers: $\log_{10} 4.52$, which is a number between 0 and 1, and $\log_{10} 10^1$, which is an integer, 1. In general, $\log_{10} x$, for any positive real number x, can be thought of as consisting of two parts. One part is a nonnegative decimal fraction that represents the logarithm of a number between 1 and 10, and is called the **mantissa** of the logarithm. The other part is an integer that represents the logarithm of a power of 10, and is called the **characteristic** of the logarithm.

Thus, for any positive real number a, we have

$$\log_{10} a = \underbrace{\log_{10} m}_{\text{mantissa}} + \underbrace{c}_{\text{characteristic}} ,$$

where $1 \le m < 10, 0 \le \log_{10} m < 1$, and c is an integer. If the characteristic of $\log_{10} a$ is negative, that is, if $c < 0$, then we usually write $\log_{10} a$ in the form

$$\log_{10} a = \underbrace{\log_{10} m}_{\text{mantissa}} + \underbrace{b - d}_{\text{characteristic}} ,$$

where $0 < b < d$, and $b - d = c$.

Example 2 Find $\log_{10} 0.0452$

Solution   We have $0.0452 = 4.52 \times 10^{-2}$. Then

$$\log_{10} 0.0452 = \log_{10} 4.52 + \log_{10} 10^{-2} = 0.6551 + (-2)$$
$$= 0.6551 + 8 - 10 = 8.6551 - 10.$$

The use of the -10 in Example 2 is customary. However, it may sometimes be more convenient to use another form. In such a case we could use $1 - 3, 18 - 20$, or any other pair of nonnegative integers b and d such that $b - d = -2$.

Now suppose we wish to find the logarithm to the base 10 of a number having four significant digits. Although this presents no problem on a calculator, Table 5 contains only numbers having three significant digits. This problem can be overcome by using linear interpolation (page 350) as shown in Example 3.

Example 3 Find $\log_{10} 3.824$.

Solution

	x		$\log_{10} x$	
	⎡3.820	0.5821	⎤	
0.010 ⎡0.004				⎤ 0.0011
	⎣3.824	$\log_{10} 3.824$	⎦ y	
	3.830	0.5832		

$$\frac{y}{0.0011} = \frac{0.004}{0.010}$$

$$y = \frac{4}{10}(0.0011) = 0.00044 = 0.0004$$

Therefore,

$$\log_{10} 3.824 = 0.5821 + 0.0004 = 0.5825.$$

If $\log_{10} a = b$, then a is called the **antilogarithm** of b, and we write

$$\text{antilog}_{10} b = a.$$

You can use Table 5 to find an approximation for $\text{antilog}_{10} b$ for any positive real number b having four significant digits to the right of the decimal point.

Example 4 Find $\text{antilog}_{10} 3.7324$.

Solution If $a = \text{antilog}_{10} 3.7324$, then $\log_{10} a = 3.7324 = 0.7324 + 3$. Now we locate the mantissa, 0.7324, in the body of Table 5 and record the corresponding value for a, 5.40. Since the characteristic of $\log_{10} a$ is 3, $a = 5.40 \times 10^3 = 5400$.

You can also use linear interpolation to find $\text{antilog}_{10} b$ when the mantissa of b has four significant digits, but is not an entry in the table. Note, however, that when $\text{antilog}_{10} b$ is found by interpolation in a four-place table, such as Table 5, the result may not be correct to four significant digits, even though the mantissa of b has four significant digits.

Example 5 Find $\text{antilog}_{10} 2.3155$.

Solution If $a = \text{antilog}_{10} 2.3155$, then $\log_{10} a = 2.3155 = 0.3155 + 2$. Noting that the characteristic is 2, we can turn our attention to finding $\text{antilog}_{10} 0.3155$ and multiply the result by 10^2, or 100. We have:

	x	$\log_{10} x$	
	2.060	0.3139	
$0.010\begin{bmatrix} y\begin{bmatrix} \\ x \end{bmatrix} \end{bmatrix}$		0.3155	$\end{bmatrix}0.0016\Big]0.0021$
	2.070	0.3160	

$$\frac{y}{0.010} = \frac{0.0016}{0.0021} \text{ or } y = 0.01\left(\frac{16}{21}\right) = 0.008$$

Then $x = 2.060 + 0.008 = 2.068$, and since $a = x \times 10^2$, we find that $a = 2.068 \times 10^2 = 206.8$.

Exercises A-2

Find $\log_{10} x$.

A 1. $x = 6.81$ 2. $x = 5.22$ 3. $x = 40$ 4. $x = 57$

5. $x = 61.3$ 6. $x = 22.73$ 7. $x = 7490$ 8. $x = 1547$

9. $x = 0.023$ 10. $x = 0.3603$ 11. $x = 0.00392$ 12. $x = 0.0009396$

Find antilog$_{10}$ y.

13. $y = 1.3096$ 14. $y = 2.6561$ 15. $y = 9.7451 - 10$

16. $y = 2.5551$ 17. $y = 4.8762 - 5$ 18. $y = 8.4843 - 10$

19. $y = 9.9800 - 10$ 20. $y = 0.5520$ 21. $y = 3.1887$

22. $y = 8.5482 - 10$ 23. $y = 1.7787 - 4$ 24. $y = 5.1220$

A-3   Computations

The properties of logarithms listed on page 341 can be used to compute products, quotients, powers, roots, or any combination of these.

Example 1   Compute $\dfrac{(2.24)^2(8.2)}{(2.12)^3}$.

Solution   We let $N = \dfrac{(2.24)^2(8.2)}{(2.12)^3}$. Then, by Property 4,

$$\log_{10} N = \log_{10} \frac{(2.24)^2(8.2)}{(2.12)^3}.$$

Using Properties 1, 2, and 3, we have

$$\log_{10} N = 2 \log_{10} 2.24 + \log_{10} 8.2 - 3 \log_{10} 2.12.$$

From Table 5,

$$2 \log_{10} 2.24 = 2(0.3502) = 0.7004$$
$$\log_{10} 8.2 \qquad\qquad\quad = 0.9138 \; (+)$$
$$\overline{\qquad\qquad\qquad\qquad\qquad 1.6142}$$
$$3 \log_{10} 2.12 = 3(0.3263) = 0.9789 \; (-)$$
$$\overline{\qquad\qquad\qquad\qquad\qquad 0.6353}$$

Therefore, $\log_{10} N = 0.6353$, and

$$N = \text{antilog}_{10} \, 0.6353 = 4.32.$$

Example 2   Compute $\sqrt[7]{0.862}$.

Solution
$$N = \sqrt[7]{0.862} = (0.862)^{\frac{1}{7}}$$

$$\log_{10} N = \log_{10} (0.862)^{\frac{1}{7}} = \frac{1}{7} \log_{10} 0.862$$

From Table 5, $\log_{10} 0.862 = 6.9355 - 7$. Notice that we use $6 - 7$ rather than $9 - 10$ for the characteristic of $\log_{10} 0.862$ because we wish to divide this logarithm by 7.

$$\log_{10} N = \frac{1}{7}(6.9355 - 7) = 0.9908 - 1$$

$$N = \text{antilog}_{10} (0.9908 - 1) = 0.979$$

When sums or differences are involved in numerical expressions, logarithms can be of help only if some of the terms involve products, quotients, powers, or roots. Calculators offer the best means of evaluating such expressions. However, you may find logarithms helpful if one or more terms in an expression are powers or roots.

Example 3   Compute $(4.13)^3 - \sqrt[4]{(20.5)^3}$.

Solution
Although logarithms are of no help in computing the difference, they can be of help in simplifying each term separately. We can then find the difference directly. Let $N = (4.13)^3$ and $M = \sqrt[4]{(20.5)^3}$. Then

$$\log_{10} N = 3 \log_{10} 4.13 \qquad \log_{10} M = \frac{3}{4} \log_{10} 20.5$$

$$\log_{10} N = 3(0.6160) \qquad \log_{10} M = \frac{3}{4}(1.3118)$$

$$\log_{10} N = 1.8480 \qquad \log_{10} M = 0.9839$$

$$N = \text{antilog}_{10} 1.8480 \qquad M = \text{antilog}_{10} 0.9839$$

$$= 70.47 \qquad\qquad\quad = 9.64$$

Therefore, $N - M = 70.47 - 9.64 = 60.83$.

If any product, quotient, power, or root involves negative numbers, we can still use logarithms for computations on the absolute values of these numbers, and then make appropriate adjustments for signs.

Exercises A-3

Write the logarithmic equation you would use to compute the expression in each of Exercises 1–16.

A   1. $(31.4)(253)$ 2. $(0.541)(2.540)$ 3. $\dfrac{517}{37.1}$ 4. $\dfrac{0.891}{0.0357}$

5. $\dfrac{(22.4)^2(12.4)}{(8.54)}$

6. $\dfrac{(16.5)(5.92)^2}{11.7}$

7. $\sqrt[6]{0.0598}$

8. $\dfrac{3}{\sqrt[4]{0.725}}$

9. $\sqrt[3]{\dfrac{(51.2)(7.81)}{2.93}}$

10. $\sqrt{\dfrac{258}{(2.17)^2(9.32)}}$

11. $\sqrt[5]{\dfrac{(6)(3.91)}{(2)(3.02)}}$

12. $\sqrt[4]{\dfrac{(3.52)(12.8)}{(5.14)(1.97)}}$

13. $\sqrt[8]{\sqrt[3]{(16.4)^2}}$

14. $\sqrt[7]{\sqrt{(2.95)^3}}$

15. $3.251\sqrt{\dfrac{(64.9)(21.2)}{15.7}}$

16. $\dfrac{(12.2)^3(4193)(5.271)}{17.83\sqrt[3]{0.934}}$

17–32. Perform the computations in Exercises 1–16 by using logarithms.

A-4 *Computations Involving Values of Trigonometric Functions*

Logarithms can be used to make computations in trigonometric problems.

Example 1 Solve the right triangle in which $a = 24.80$ and $b = 39.70$.

Solution We make a sketch. By inspection, $\tan A = \dfrac{a}{b} = \dfrac{24.80}{30.70}$.

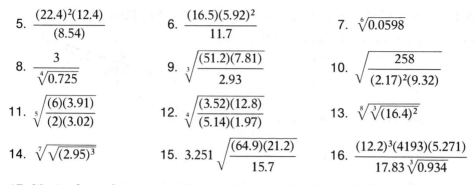

Therefore,

$\log_{10} \tan A = \log_{10} 24.80 - \log_{10} 39.70$

$\qquad = 1.3945 - 1.5988$

$\qquad = (11.3945 - 10) - 1.5988$

$\log_{10} \tan A = 9.7957 - 10$

From Table 2, $\tan A = 0.625$.
Therefore, from Table 2, $\angle A = 32°$.
Then $\angle B = 90° - \angle A = 90° - 32° = 58°$.

To find c, we note that $c = \dfrac{a}{\sin A}$,

so that $\log_{10} c = \log_{10} a - \log_{10} \sin A$, or

$\log_{10} c = \log_{10} 24.80 - \log_{10} \sin 32°$

$\qquad = \log_{10} 24.80 - \log_{10} 0.5299$

$\qquad = 1.3945 - (9.7242 - 10)$

$\qquad = (11.3945 - 10) - (9.7242 - 10) = 1.6703$

Therefore, $c = \text{antilog}_{10} 1.6703 = 46.81$.
Then $\angle A = 32°$, $\angle B = 58°$, and $c = 46.81$.

Example 2 In $\triangle ABC$, $b = 2780$, $\angle A = 34°$, and $\angle B = 14°$. Find a.

Solution We make a sketch. Using the law of sines, we have

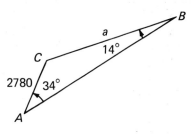

$$\frac{a}{\sin A} = \frac{b}{\sin B}$$

from which we obtain

$$a = \frac{b \sin A}{\sin B}.$$

Then $\log_{10} a = \log_{10} b + \log_{10} \sin A - \log_{10} \sin B$, or

$$\log_{10} a = \log_{10} 2780 + \log_{10} \sin 34° - \log_{10} \sin 14°$$

$$= \log_{10} 2780 + \log_{10} 0.5592 - \log_{10} 0.2419$$

From Table 5,

$$\log_{10} 2780 = 3.4440$$

$$\log_{10} 0.5590 = 9.7474 - 10$$

$$\log_{10} 0.2420 = 9.3838 - 10$$

Thus, $\log_{10} a = 3.4440 + (9.7474 - 10) - (9.3838 - 10)$

$$= (13.1914 - 10) - (9.3838 - 10)$$

$$= 3.8076$$

Therefore, $a = \text{antilog}_{10} 3.8076 = 6420$.

A table of logarithmic values of trigonometric functions is given on page 372 (Table 6). See page 351 for an example using this table.

Exercises A-4

Use logarithms to solve the following right triangles. In each case, make a sketch. $\angle C = 90°$.

A 1. $b = 36.4$, $\angle A = 43°$ 2. $a = 5.07$, $\angle A = 51°$

3. $a = 16.8$, $c = 24.2$ 4. $a = 11.9$, $b = 6.83$

5. $a = 14.23$, $c = 38.70$ 6. $a = 6.492$, $b = 9.831$

Solve the following oblique triangles, using logarithms as convenient. In each case, make a sketch.

B 7. $\angle C = 76°$, $\angle B = 41°$, $a = 7.08$ 8. $c = 27$, $\angle A = 25°30'$, $\angle B = 95°30'$

9. $\angle A = 71°$, $\angle C = 77°$, $b = 113$ 10. $b = 128$, $c = 57.1$, $\angle B = 110°$

Linear Interpolation

When using the trigonometric or logarithmic tables, it may be necessary to use linear interpolation to approximate values for numbers between those listed in the table. The given diagram shows part of the graph of a function f, where x is a number between the numbers a and b listed in the table for f. To find $f(x)$, consider the line g joining the points A and B. $\overline{AB}$ can be assumed to be relatively close to the curve joining A and B. (The diagram is exaggerated to make a clearer picture.) The second coordinate of point M on $\overline{AB}$, $g(x)$, is, then, a very good approximation of $f(x)$. Moreover, we can compute the value of $g(x)$ by using the formula for the slope of $\overline{AB}$. Recall from algebra that if (x_1, y_1) and (x_2, y_2) are any two points on a non-vertical line, then the slope m of the line is given by $m = \dfrac{y_2 - y_1}{x_2 - x_1}$. Since the slope of $\overline{AM}$ equals the slope of $\overline{AB}$ then,

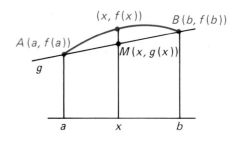

$$\frac{g(x) - f(a)}{x - a} = \frac{f(b) - f(a)}{b - a}.$$

Being given a, b, and x, we can find $f(a)$ and $f(b)$ in the table and then solve for $g(x)$ in the equation above to obtain the desired approximation for $f(x)$.

Example 1   Find $\log_{10} 1.357$.

Solution   Entries for $\log_{10} 1.350$ and $\log_{10} 1.360$ can be found in Table 5, but not for $\log_{10} 1.357$. We know, however, that $\log_{10} 1.350 <$ $\log_{10} 1.357 < \log_{10} 1.360$. Therefore, $0.1303 < \log_{10} 1.357 < 0.1335$. We can use linear interpolation and arrange the facts as shown.

	x	$\log_{10} x$	
	1.350	0.1303	
0.007 [	1.357	$\log_{10} 1.357$] y	0.0032
0.010	1.360	0.1335	

$$\frac{0.007}{0.010} = \frac{y}{0.0032} \quad \text{or} \quad \frac{7}{10} = \frac{y}{0.0032}$$

$$y = \frac{7}{10}(0.0032) = 0.0022$$

Therefore, $\log_{10} 1.357 \approx 0.1303 + 0.0022 = 0.1325$

Example 2 Find cos 35° 53′.

Solution Entries for cos 35° 50′ and cos 36° 00′ can be found in Table 2, but not for cos 35° 53′. We can use linear interpolation and arrange the facts as shown.

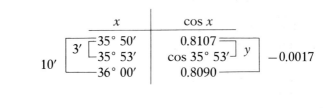

$$\frac{3}{10} = \frac{y}{-0.0017}$$

$$y = \frac{3}{10}(-0.0017) = -0.0005$$

Therefore, cos 35° 53′ $\approx -0.0005 + 0.8107 = 0.8102$.

Table 6 on page 372 gives logarithmic values of trigonometric functions. To make linear interpolation easier, the difference between successive logarithms are given in the columns headed d and cd. The differences in the cd column apply to both the tangent logarithm (Ltan) and the cotangent logarithm (Lcot). A difference of 27, for example, means either 0.0027 or -0.0027, as the case may be.

Example 3 Find $\log_{10} \cos 34°22′$.

Solution We can use linear interpolation and arrange the facts as shown.

θ	$\log_{10} \cos \theta$
┌──34° 20′	9.9169 − 10 ──
10′ 8′┌─34° 22′	$\log_{10}\cos 34°22′$ ┐y −0.0009
└─34° 30′	9.9160 − 10 ══

$$\frac{8}{10} = \frac{y}{-0.0009}$$

$$y = \frac{8}{10}(-0.0009) = -0.0007$$

Therefore, $\log_{10} \cos 34°22′ \approx -0.0007 + 9.9169 - 10 = 9.9162 - 10$.

Table 1 Trigonometric Functions of θ (θ in decimal degrees)

θ Degrees	θ Radians	$\sin \theta$	$\cos \theta$	$\tan \theta$	$\cot \theta$	$\sec \theta$	$\csc \theta$		
0.0	.0000	.0000	1.0000	.0000	undefined	1.000	undefined	1.5708	**90.0**
0.1	.0017	.0017	1.0000	.0017	573.0	1.000	573.0	1.5691	89.9
0.2	.0035	.0035	1.0000	.0035	286.5	1.000	286.5	1.5673	89.8
0.3	.0052	.0052	1.0000	.0052	191.0	1.000	191.0	1.5656	89.7
0.4	.0070	.0070	1.0000	.0070	143.2	1.000	143.2	1.5638	89.6
0.5	.0087	.0087	1.0000	.0087	114.6	1.000	114.6	1.5621	89.5
0.6	.0105	.0105	.9999	.0105	95.49	1.000	95.49	1.5603	89.4
0.7	.0122	.0122	.9999	.0122	81.85	1.000	81.85	1.5586	89.3
0.8	.0140	.0140	.9999	.0140	71.62	1.000	71.62	1.5568	89.2
0.9	.0157	.0157	.9999	.0157	63.66	1.000	63.66	1.5551	89.1
1.0	.0175	.0175	.9998	.0175	57.29	1.000	57.30	1.5533	**89.0**
1.1	.0192	.0192	.9998	.0192	52.08	1.000	52.09	1.5516	88.9
1.2	.0209	.0209	.9998	.0209	47.74	1.000	47.75	1.5499	88.8
1.3	.0227	.0227	.9997	.0227	44.07	1.000	44.08	1.5481	88.7
1.4	.0244	.0244	.9997	.0244	40.92	1.000	40.93	1.5464	88.6
1.5	.0262	.0262	.9997	.0262	38.19	1.000	38.20	1.5446	88.5
1.6	.0279	.0279	.9996	.0279	35.80	1.000	35.81	1.5429	88.4
1.7	.0297	.0297	.9996	.0297	33.69	1.000	33.71	1.5411	88.3
1.8	.0314	.0314	.9995	.0314	31.82	1.000	31.84	1.5394	88.2
1.9	.0332	.0332	.9995	.0332	30.14	1.001	30.16	1.5376	88.1
2.0	.0349	.0349	.9994	.0349	28.64	1.001	28.65	1.5359	**88.0**
2.1	.0367	.0366	.9993	.0367	27.27	1.001	27.29	1.5341	87.9
2.2	.0384	.0384	.9993	.0384	26.03	1.001	26.05	1.5324	87.8
2.3	.0401	.0401	.9992	.0402	24.90	1.001	24.92	1.5307	87.7
2.4	.0419	.0419	.9991	.0419	23.86	1.001	23.88	1.5289	87.6
2.5	.0436	.0436	.9990	.0437	22.90	1.001	22.93	1.5272	87.5
2.6	.0454	.0454	.9990	.0454	22.02	1.001	22.04	1.5254	87.4
2.7	.0471	.0471	.9989	.0472	21.20	1.001	21.23	1.5237	87.3
2.8	.0489	.0488	.9988	.0489	20.45	1.001	20.47	1.5219	87.2
2.9	.0506	.0506	.9987	.0507	19.74	1.001	19.77	1.5202	87.1
3.0	.0524	.0523	.9986	.0524	19.08	1.001	19.11	1.5184	**87.0**
3.1	.0541	.0541	.9985	.0542	18.46	1.001	18.49	1.5167	86.9
3.2	.0559	.0558	.9984	.0559	17.89	1.002	17.91	1.5149	86.8
3.3	.0576	.0576	.9983	.0577	17.34	1.002	17.37	1.5132	86.7
3.4	.0593	.0593	.9982	.0594	16.83	1.002	16.86	1.5115	86.6
3.5	.0611	.0610	.9981	.0612	16.35	1.002	16.38	1.5097	86.5
3.6	.0628	.0628	.9980	.0629	15.89	1.002	15.93	1.5080	86.4
3.7	.0646	.0645	.9979	.0647	15.46	1.002	15.50	1.5062	86.3
3.8	.0663	.0663	.9978	.0664	15.06	1.002	15.09	1.5045	86.2
3.9	.0681	.0680	.9977	.0682	14.67	1.002	14.70	1.5027	86.1
4.0	.0698	.0698	.9976	.0699	14.30	1.002	14.34	1.5010	**86.0**
4.1	.0716	.0715	.9974	.0717	13.95	1.003	13.99	1.4992	85.9
4.2	.0733	.0732	.9973	.0734	13.62	1.003	13.65	1.4975	85.8
4.3	.0750	.0750	.9972	.0752	13.30	1.003	13.34	1.4957	85.7
4.4	.0768	.0767	.9971	.0769	13.00	1.003	13.03	1.4940	85.6
4.5	.0785	.0785	.9969	.0787	12.71	1.003	12.75	1.4923	85.5
4.6	.0803	.0802	.9968	.0805	12.43	1.003	12.47	1.4905	85.4
4.7	.0820	.0819	.9966	.0822	12.16	1.003	12.20	1.4888	85.3
4.8	.0838	.0837	.9965	.0840	11.91	1.004	11.95	1.4870	85.2
4.9	.0855	.0854	.9963	.0857	11.66	1.004	11.71	1.4853	85.1
5.0	.0873	.0872	.9962	.0875	11.43	1.004	11.47	1.4835	**85.0**
5.1	.0890	.0889	.9960	.0892	11.20	1.004	11.25	1.4818	84.9
5.2	.0908	.0906	.9959	.0910	10.99	1.004	11.03	1.4800	84.8
5.3	.0925	.0924	.9957	.0928	10.78	1.004	10.83	1.4783	84.7
5.4	.0942	.0941	.9956	.0945	10.58	1.004	10.63	1.4765	84.6
5.5	.0960	.0958	.9954	.0963	10.39	1.005	10.43	1.4748	84.5
5.6	.0977	.0976	.9952	.0981	10.20	1.005	10.25	1.4731	84.4
5.7	.0995	.0993	.9951	.0998	10.02	1.005	10.07	1.4713	84.3
5.8	.1012	.1011	.9949	.1016	9.845	1.005	9.895	1.4696	84.2
5.9	.1030	.1028	.9947	.1033	9.677	1.005	9.728	1.4678	84.1
6.0	.1047	.1045	.9945	.1051	9.514	1.006	9.567	1.4661	**84.0**
		$\cos \theta$	$\sin \theta$	$\cot \theta$	$\tan \theta$	$\csc \theta$	$\sec \theta$	θ Radians	θ Degrees

Table 1 Trigonometric Functions of θ (θ in decimal degrees)

θ Degrees	θ Radians	$\sin \theta$	$\cos \theta$	$\tan \theta$	$\cot \theta$	$\sec \theta$	$\csc \theta$		
6.0	.1047	.1045	.9945	.1051	9.514	1.006	9.567	1.4661	**84.0**
6.1	.1065	.1063	.9943	.1069	9.357	1.006	9.411	1.4643	83.9
6.2	.1082	.1080	.9942	.1086	9.205	1.006	9.259	1.4626	83.8
6.3	.1100	.1097	.9940	.1104	9.058	1.006	9.113	1.4608	83.7
6.4	.1117	.1115	.9938	.1122	8.915	1.006	8.971	1.4591	83.6
6.5	.1134	.1132	.9936	.1139	8.777	1.006	8.834	1.4574	83.5
6.6	.1152	.1149	.9934	.1157	8.643	1.007	8.700	1.4556	83.4
6.7	.1169	.1167	.9932	.1175	8.513	1.007	8.571	1.4539	83.3
6.8	.1187	.1184	.9930	.1192	8.386	1.007	8.446	1.4521	83.2
6.9	.1204	.1201	.9928	.1210	8.264	1.007	8.324	1.4504	83.1
7.0	.1222	.1219	.9925	.1228	8.144	1.008	8.206	1.4486	**83.0**
7.1	.1239	.1236	.9923	.1246	8.028	1.008	8.091	1.4469	82.9
7.2	.1257	.1253	.9921	.1263	7.916	1.008	7.979	1.4451	82.8
7.3	.1274	.1271	.9919	.1281	7.806	1.008	7.870	1.4434	82.7
7.4	.1292	.1288	.9917	.1299	7.700	1.008	7.764	1.4416	82.6
7.5	.1309	.1305	.9914	.1317	7.596	1.009	7.661	1.4399	82.5
7.6	.1326	.1323	.9912	.1334	7.495	1.009	7.561	1.4382	82.4
7.7	.1344	.1340	.9910	.1352	7.396	1.009	7.463	1.4364	82.3
7.8	.1361	.1357	.9907	.1370	7.300	1.009	7.368	1.4347	82.2
7.9	.1379	.1374	.9905	.1388	7.207	1.010	7.276	1.4329	82.1
8.0	.1396	.1392	.9903	.1405	7.115	1.010	7.185	1.4312	**82.0**
8.1	.1414	.1409	.9900	.1423	7.026	1.010	7.097	1.4294	81.9
8.2	.1431	.1426	.9898	.1441	6.940	1.010	7.011	1.4277	81.8
8.3	.1449	.1444	.9895	.1459	6.855	1.011	6.927	1.4259	81.7
8.4	.1466	.1461	.9893	.1477	6.772	1.011	6.845	1.4242	81.6
8.5	.1484	.1478	.9890	.1495	6.691	1.011	6.765	1.4224	81.5
8.6	.1501	.1495	.9888	.1512	6.612	1.011	6.687	1.4207	81.4
8.7	.1518	.1513	.9885	.1530	6.535	1.012	6.611	1.4190	81.3
8.8	.1536	.1530	.9882	.1548	6.460	1.012	6.537	1.4172	81.2
8.9	.1553	.1547	.9880	.1566	6.386	1.012	6.464	1.4155	81.1
9.0	.1571	.1564	.9877	.1584	6.314	1.012	6.392	1.4137	**81.0**
9.1	.1588	.1582	.9874	.1602	6.243	1.013	6.323	1.4120	80.9
9.2	.1606	.1599	.9871	.1620	6.174	1.013	6.255	1.4102	80.8
9.3	.1623	.1616	.9869	.1638	6.107	1.013	6.188	1.4085	80.7
9.4	.1641	.1633	.9866	.1655	6.041	1.014	6.123	1.4067	80.6
9.5	.1658	.1650	.9863	.1673	5.976	1.014	6.059	1.4050	80.5
9.6	.1676	.1668	.9860	.1691	5.912	1.014	5.996	1.4032	80.4
9.7	.1693	.1685	.9857	.1709	5.850	1.015	5.935	1.4015	80.3
9.8	.1710	.1702	.9854	.1727	5.789	1.015	5.875	1.3998	80.2
9.9	.1728	.1719	.9851	.1745	5.730	1.015	5.816	1.3980	80.1
10.0	.1745	.1736	.9848	.1763	5.671	1.015	5.759	1.3963	**80.0**
10.1	.1763	.1754	.9845	.1781	5.614	1.016	5.702	1.3945	79.9
10.2	.1780	.1771	.9842	.1799	5.558	1.016	5.647	1.3928	79.8
10.3	.1798	.1788	.9839	.1817	5.503	1.016	5.593	1.3910	79.7
10.4	.1815	.1805	.9836	.1835	5.449	1.017	5.540	1.3893	79.6
10.5	.1833	.1822	.9833	.1853	5.396	1.017	5.487	1.3875	79.5
10.6	.1850	.1840	.9829	.1871	5.343	1.017	5.436	1.3858	79.4
10.7	.1868	.1857	.9826	.1890	5.292	1.018	5.386	1.3840	79.3
10.8	.1885	.1874	.9823	.1908	5.242	1.018	5.337	1.3823	79.2
10.9	.1902	.1891	.9820	.1926	5.193	1.018	5.288	1.3806	79.1
11.0	.1920	.1908	.9816	.1944	5.145	1.019	5.241	1.3788	**79.0**
11.1	.1937	.1925	.9813	.1962	5.097	1.019	5.194	1.3771	78.9
11.2	.1955	.1942	.9810	.1980	5.050	1.019	5.148	1.3753	78.8
11.3	.1972	.1959	.9806	.1998	5.005	1.020	5.103	1.3736	78.7
11.4	.1990	.1977	.9803	.2016	4.959	1.020	5.059	1.3718	78.6
11.5	.2007	.1994	.9799	.2035	4.915	1.020	5.016	1.3701	78.5
11.6	.2025	.2011	.9796	.2053	4.872	1.021	4.973	1.3683	78.4
11.7	.2042	.2028	.9792	.2071	4.829	1.021	4.931	1.3666	78.3
11.8	.2059	.2045	.9789	.2089	4.787	1.022	4.890	1.3648	78.2
11.9	.2077	.2062	.9785	.2107	4.745	1.022	4.850	1.3631	78.1
12.0	.2094	.2079	.9781	.2126	4.705	1.022	4.810	1.3614	**78.0**
		$\cos \theta$	$\sin \theta$	$\cot \theta$	$\tan \theta$	$\csc \theta$	$\sec \theta$	θ Radians	θ Degrees

Table 1 Trigonometric Functions of θ (θ in decimal degrees)

θ Degrees	θ Radians	$\sin \theta$	$\cos \theta$	$\tan \theta$	$\cot \theta$	$\sec \theta$	$\csc \theta$		
12.0	.2094	.2079	.9781	.2126	4.705	1.022	4.810	1.3614	**78.0**
12.1	.2112	.2096	.9778	.2144	4.665	1.023	4.771	1.3596	77.9
12.2	.2129	.2113	.9774	.2162	4.625	1.023	4.732	1.3579	77.8
12.3	.2147	.2130	.9770	.2180	4.586	1.023	4.694	1.3561	77.7
12.4	.2164	.2147	.9767	.2199	4.548	1.024	4.657	1.3544	77.6
12.5	.2182	.2164	.9763	.2217	4.511	1.024	4.620	1.3526	77.5
12.6	.2199	.2181	.9759	.2235	4.474	1.025	4.584	1.3509	77.4
12.7	.2217	.2198	.9755	.2254	4.437	1.025	4.549	1.3491	77.3
12.8	.2234	.2215	.9751	.2272	4.402	1.025	4.514	1.3474	77.2
12.9	.2251	.2233	.9748	.2290	4.366	1.026	4.479	1.3456	77.1
13.0	.2269	.2250	.9744	.2309	4.331	1.026	4.445	1.3439	**77.0**
13.1	.2286	.2267	.9740	.2327	4.297	1.027	4.412	1.3422	76.9
13.2	.2304	.2284	.9736	.2345	4.264	1.027	4.379	1.3404	76.8
13.3	.2321	.2300	.9732	.2364	4.230	1.028	4.347	1.3387	76.7
13.4	.2339	.2317	.9728	.2382	4.198	1.028	4.315	1.3369	76.6
13.5	.2356	.2334	.9724	.2401	4.165	1.028	4.284	1.3352	76.5
13.6	.2374	.2351	.9720	.2419	4.134	1.029	4.253	1.3334	76.4
13.7	.2391	.2368	.9715	.2438	4.102	1.029	4.222	1.3317	76.3
13.8	.2409	.2385	.9711	.2456	4.071	1.030	4.192	1.3299	76.2
13.9	.2426	.2402	.9707	.2475	4.041	1.030	4.163	1.3282	76.1
14.0	.2443	.2419	.9703	.2493	4.011	1.031	4.134	1.3265	**76.0**
14.1	.2461	.2436	.9699	.2512	3.981	1.031	4.105	1.3247	75.9
14.2	.2478	.2453	.9694	.2530	3.952	1.032	4.077	1.3230	75.8
14.3	.2496	.2470	.9690	.2549	3.923	1.032	4.049	1.3212	75.7
14.4	.2513	.2487	.9686	.2568	3.895	1.032	4.021	1.3195	75.6
14.5	.2531	.2504	.9681	.2586	3.867	1.033	3.994	1.3177	75.5
14.6	.2548	.2521	.9677	.2605	3.839	1.033	3.967	1.3160	75.4
14.7	.2566	.2538	.9673	.2623	3.812	1.034	3.941	1.3142	75.3
14.8	.2583	.2554	.9668	.2642	3.785	1.034	3.915	1.3125	75.2
14.9	.2601	.2571	.9664	.2661	3.758	1.035	3.889	1.3107	75.1
15.0	.2618	.2588	.9659	.2679	3.732	1.035	3.864	1.3090	**75.0**
15.1	.2635	.2605	.9655	.2698	3.706	1.036	3.839	1.3073	74.9
15.2	.2653	.2622	.9650	.2717	3.681	1.036	3.814	1.3055	74.8
15.3	.2670	.2639	.9646	.2736	3.655	1.037	3.790	1.3038	74.7
15.4	.2688	.2656	.9641	.2754	3.630	1.037	3.766	1.3020	74.6
15.5	.2705	.2672	.9636	.2773	3.606	1.038	3.742	1.3003	74.5
15.6	.2723	.2689	.9632	.2792	3.582	1.038	3.719	1.2985	74.4
15.7	.2740	.2706	.9627	.2811	3.558	1.039	3.695	1.2968	74.3
15.8	.2758	.2723	.9622	.2830	3.534	1.039	3.673	1.2950	74.2
15.9	.2775	.2740	.9617	.2849	3.511	1.040	3.650	1.2933	74.1
16.0	.2793	.2756	.9613	.2867	3.487	1.040	3.628	1.2915	**74.0**
16.1	.2810	.2773	.9608	.2886	3.465	1.041	3.606	1.2898	73.9
16.2	.2827	.2790	.9603	.2905	3.442	1.041	3.584	1.2881	73.8
16.3	.2845	.2807	.9598	.2924	3.420	1.042	3.563	1.2863	73.7
16.4	.2862	.2823	.9593	.2943	3.398	1.042	3.542	1.2846	73.6
16.5	.2880	.2840	.9588	.2962	3.376	1.043	3.521	1.2828	73.5
16.6	.2897	.2857	.9583	.2981	3.354	1.043	3.500	1.2811	73.4
16.7	.2915	.2874	.9578	.3000	3.333	1.044	3.480	1.2793	73.3
16.8	.2932	.2890	.9573	.3019	3.312	1.045	3.460	1.2776	73.2
16.9	.2950	.2907	.9568	.3038	3.291	1.045	3.440	1.2758	73.1
17.0	.2967	.2924	.9563	.3057	3.271	1.046	3.420	1.2741	**73.0**
17.1	.2985	.2940	.9558	.3076	3.251	1.046	3.401	1.2723	72.9
17.2	.3002	.2957	.9553	.3096	3.230	1.047	3.382	1.2706	72.8
17.3	.3019	.2974	.9548	.3115	3.211	1.047	3.363	1.2689	72.7
17.4	.3037	.2990	.9542	.3134	3.191	1.048	3.344	1.2671	72.6
17.5	.3054	.3007	.9537	.3153	3.172	1.049	3.326	1.2654	72.5
17.6	.3072	.3024	.9532	.3172	3.152	1.049	3.307	1.2636	72.4
17.7	.3089	.3040	.9527	.3191	3.133	1.050	3.289	1.2619	72.3
17.8	.3107	.3057	.9521	.3211	3.115	1.050	3.271	1.2601	72.2
17.9	.3124	.3074	.9516	.3230	3.096	1.051	3.254	1.2584	72.1
18.0	.3142	.3090	.9511	.3249	3.078	1.051	3.236	1.2566	**72.0**
		$\cos \theta$	$\sin \theta$	$\cot \theta$	$\tan \theta$	$\csc \theta$	$\sec \theta$	θ Radians	θ Degrees

Table 1 Trigonometric Functions of θ (θ in decimal degrees)

θ Degrees	θ Radians	$\sin \theta$	$\cos \theta$	$\tan \theta$	$\cot \theta$	$\sec \theta$	$\csc \theta$		
18.0	.3142	.3090	.9511	.3249	3.078	1.051	3.236	1.2566	**72.0**
18.1	.3159	.3107	.9505	.3269	3.060	1.052	3.219	1.2549	71.9
18.2	.3177	.3123	.9500	.3288	3.042	1.053	3.202	1.2531	71.8
18.3	.3194	.3140	.9494	.3307	3.024	1.053	3.185	1.2514	71.7
18.4	.3211	.3156	.9489	.3327	3.006	1.054	3.168	1.2497	71.6
18.5	.3229	.3173	.9483	.3346	2.989	1.054	3.152	1.2479	71.5
18.6	.3246	.3190	.9478	.3365	2.971	1.055	3.135	1.2462	71.4
18.7	.3264	.3206	.9472	.3385	2.954	1.056	3.119	1.2444	71.3
18.8	.3281	.3223	.9466	.3404	2.937	1.056	3.103	1.2427	71.2
18.9	.3299	.3239	.9461	.3424	2.921	1.057	3.087	1.2409	71.1
19.0	.3316	.3256	.9455	.3443	2.904	1.058	3.072	1.2392	**71.0**
19.1	.3334	.3272	.9449	.3463	2.888	1.058	3.056	1.2374	70.9
19.2	.3351	.3289	.9444	.3482	2.872	1.059	3.041	1.2357	70.8
19.3	.3368	.3305	.9438	.3502	2.856	1.060	3.026	1.2339	70.7
19.4	.3386	.3322	.9432	.3522	2.840	1.060	3.011	1.2322	70.6
19.5	.3403	.3338	.9426	.3541	2.824	1.061	2.996	1.2305	70.5
19.6	.3421	.3355	.9421	.3561	2.808	1.062	2.981	1.2287	70.4
19.7	.3438	.3371	.9415	.3581	2.793	1.062	2.967	1.2270	70.3
19.8	.3456	.3387	.9409	.3600	2.778	1.063	2.952	1.2252	70.2
19.9	.3473	.3404	.9403	.3620	2.762	1.064	2.938	1.2235	70.1
20.0	.3491	.3420	.9397	.3640	2.747	1.064	2.924	1.2217	**70.0**
20.1	.3508	.3437	.9391	.3659	2.733	1.065	2.910	1.2200	69.9
20.2	.3526	.3453	.9385	.3679	2.718	1.066	2.896	1.2182	69.8
20.3	.3543	.3469	.9379	.3699	2.703	1.066	2.882	1.2165	69.7
20.4	.3560	.3486	.9373	.3719	2.689	1.067	2.869	1.2147	69.6
20.5	.3578	.3502	.9367	.3739	2.675	1.068	2.855	1.2130	69.5
20.6	.3595	.3518	.9361	.3759	2.660	1.068	2.842	1.2113	69.4
20.7	.3613	.3535	.9354	.3779	2.646	1.069	2.829	1.2095	69.3
20.8	.3630	.3551	.9348	.3799	2.633	1.070	2.816	1.2078	69.2
20.9	.3648	.3567	.9342	.3819	2.619	1.070	2.803	1.2060	69.1
21.0	.3665	.3584	.9336	.3839	2.605	1.071	2.790	1.2043	**69.0**
21.1	.3683	.3600	.9330	.3859	2.592	1.072	2.778	1.2025	68.9
21.2	.3700	.3616	.9323	.3879	2.578	1.073	2.765	1.2008	68.8
21.3	.3718	.3633	.9317	.3899	2.565	1.073	2.753	1.1991	68.7
21.4	.3735	.3649	.9311	.3919	2.552	1.074	2.741	1.1973	68.6
21.5	.3752	.3665	.9304	.3939	2.539	1.075	2.729	1.1956	68.5
21.6	.3770	.3681	.9298	.3959	2.526	1.076	2.716	1.1938	68.4
21.7	.3787	.3697	.9291	.3979	2.513	1.076	2.705	1.1921	68.3
21.8	.3805	.3714	.9285	.4000	2.500	1.077	2.693	1.1903	68.2
21.9	.3822	.3730	.9278	.4020	2.488	1.078	2.681	1.1886	68.1
22.0	.3840	.3746	.9272	.4040	2.475	1.079	2.669	1.1868	**68.0**
22.1	.3857	.3762	.9265	.4061	2.463	1.079	2.658	1.1851	67.9
22.2	.3875	.3778	.9259	.4081	2.450	1.080	2.647	1.1833	67.8
22.3	.3892	.3795	.9252	.4101	2.438	1.081	2.635	1.1816	67.7
22.4	.3910	.3811	.9245	.4122	2.426	1.082	2.624	1.1798	67.6
22.5	.3927	.3827	.9239	.4142	2.414	1.082	2.613	1.1781	67.5
22.6	.3944	.3843	.9232	.4163	2.402	1.083	2.602	1.1764	67.4
22.7	.3962	.3859	.9225	.4183	2.391	1.084	2.591	1.1746	67.3
22.8	.3979	.3875	.9219	.4204	2.379	1.085	2.581	1.1729	67.2
22.9	.3997	.3891	.9212	.4224	2.367	1.086	2.570	1.1711	67.1
23.0	.4014	.3907	.9205	.4245	2.356	1.086	2.559	1.1694	**67.0**
23.1	.4032	.3923	.9198	.4265	2.344	1.087	2.549	1.1676	66.9
23.2	.4049	.3939	.9191	.4286	2.333	1.088	2.538	1.1659	66.8
23.3	.4067	.3955	.9184	.4307	2.322	1.089	2.528	1.1641	66.7
23.4	.4084	.3971	.9178	.4327	2.311	1.090	2.518	1.1624	66.6
23.5	.4102	.3987	.9171	.4348	2.300	1.090	2.508	1.1606	66.5
23.6	.4119	.4003	.9164	.4369	2.289	1.091	2.498	1.1589	66.4
23.7	.4136	.4019	.9157	.4390	2.278	1.092	2.488	1.1572	66.3
23.8	.4154	.4035	.9150	.4411	2.267	1.093	2.478	1.1554	66.2
23.9	.4171	.4051	.9143	.4431	2.257	1.094	2.468	1.1537	66.1
24.0	.4189	.4067	.9135	.4452	2.246	1.095	2.459	1.1519	**66.0**
		$\cos \theta$	$\sin \theta$	$\cot \theta$	$\tan \theta$	$\csc \theta$	$\sec \theta$	θ Radians	θ Degrees

Table 1 Trigonometric Functions of θ (θ in decimal degrees)

θ Degrees	θ Radians	$\sin \theta$	$\cos \theta$	$\tan \theta$	$\cot \theta$	$\sec \theta$	$\csc \theta$		
24.0	.4189	.4067	.9135	.4452	2.246	1.095	2.459	1.1519	**66.0**
24.1	.4206	.4083	.9128	.4473	2.236	1.095	2.449	1.1502	65.9
24.2	.4224	.4099	.9121	.4494	2.225	1.096	2.439	1.1484	65.8
24.3	.4241	.4115	.9114	.4515	2.215	1.097	2.430	1.1467	65.7
24.4	.4259	.4131	.9107	.4536	2.204	1.098	2.421	1.1449	65.6
24.5	.4276	.4147	.9100	.4557	2.194	1.099	2.411	1.1432	65.5
24.6	.4294	.4163	.9092	.4578	2.184	1.100	2.402	1.1414	65.4
24.7	.4311	.4179	.9085	.4599	2.174	1.101	2.393	1.1397	65.3
24.8	.4328	.4195	.9078	.4621	2.164	1.102	2.384	1.1380	65.2
24.9	.4346	.4210	.9070	.4642	2.154	1.102	2.375	1.1362	65.1
25.0	.4363	.4226	.9063	.4663	2.145	1.103	2.366	1.1345	**65.0**
25.1	.4381	.4242	.9056	.4684	2.135	1.104	2.357	1.1327	64.9
25.2	.4398	.4258	.9048	.4706	2.125	1.105	2.349	1.1310	64.8
25.3	.4416	.4274	.9041	.4727	2.116	1.106	2.340	1.1292	64.7
25.4	.4433	.4289	.9033	.4748	2.106	1.107	2.331	1.1275	64.6
25.5	.4451	.4305	.9026	.4770	2.097	1.108	2.323	1.1257	64.5
25.6	.4468	.4321	.9018	.4791	2.087	1.109	2.314	1.1240	64.4
25.7	.4485	.4337	.9011	.4813	2.078	1.110	2.306	1.1222	64.3
25.8	.4503	.4352	.9003	.4834	2.069	1.111	2.298	1.1205	64.2
25.9	.4520	.4368	.8996	.4856	2.059	1.112	2.289	1.1188	64.1
26.0	.4538	.4384	.8988	.4877	2.050	1.113	2.281	1.1170	**64.0**
26.1	.4555	.4399	.8980	.4899	2.041	1.114	2.273	1.1153	63.9
26.2	.4573	.4415	.8973	.4921	2.032	1.115	2.265	1.1135	63.8
26.3	.4590	.4431	.8965	.4942	2.023	1.115	2.257	1.1118	63.7
26.4	.4608	.4446	.8957	.4964	2.014	1.116	2.249	1.1100	63.6
26.5	.4625	.4462	.8949	.4986	2.006	1.117	2.241	1.1083	63.5
26.6	.4643	.4478	.8942	.5008	1.997	1.118	2.233	1.1065	63.4
26.7	.4660	.4493	.8934	.5029	1.988	1.119	2.226	1.1048	63.3
26.8	.4677	.4509	.8926	.5051	1.980	1.120	2.218	1.1030	63.2
26.9	.4695	.4524	.8918	.5073	1.971	1.121	2.210	1.1013	63.1
27.0	.4712	.4540	.8910	.5095	1.963	1.122	2.203	1.0996	**63.0**
27.1	.4730	.4555	.8902	.5117	1.954	1.123	2.195	1.0978	62.9
27.2	.4747	.4571	.8894	.5139	1.946	1.124	2.188	1.0961	62.8
27.3	.4765	.4586	.8886	.5161	1.937	1.125	2.180	1.0943	62.7
27.4	.4782	.4602	.8878	.5184	1.929	1.126	2.173	1.0926	62.6
27.5	.4800	.4617	.8870	.5206	1.921	1.127	2.166	1.0908	62.5
27.6	.4817	.4633	.8862	.5228	1.913	1.128	2.158	1.0891	62.4
27.7	.4835	.4648	.8854	.5250	1.905	1.129	2.151	1.0873	62.3
27.8	.4852	.4664	.8846	.5272	1.897	1.130	2.144	1.0856	62.2
27.9	.4869	.4679	.8838	.5295	1.889	1.132	2.137	1.0838	62.1
28.0	.4887	.4695	.8829	.5317	1.881	1.133	2.130	1.0821	**62.0**
28.1	.4904	.4710	.8821	.5340	1.873	1.134	2.123	1.0804	61.9
28.2	.4922	.4726	.8813	.5362	1.865	1.135	2.116	1.0786	61.8
28.3	.4939	.4741	.8805	.5384	1.857	1.136	2.109	1.0769	61.7
28.4	.4957	.4756	.8796	.5407	1.849	1.137	2.103	1.0751	61.6
28.5	.4974	.4772	.8788	.5430	1.842	1.138	2.096	1.0734	61.5
28.6	.4992	.4787	.8780	.5452	1.834	1.139	2.089	1.0716	61.4
28.7	.5009	.4802	.8771	.5475	1.827	1.140	2.082	1.0699	61.3
28.8	.5027	.4818	.8763	.5498	1.819	1.141	2.076	1.0681	61.2
28.9	.5044	.4833	.8755	.5520	1.811	1.142	2.069	1.0664	61.1
29.0	.5061	.4848	.8746	.5543	1.804	1.143	2.063	1.0647	**61.0**
29.1	.5079	.4863	.8738	.5566	1.797	1.144	2.056	1.0629	60.9
29.2	.5096	.4879	.8729	.5589	1.789	1.146	2.050	1.0612	60.8
29.3	.5114	.4894	.8721	.5612	1.782	1.147	2.043	1.0594	60.7
29.4	.5131	.4909	.8712	.5635	1.775	1.148	2.037	1.0577	60.6
29.5	.5149	.4924	.8704	.5658	1.767	1.149	2.031	1.0559	60.5
29.6	.5166	.4939	.8695	.5681	1.760	1.150	2.025	1.0542	60.4
29.7	.5184	.4955	.8686	.5704	1.753	1.151	2.018	1.0524	60.3
29.8	.5201	.4970	.8678	.5727	1.746	1.152	2.012	1.0507	60.2
29.9	.5219	.4985	.8669	.5750	1.739	1.154	2.006	1.0489	60.1
30.0	.5236	.5000	.8660	.5774	1.732	1.155	2.000	1.0472	**60.0**
		$\cos \theta$	$\sin \theta$	$\cot \theta$	$\tan \theta$	$\csc \theta$	$\sec \theta$	θ Radians	θ Degrees

Table 1 Trigonometric Functions of θ (θ in decimal degrees)

θ Degrees	θ Radians	$\sin \theta$	$\cos \theta$	$\tan \theta$	$\cot \theta$	$\sec \theta$	$\csc \theta$		
30.0	.5236	.5000	.8660	.5774	1.732	1.155	2.000	1.0472	**60.0**
30.1	.5253	.5015	.8652	.5797	1.725	1.156	1.994	1.0455	59.9
30.2	.5271	.5030	.8643	.5820	1.718	1.157	1.988	1.0437	59.8
30.3	.5288	.5045	.8634	.5844	1.711	1.158	1.982	1.0420	59.7
30.4	.5306	.5060	.8625	.5867	1.704	1.159	1.976	1.0402	59.6
30.5	.5323	.5075	.8616	.5890	1.698	1.161	1.970	1.0385	59.5
30.6	.5341	.5090	.8607	.5914	1.691	1.162	1.964	1.0367	59.4
30.7	.5358	.5105	.8599	.5938	1.684	1.163	1.959	1.0350	59.3
30.8	.5376	.5120	.8590	.5961	1.678	1.164	1.953	1.0332	59.2
30.9	.5393	.5135	.8581	.5985	1.671	1.165	1.947	1.0315	59.1
31.0	.5411	.5150	.8572	.6009	1.664	1.167	1.942	1.0297	**59.0**
31.1	.5428	.5165	.8563	.6032	1.658	1.168	1.936	1.0280	58.9
31.2	.5445	.5180	.8554	.6056	1.651	1.169	1.930	1.0263	58.8
31.3	.5463	.5195	.8545	.6080	1.645	1.170	1.925	1.0245	58.7
31.4	.5480	.5210	.8535	.6104	1.638	1.172	1.919	1.0228	58.6
31.5	.5498	.5225	.8526	.6128	1.632	1.173	1.914	1.0210	58.5
31.6	.5515	.5240	.8517	.6152	1.625	1.174	1.908	1.0193	58.4
31.7	.5533	.5255	.8508	.6176	1.619	1.175	1.903	1.0175	58.3
31.8	.5550	.5270	.8499	.6200	1.613	1.177	1.898	1.0158	58.2
31.9	.5568	.5284	.8490	.6224	1.607	1.178	1.892	1.0140	58.1
32.0	.5585	.5299	.8480	.6249	1.600	1.179	1.887	1.0123	**58.0**
32.1	.5603	.5314	.8471	.6273	1.594	1.180	1.882	1.0105	57.9
32.2	.5620	.5329	.8462	.6297	1.588	1.182	1.877	1.0088	57.8
32.3	.5637	.5344	.8453	.6322	1.582	1.183	1.871	1.0071	57.7
32.4	.5655	.5358	.8443	.6346	1.576	1.184	1.866	1.0053	57.6
32.5	.5672	.5373	.8434	.6371	1.570	1.186	1.861	1.0036	57.5
32.6	.5690	.5388	.8425	.6395	1.564	1.187	1.856	1.0018	57.4
32.7	.5707	.5402	.8415	.6420	1.558	1.188	1.851	1.0001	57.3
32.8	.5725	.5417	.8406	.6445	1.552	1.190	1.846	.9983	57.2
32.9	.5742	.5432	.8396	.6469	1.546	1.191	1.841	.9966	57.1
33.0	.5760	.5446	.8387	.6494	1.540	1.192	1.836	.9948	**57.0**
33.1	.5777	.5461	.8377	.6519	1.534	1.194	1.831	.9931	56.9
33.2	.5794	.5476	.8368	.6544	1.528	1.195	1.826	.9913	56.8
33.3	.5812	.5490	.8358	.6569	1.522	1.196	1.821	.9896	56.7
33.4	.5829	.5505	.8348	.6594	1.517	1.198	1.817	.9879	56.6
33.5	.5847	.5519	.8339	.6619	1.511	1.199	1.812	.9861	56.5
33.6	.5864	.5534	.8329	.6644	1.505	1.201	1.807	.9844	56.4
33.7	.5882	.5548	.8320	.6669	1.499	1.202	1.802	.9826	56.3
33.8	.5899	.5563	.8310	.6694	1.494	1.203	1.798	.9809	56.2
33.9	.5917	.5577	.8300	.6720	1.488	1.205	1.793	.9791	56.1
34.0	.5934	.5592	.8290	.6745	1.483	1.206	1.788	.9774	**56.0**
34.1	.5952	.5606	.8281	.6771	1.477	1.208	1.784	.9756	55.9
34.2	.5969	.5621	.8271	.6796	1.471	1.209	1.779	.9739	55.8
34.3	.5986	.5635	.8261	.6822	1.466	1.211	1.775	.9721	55.7
34.4	.6004	.5650	.8251	.6847	1.460	1.212	1.770	.9704	55.6
34.5	.6021	.5664	.8241	.6873	1.455	1.213	1.766	.9687	55.5
34.6	.6039	.5678	.8231	.6899	1.450	1.215	1.761	.9669	55.4
34.7	.6056	.5693	.8221	.6924	1.444	1.216	1.757	.9652	55.3
34.8	.6074	.5707	.8211	.6950	1.439	1.218	1.752	.9634	55.2
34.9	.6091	.5721	.8202	.6976	1.433	1.219	1.748	.9617	55.1
35.0	.6109	.5736	.8192	.7002	1.428	1.221	1.743	.9599	**55.0**
35.1	.6126	.5750	.8181	.7028	1.423	1.222	1.739	.9582	54.9
35.2	.6144	.5764	.8171	.7054	1.418	1.224	1.735	.9564	54.8
35.3	.6161	.5779	.8161	.7080	1.412	1.225	1.731	.9547	54.7
35.4	.6178	.5793	.8151	.7107	1.407	1.227	1.726	.9529	54.6
35.5	.6196	.5807	.8141	.7133	1.402	1.228	1.722	.9512	54.5
35.6	.6213	.5821	.8131	.7159	1.397	1.230	1.718	.9495	54.4
35.7	.6231	.5835	.8121	.7186	1.392	1.231	1.714	.9477	54.3
35.8	.6248	.5850	.8111	.7212	1.387	1.233	1.710	.9460	54.2
35.9	.6266	.5864	.8100	.7239	1.381	1.235	1.705	.9442	54.1
36.0	.6283	.5878	.8090	.7265	1.376	1.236	1.701	.9425	**54.0**
		$\cos \theta$	$\sin \theta$	$\cot \theta$	$\tan \theta$	$\csc \theta$	$\sec \theta$	θ Radians	θ Degrees

Table 1 Trigonometric Functions of θ (θ in decimal degrees)

θ Degrees	θ Radians	$\sin\theta$	$\cos\theta$	$\tan\theta$	$\cot\theta$	$\sec\theta$	$\csc\theta$		
36.0	.6283	.5878	.8090	.7265	1.376	1.236	1.701	.9425	**54.0**
36.1	.6301	.5892	.8080	.7292	1.371	1.238	1.697	.9407	53.9
36.2	.6318	.5906	.8070	.7319	1.366	1.239	1.693	.9390	53.8
36.3	.6336	.5920	.8059	.7346	1.361	1.241	1.689	.9372	53.7
36.4	.6353	.5934	.8049	.7373	1.356	1.242	1.685	.9355	53.6
36.5	.6370	.5948	.8039	.7400	1.351	1.244	1.681	.9338	53.5
36.6	.6388	.5962	.8028	.7427	1.347	1.246	1.677	.9320	53.4
36.7	.6405	.5976	.8018	.7454	1.342	1.247	1.673	.9303	53.3
36.8	.6423	.5990	.8007	.7481	1.337	1.249	1.669	.9285	53.2
36.9	.6440	.6004	.7997	.7508	1.332	1.250	1.666	.9268	53.1
37.0	.6458	.6018	.7986	.7536	1.327	1.252	1.662	.9250	**53.0**
37.1	.6475	.6032	.7976	.7563	1.322	1.254	1.658	.9233	52.9
37.2	.6493	.6046	.7965	.7590	1.317	1.255	1.654	.9215	52.8
37.3	.6510	.6060	.7955	.7618	1.313	1.257	1.650	.9198	52.7
37.4	.6528	.6074	.7944	.7646	1.308	1.259	1.646	.9180	52.6
37.5	.6545	.6088	.7934	.7673	1.303	1.260	1.643	.9163	52.5
37.6	.6562	.6101	.7923	.7701	1.299	1.262	1.639	.9146	52.4
37.7	.6580	.6115	.7912	.7729	1.294	1.264	1.635	.9128	52.3
37.8	.6597	.6129	.7902	.7757	1.289	1.266	1.632	.9111	52.2
37.9	.6615	.6143	.7891	.7785	1.285	1.267	1.628	.9093	52.1
38.0	.6632	.6157	.7880	.7813	1.280	1.269	1.624	.9076	**52.0**
38.1	.6650	.6170	.7869	.7841	1.275	1.271	1.621	.9058	51.9
38.2	.6667	.6184	.7859	.7869	1.271	1.272	1.617	.9041	51.8
38.3	.6685	.6198	.7848	.7898	1.266	1.274	1.613	.9023	51.7
38.4	.6702	.6211	.7837	.7926	1.262	1.276	1.610	.9006	51.6
38.5	.6720	.6225	.7826	.7954	1.257	1.278	1.606	.8988	51.5
38.6	.6737	.6239	.7815	.7983	1.253	1.280	1.603	.8971	51.4
38.7	.6754	.6252	.7804	.8012	1.248	1.281	1.599	.8954	51.3
38.8	.6772	.6266	.7793	.8040	1.244	1.283	1.596	.8936	51.2
38.9	.6789	.6280	.7782	.8069	1.239	1.285	1.592	.8919	51.1
39.0	.6807	.6293	.7771	.8098	1.235	1.287	1.589	.8901	**51.0**
39.1	.6824	.6307	.7760	.8127	1.230	1.289	1.586	.8884	50.9
39.2	.6842	.6320	.7749	.8156	1.226	1.290	1.582	.8866	50.8
39.3	.6859	.6334	.7738	.8185	1.222	1.292	1.579	.8849	50.7
39.4	.6877	.6347	.7727	.8214	1.217	1.294	1.575	.8831	50.6
39.5	.6894	.6361	.7716	.8243	1.213	1.296	1.572	.8814	50.5
39.6	.6912	.6374	.7705	.8273	1.209	1.298	1.569	.8796	50.4
39.7	.6929	.6388	.7694	.8302	1.205	1.300	1.566	.8779	50.3
39.8	.6946	.6401	.7683	.8332	1.200	1.302	1.562	.8762	50.2
39.9	.6964	.6414	.7672	.8361	1.196	1.304	1.559	.8744	50.1
40.0	.6981	.6428	.7660	.8391	1.192	1.305	1.556	.8727	**50.0**
40.1	.6999	.6441	.7649	.8421	1.188	1.307	1.552	.8709	49.9
40.2	.7016	.6455	.7638	.8451	1.183	1.309	1.549	.8692	49.8
40.3	.7034	.6468	.7627	.8481	1.179	1.311	1.546	.8674	49.7
40.4	.7051	.6481	.7615	.8511	1.175	1.313	1.543	.8657	49.6
40.5	.7069	.6494	.7604	.8541	1.171	1.315	1.540	.8639	49.5
40.6	.7086	.6508	.7593	.8571	1.167	1.317	1.537	.8622	49.4
40.7	.7103	.6521	.7581	.8601	1.163	1.319	1.534	.8604	49.3
40.8	.7121	.6534	.7570	.8632	1.159	1.321	1.530	.8587	49.2
40.9	.7138	.6547	.7559	.8662	1.154	1.323	1.527	.8570	49.1
41.0	.7156	.6561	.7547	.8693	1.150	1.325	1.524	.8552	**49.0**
41.1	.7173	.6574	.7536	.8724	1.146	1.327	1.521	.8535	48.9
41.2	.7191	.6587	.7524	.8754	1.142	1.329	1.518	.8517	48.8
41.3	.7208	.6600	.7513	.8785	1.138	1.331	1.515	.8500	48.7
41.4	.7226	.6613	.7501	.8816	1.134	1.333	1.512	.8482	48.6
41.5	.7243	.6626	.7490	.8847	1.130	1.335	1.509	.8465	48.5
41.6	.7261	.6639	.7478	.8878	1.126	1.337	1.506	.8447	48.4
41.7	.7278	.6652	.7466	.8910	1.122	1.339	1.503	.8430	48.3
41.8	.7295	.6665	.7455	.8941	1.118	1.341	1.500	.8412	48.2
41.9	.7313	.6678	.7443	.8972	1.115	1.344	1.497	.8395	48.1
42.0	.7330	.6691	.7431	.9004	1.111	1.346	1.494	.8378	**48.0**
		$\cos\theta$	$\sin\theta$	$\cot\theta$	$\tan\theta$	$\csc\theta$	$\sec\theta$	θ Radians	θ Degrees

Table 1 Trigonometric Functions of θ (θ in decimal degrees)

θ Degrees	θ Radians	$\sin \theta$	$\cos \theta$	$\tan \theta$	$\cot \theta$	$\sec \theta$	$\csc \theta$		
42.0	.7330	.6691	.7431	.9004	1.111	1.346	1.494	.8378	**48.0**
42.1	.7348	.6704	.7420	.9036	1.107	1.348	1.492	.8360	47.9
42.2	.7365	.6717	.7408	.9067	1.103	1.350	1.489	.8343	47.8
42.3	.7383	.6730	.7396	.9099	1.099	1.352	1.486	.8325	47.7
42.4	.7400	.6743	.7385	.9131	1.095	1.354	1.483	.8308	47.6
42.5	.7418	.6756	.7373	.9163	1.091	1.356	1.480	.8290	47.5
42.6	.7435	.6769	.7361	.9195	1.087	1.359	1.477	.8273	47.4
42.7	.7453	.6782	.7349	.9228	1.084	1.361	1.475	.8255	47.3
42.8	.7470	.6794	.7337	.9260	1.080	1.363	1.472	.8238	47.2
42.9	.7487	.6807	.7325	.9293	1.076	1.365	1.469	.8221	47.1
43.0	.7505	.6820	.7314	.9325	1.072	1.367	1.466	.8203	**47.0**
43.1	.7522	.6833	.7302	.9358	1.069	1.370	1.464	.8186	46.9
43.2	.7540	.6845	.7290	.9391	1.065	1.372	1.461	.8168	46.8
43.3	.7557	.6858	.7278	.9424	1.061	1.374	1.458	.8151	46.7
43.4	.7575	.6871	.7266	.9457	1.057	1.376	1.455	.8133	46.6
43.5	.7592	.6884	.7254	.9490	1.054	1.379	1.453	.8116	46.5
43.6	.7610	.6896	.7242	.9523	1.050	1.381	1.450	.8098	46.4
43.7	.7627	.6909	.7230	.9556	1.046	1.383	1.447	.8081	46.3
43.8	.7645	.6921	.7218	.9590	1.043	1.386	1.445	.8063	46.2
43.9	.7662	.6934	.7206	.9623	1.039	1.388	1.442	.8046	46.1
44.0	.7679	.6947	.7193	.9657	1.036	1.390	1.440	.8029	**46.0**
44.1	.7697	.6959	.7181	.9691	1.032	1.393	1.437	.8011	45.9
44.2	.7714	.6972	.7169	.9725	1.028	1.395	1.434	.7994	45.8
44.3	.7732	.6984	.7157	.9759	1.025	1.397	1.432	.7976	45.7
44.4	.7749	.6997	.7145	.9793	1.021	1.400	1.429	.7959	45.6
44.5	.7767	.7009	.7133	.9827	1.018	1.402	1.427	.7941	45.5
44.6	.7784	.7022	.7120	.9861	1.014	1.404	1.424	.7924	45.4
44.7	.7802	.7034	.7108	.9896	1.011	1.407	1.422	.7906	45.3
44.8	.7819	.7046	.7096	.9930	1.007	1.409	1.419	.7889	45.2
44.9	.7837	.7059	.7083	.9965	1.003	1.412	1.417	.7871	45.1
45.0	.7854	.7071	.7071	1.0000	1.000	1.414	1.414	.7854	**45.0**
		$\cos \theta$	$\sin \theta$	$\cot \theta$	$\tan \theta$	$\csc \theta$	$\sec \theta$	θ Radians	θ Degrees

Table 2 Trigonometric Functions of θ (θ in degrees and minutes)

m (θ) Degrees	Radians	sin θ	csc θ	tan θ	cot θ	sec θ	cos θ		
0° 00′	.0000	.0000	Undefined	.0000	Undefined	1.000	1.0000	1.5708	90° 00′
10′	.0029	.0029	343.8	.0029	343.8	1.000	1.0000	1.5679	50′
20′	.0058	.0058	171.9	.0058	171.9	1.000	1.0000	1.5650	40′
30′	.0087	.0087	114.6	.0087	114.6	1.000	1.0000	1.5621	30′
40′	.0116	.0116	85.95	.0116	85.94	1.000	.9999	1.5592	20′
50′	.0145	.0145	68.76	.0145	68.75	1.000	.9999	1.5563	10′
1° 00′	.0175	.0175	57.30	.0175	57.29	1.000	.9998	1.5533	89° 00′
10′	.0204	.0204	49.11	.0204	49.10	1.000	.9998	1.5504	50′
20′	.0233	.0233	42.98	.0233	42.96	1.000	.9997	1.5475	40′
30′	.0262	.0262	38.20	.0262	38.19	1.000	.9997	1.5446	30′
40′	.0291	.0291	34.38	.0291	34.37	1.000	.9996	1.5417	20′
50′	.0320	.0320	31.26	.0320	31.24	1.001	.9995	1.5388	10′
2° 00′	.0349	.0349	28.65	.0349	28.64	1.001	.9994	1.5359	88° 00′
10′	.0378	.0378	26.45	.0378	26.43	1.001	.9993	1.5330	50′
20′	.0407	.0407	24.56	.0407	24.54	1.001	.9992	1.5301	40′
30′	.0436	.0436	22.93	.0437	22.90	1.001	.9990	1.5272	30′
40′	.0465	.0465	21.49	.0466	21.47	1.001	.9989	1.5243	20′
50′	.0495	.0494	20.23	.0495	20.21	1.001	.9988	1.5213	10′
3° 00′	.0524	.0523	19.11	.0524	19.08	1.001	.9986	1.5184	87° 00′
10′	.0553	.0552	18.10	.0553	18.07	1.002	.9985	1.5155	50′
20′	.0582	.0581	17.20	.0582	17.17	1.002	.9983	1.5126	40′
30′	.0611	.0610	16.38	.0612	16.35	1.002	.9981	1.5097	30′
40′	.0640	.0640	15.64	.0641	15.60	1.002	.9980	1.5068	20′
50′	.0669	.0669	14.96	.0670	14.92	1.002	.9978	1.5039	10′
4° 00′	.0698	.0698	14.34	.0699	14.30	1.002	.9976	1.5010	86° 00′
10′	.0727	.0727	13.76	.0729	13.73	1.003	.9974	1.4981	50′
20′	.0756	.0756	13.23	.0758	13.20	1.003	.9971	1.4952	40′
30′	.0785	.0785	12.75	.0787	12.71	1.003	.9969	1.4923	30′
40′	.0814	.0814	12.29	.0816	12.25	1.003	.9967	1.4893	20′
50′	.0844	.0843	11.87	.0846	11.83	1.004	.9964	1.4864	10′
5° 00′	.0873	.0872	11.47	.0875	11.43	1.004	.9962	1.4835	85° 00′
10′	.0902	.0901	11.10	.0904	11.06	1.004	.9959	1.4806	50′
20′	.0931	.0929	10.76	.0934	10.71	1.004	.9957	1.4777	40′
30′	.0960	.0958	10.43	.0963	10.39	1.005	.9954	1.4748	30′
40′	.0989	.0987	10.13	.0992	10.08	1.005	.9951	1.4719	20′
50′	.1018	.1016	9.839	.1022	9.788	1.005	.9948	1.4690	10′
6° 00′	.1047	.1045	9.567	.1051	9.514	1.006	.9945	1.4661	84° 00′
10′	.1076	.1074	9.309	.1080	9.255	1.006	.9942	1.4632	50′
20′	.1105	.1103	9.065	.1110	9.010	1.006	.9939	1.4603	40′
30′	.1134	.1132	8.834	.1139	8.777	1.006	.9936	1.4573	30′
40′	.1164	.1161	8.614	.1169	8.556	1.007	.9932	1.4544	20′
50′	.1193	.1190	8.405	.1198	8.345	1.007	.9929	1.4515	10′
7° 00′	.1222	.1219	8.206	.1228	8.144	1.008	.9925	1.4486	83° 00′
10′	.1251	.1248	8.016	.1257	7.953	1.008	.9922	1.4457	50′
20′	.1280	.1276	7.834	.1287	7.770	1.008	.9918	1.4428	40′
30′	.1309	.1305	7.661	.1317	7.596	1.009	.9914	1.4399	30′
40′	.1338	.1334	7.496	.1346	7.429	1.009	.9911	1.4370	20′
50′	.1367	.1363	7.337	.1376	7.269	1.009	.9907	1.4341	10′
8° 00′	.1396	.1392	7.185	.1405	7.115	1.010	.9903	1.4312	82° 00′
10′	.1425	.1421	7.040	.1435	6.968	1.010	.9899	1.4283	50′
20′	.1454	.1449	6.900	.1465	6.827	1.011	.9894	1.4254	40′
30′	.1484	.1478	6.765	.1495	6.691	1.011	.9890	1.4224	30′
40′	.1513	.1507	6.636	.1524	6.561	1.012	.9886	1.4195	20′
50′	.1542	.1536	6.512	.1554	6.435	1.012	.9881	1.4166	10′
9° 00′	.1571	.1564	6.392	.1584	6.314	1.012	.9877	1.4137	81° 00′
		cos θ	sec θ	cot θ	tan θ	csc θ	sin θ	Radians	Degrees m (θ)

Table 2 Trigonometric Functions of θ (θ in degrees and minutes)

$m(\theta)$ Degrees	Radians	$\sin\theta$	$\csc\theta$	$\tan\theta$	$\cot\theta$	$\sec\theta$	$\cos\theta$		
9° 00′	.1571	.1564	6.392	.1584	6.314	1.012	.9877	1.4137	**81° 00′**
10′	.1600	.1593	6.277	.1614	6.197	1.013	.9872	1.4108	50′
20′	.1629	.1622	6.166	.1644	6.084	1.013	.9868	1.4079	40′
30′	.1658	.1650	6.059	.1673	5.976	1.014	.9863	1.4050	30′
40′	.1687	.1679	5.955	.1703	5.871	1.014	.9858	1.4021	20′
50′	.1716	.1708	5.855	.1733	5.769	1.015	.9853	1.3992	10′
10° 00′	.1745	.1736	5.759	.1763	5.671	1.015	.9848	1.3963	**80° 00′**
10′	.1774	.1765	5.665	.1793	5.576	1.016	.9843	1.3934	50′
20′	.1804	.1794	5.575	.1823	5.485	1.016	.9838	1.3904	40′
30′	.1833	.1822	5.487	.1853	5.396	1.017	.9833	1.3875	30′
40′	.1862	.1851	5.403	.1883	5.309	1.018	.9827	1.3846	20′
50′	.1891	.1880	5.320	.1914	5.226	1.018	.9822	1.3817	10′
11° 00′	.1920	.1908	5.241	.1944	5.145	1.019	.9816	1.3788	**79° 00′**
10′	.1949	.1937	5.164	.1974	5.066	1.019	.9811	1.3759	50′
20′	.1978	.1965	5.089	.2004	4.989	1.020	.9805	1.3730	40′
30′	.2007	.1994	5.016	.2035	4.915	1.020	.9799	1.3701	30′
40′	.2036	.2022	4.945	.2065	4.843	1.021	.9793	1.3672	20′
50′	.2065	.2051	4.876	.2095	4.773	1.022	.9787	1.3643	10′
12° 00′	.2094	.2079	4.810	.2126	4.705	1.022	.9781	1.3614	**78° 00′**
10′	.2123	.2108	4.745	.2156	4.638	1.023	.9775	1.3584	50′
20′	.2153	.2136	4.682	.2186	4.574	1.024	.9769	1.3555	40′
30′	.2182	.2164	4.620	.2217	4.511	1.024	.9763	1.3526	30′
40′	.2211	.2193	4.560	.2247	4.449	1.025	.9757	1.3497	20′
50′	.2240	.2221	4.502	.2278	4.390	1.026	.9750	1.3468	10′
13° 00′	.2269	.2250	4.445	.2309	4.331	1.026	.9744	1.3439	**77° 00′**
10′	.2298	.2278	4.390	.2339	4.275	1.027	.9737	1.3410	50′
20′	.2327	.2306	4.336	.2370	4.219	1.028	.9730	1.3381	40′
30′	.2356	.2334	4.284	.2401	4.165	1.028	.9724	1.3352	30′
40′	.2385	.2363	4.232	.2432	4.113	1.029	.9717	1.3323	20′
50′	.2414	.2391	4.182	.2462	4.061	1.030	.9710	1.3294	10′
14° 00′	.2443	.2419	4.134	.2493	4.011	1.031	.9703	1.3265	**76° 00′**
10′	.2473	.2447	4.086	.2524	3.962	1.031	.9696	1.3235	50′
20′	.2502	.2476	4.039	.2555	3.914	1.032	.9689	1.3206	40′
30′	.2531	.2504	3.994	.2586	3.867	1.033	.9681	1.3177	30′
40′	.2560	.2532	3.950	.2617	3.821	1.034	.9674	1.3148	20′
50′	.2589	.2560	3.906	.2648	3.776	1.034	.9667	1.3119	10′
15° 00′	.2618	.2588	3.864	.2679	3.732	1.035	.9659	1.3090	**75° 00′**
10′	.2647	.2616	3.822	.2711	3.689	1.036	.9652	1.3061	50′
20′	.2676	.2644	3.782	.2742	3.647	1.037	.9644	1.3032	40′
30′	.2705	.2672	3.742	.2773	3.606	1.038	.9636	1.3003	30′
40′	.2734	.2700	3.703	.2805	3.566	1.039	.9628	1.2974	20′
50′	.2763	.2728	3.665	.2836	3.526	1.039	.9621	1.2945	10′
16° 00′	.2793	.2756	3.628	.2867	3.487	1.040	.9613	1.2915	**74° 00′**
10′	.2822	.2784	3.592	.2899	3.450	1.041	.9605	1.2886	50′
20′	.2851	.2812	3.556	.2931	3.412	1.042	.9596	1.2857	40′
30′	.2880	.2840	3.521	.2962	3.376	1.043	.9588	1.2828	30′
40′	.2909	.2868	3.487	.2994	3.340	1.044	.9580	1.2799	20′
50′	.2938	.2896	3.453	.3026	3.305	1.045	.9572	1.2770	10′
17° 00′	.2967	.2924	3.420	.3057	3.271	1.046	.9563	1.2741	**73° 00′**
10′	.2996	.2952	3.388	.3089	3.237	1.047	.9555	1.2712	50′
20′	.3025	.2979	3.357	.3121	3.204	1.048	.9546	1.2683	40′
30′	.3054	.3007	3.326	.3153	3.172	1.049	.9537	1.2654	30′
40′	.3083	.3035	3.295	.3185	3.140	1.049	.9528	1.2625	20′
50′	.3113	.3062	3.265	.3217	3.108	1.050	.9520	1.2595	10′
18° 00′	.3142	.3090	3.236	.3249	3.078	1.051	.9511	1.2566	**72° 00′**
		$\cos\theta$	$\sec\theta$	$\cot\theta$	$\tan\theta$	$\csc\theta$	$\sin\theta$	Radians	Degrees $m(\theta)$

Table 2   Trigonometric Functions of θ (θ in degrees and minutes)

m (θ) Degrees	Radians	sin θ	csc θ	tan θ	cot θ	sec θ	cos θ		
18° 00′	.3142	.3090	3.236	.3249	3.078	1.051	.9511	1.2566	72° 00′
10′	.3171	.3118	3.207	.3281	3.047	1.052	.9502	1.2537	50′
20′	.3200	.3145	3.179	.3314	3.018	1.053	.9492	1.2508	40′
30′	.3229	.3173	3.152	.3346	2.989	1.054	.9483	1.2479	30′
40′	.3258	.3201	3.124	.3378	2.960	1.056	.9474	1.2450	20′
50′	.3287	.3228	3.098	.3411	2.932	1.057	.9465	1.2421	10′
19° 00′	.3316	.3256	3.072	.3443	2.904	1.058	.9455	1.2392	71° 00′
10′	.3345	.3283	3.046	.3476	2.877	1.059	.9446	1.2363	50′
20′	.3374	.3311	3.021	.3508	2.850	1.060	.9436	1.2334	40′
30′	.3403	.3338	2.996	.3541	2.824	1.061	.9426	1.2305	30′
40′	.3432	.3365	2.971	.3574	2.798	1.062	.9417	1.2275	20′
50′	.3462	.3393	2.947	.3607	2.773	1.063	.9407	1.2246	10′
20° 00′	.3491	.3420	2.924	.3640	2.747	1.064	.9397	1.2217	70° 00′
10′	.3520	.3448	2.901	.3673	2.723	1.065	.9387	1.2188	50′
20′	.3549	.3475	2.878	.3706	2.699	1.066	.9377	1.2159	40′
30′	.3578	.3502	2.855	.3739	2.675	1.068	.9367	1.2130	30′
40′	.3607	.3529	2.833	.3772	2.651	1.069	.9356	1.2101	20′
50′	.3636	.3557	2.812	.3805	2.628	1.070	.9346	1.2072	10′
21° 00′	.3665	.3584	2.790	.3839	2.605	1.071	.9336	1.2043	69° 00′
10′	.3694	.3611	2.769	.3872	2.583	1.072	.9325	1.2014	50′
20′	.3723	.3638	2.749	.3906	2.560	1.074	.9315	1.1985	40′
30′	.3752	.3665	2.729	.3939	2.539	1.075	.9304	1.1956	30′
40′	.3782	.3692	2.709	.3973	2.517	1.076	.9293	1.1926	20′
50′	.3811	.3719	2.689	.4006	2.496	1.077	.9283	1.1897	10′
22° 00′	.3840	.3746	2.669	.4040	2.475	1.079	.9272	1.1868	68° 00′
10′	.3869	.3773	2.650	.4074	2.455	1.080	.9261	1.1839	50′
20′	.3898	.3800	2.632	.4108	2.434	1.081	.9250	1.1810	40′
30′	.3927	.3827	2.613	.4142	2.414	1.082	.9239	1.1781	30′
40′	.3956	.3854	2.595	.4176	2.394	1.084	.9228	1.1752	20′
50′	.3985	.3881	2.577	.4210	2.375	1.085	.9216	1.1723	10′
23° 00′	.4014	.3907	2.559	.4245	2.356	1.086	.9205	1.1694	67° 00′
10′	.4043	.3934	2.542	.4279	2.337	1.088	.9194	1.1665	50′
20′	.4072	.3961	2.525	.4314	2.318	1.089	.9182	1.1636	40′
30′	.4102	.3987	2.508	.4348	2.300	1.090	.9171	1.1606	30′
40′	.4131	.4014	2.491	.4383	2.282	1.092	.9159	1.1577	20′
50′	.4160	.4041	2.475	.4417	2.264	1.093	.9147	1.1548	10′
24° 00′	.4189	.4067	2.459	.4452	2.246	1.095	.9135	1.1519	66° 00′
10′	.4218	.4094	2.443	.4487	2.229	1.096	.9124	1.1490	50′
20′	.4247	.4120	2.427	.4522	2.211	1.097	.9112	1.1461	40′
30′	.4276	.4147	2.411	.4557	2.194	1.099	.9100	1.1432	30′
40′	.4305	.4173	2.396	.4592	2.177	1.100	.9088	1.1403	20′
50′	.4334	.4200	2.381	.4628	2.161	1.102	.9075	1.1374	10′
25° 00′	.4363	.4226	2.366	.4663	2.145	1.103	.9063	1.1345	65° 00′
10′	.4392	.4253	2.352	.4699	2.128	1.105	.9051	1.1316	50′
20′	.4422	.4279	2.337	.4734	2.112	1.106	.9038	1.1286	40′
30′	.4451	.4305	2.323	.4770	2.097	1.108	.9026	1.1257	30′
40′	.4480	.4331	2.309	.4806	2.081	1.109	.9013	1.1228	20′
50′	.4509	.4358	2.295	.4841	2.066	1.111	.9001	1.1199	10′
26° 00′	.4538	.4384	2.281	.4877	2.050	1.113	.8988	1.1170	64° 00′
10′	.4567	.4410	2.268	.4913	2.035	1.114	.8975	1.1141	50′
20′	.4596	.4436	2.254	.4950	2.020	1.116	.8962	1.1112	40′
30′	.4625	.4462	2.241	.4986	2.006	1.117	.8949	1.1083	30′
40′	.4654	.4488	2.228	.5022	1.991	1.119	.8936	1.1054	20′
50′	.4683	.4514	2.215	.5059	1.977	1.121	.8923	1.1025	10′
27° 00′	.4712	.4540	2.203	.5095	1.963	1.122	.8910	1.0996	63° 00′
		cos θ	sec θ	cot θ	tan θ	csc θ	sin θ	Radians	Degrees m (θ)

Table 2 Trigonometric Functions of θ (θ in degrees and minutes)

$m(\theta)$ Degrees	Radians	$\sin \theta$	$\csc \theta$	$\tan \theta$	$\cot \theta$	$\sec \theta$	$\cos \theta$		
27° 00'	.4712	.4540	2.203	.5095	1.963	1.122	.8910	1.0996	63° 00'
10'	.4741	.4566	2.190	.5132	1.949	1.124	.8897	1.0966	50'
20'	.4771	.4592	2.178	.5169	1.935	1.126	.8884	1.0937	40'
30'	.4800	.4617	2.166	.5206	1.921	1.127	.8870	1.0908	30'
40'	.4829	.4643	2.154	.5243	1.907	1.129	.8857	1.0879	20'
50'	.4858	.4669	2.142	.5280	1.894	1.131	.8843	1.0850	10'
28° 00'	.4887	.4695	2.130	.5317	1.881	1.133	.8829	1.0821	62° 00'
10'	.4916	.4720	2.118	.5354	1.868	1.134	.8816	1.0792	50'
20'	.4945	.4746	2.107	.5392	1.855	1.136	.8802	1.0763	40'
30'	.4974	.4772	2.096	.5430	1.842	1.138	.8788	1.0734	30'
40'	.5003	.4797	2.085	.5467	1.829	1.140	.8774	1.0705	20'
50'	.5032	.4823	2.074	.5505	1.816	1.142	.8760	1.0676	10'
29° 00'	.5061	.4848	2.063	.5543	1.804	1.143	.8746	1.0647	61° 00'
10'	.5091	.4874	2.052	.5581	1.792	1.145	.8732	1.0617	50'
20'	.5120	.4899	2.041	.5619	1.780	1.147	.8718	1.0588	40'
30'	.5149	.4924	2.031	.5658	1.767	1.149	.8704	1.0559	30'
40'	.5178	.4950	2.020	.5696	1.756	1.151	.8689	1.0530	20'
50'	.5207	.4975	2.010	.5735	1.744	1.153	.8675	1.0501	10'
30° 00'	.5236	.5000	2.000	.5774	1.732	1.155	.8660	1.0472	60° 00'
10'	.5265	.5025	1.990	.5812	1.720	1.157	.8646	1.0443	50'
20'	.5294	.5050	1.980	.5851	1.709	1.159	.8631	1.0414	40'
30'	.5323	.5075	1.970	.5890	1.698	1.161	.8616	1.0385	30'
40'	.5352	.5100	1.961	.5930	1.686	1.163	.8601	1.0356	20'
50'	.5381	.5125	1.951	.5969	1.675	1.165	.8587	1.0327	10'
31° 00'	.5411	.5150	1.942	.6009	1.664	1.167	.8572	1.0297	59° 00'
10'	.5440	.5175	1.932	.6048	1.653	1.169	.8557	1.0268	50'
20'	.5469	.5200	1.923	.6088	1.643	1.171	.8542	1.0239	40'
30'	.5498	.5225	1.914	.6128	1.632	1.173	.8526	1.0210	30'
40'	.5527	.5250	1.905	.6168	1.621	1.175	.8511	1.0181	20'
50'	.5556	.5275	1.896	.6208	1.611	1.177	.8496	1.0152	10'
32° 00'	.5585	.5299	1.887	.6249	1.600	1.179	.8480	1.0123	58° 00'
10'	.5614	.5324	1.878	.6289	1.590	1.181	.8465	1.0094	50'
20'	.5643	.5348	1.870	.6330	1.580	1.184	.8450	1.0065	40'
30'	.5672	.5373	1.861	.6371	1.570	1.186	.8434	1.0036	30'
40'	.5701	.5398	1.853	.6412	1.560	1.188	.8418	1.0007	20'
50'	.5730	.5422	1.844	.6453	1.550	1.190	.8403	.9977	10'
33° 00'	.5760	.5446	1.836	.6494	1.540	1.192	.8387	.9948	57° 00'
10'	.5789	.5471	1.828	.6536	1.530	1.195	.8371	.9919	50'
20'	.5818	.5495	1.820	.6577	1.520	1.197	.8355	.9890	40'
30'	.5847	.5519	1.812	.6619	1.511	1.199	.8339	.9861	30'
40'	.5876	.5544	1.804	.6661	1.501	1.202	.8323	.9832	20'
50'	.5905	.5568	1.796	.6703	1.492	1.204	.8307	.9803	10'
34° 00'	.5934	.5592	1.788	.6745	1.483	1.206	.8290	.9774	56° 00'
10'	.5963	.5616	1.781	.6787	1.473	1.209	.8274	.9745	50'
20'	.5992	.5640	1.773	.6830	1.464	1.211	.8258	.9716	40'
30'	.6021	.5664	1.766	.6873	1.455	1.213	.8241	.9687	30'
40'	.6050	.5688	1.758	.6916	1.446	1.216	.8225	.9657	20'
50'	.6080	.5712	1.751	.6959	1.437	1.218	.8208	.9628	10'
35° 00'	.6109	.5736	1.743	.7002	1.428	1.221	.8192	.9599	55° 00'
10'	.6138	.5760	1.736	.7046	1.419	1.223	.8175	.9570	50'
20'	.6167	.5783	1.729	.7089	1.411	1.226	.8158	.9541	40'
30'	.6196	.5807	1.722	.7133	1.402	1.228	.8141	.9512	30'
40'	.6225	.5831	1.715	.7177	1.393	1.231	.8124	.9483	20'
50'	.6254	.5854	1.708	.7221	1.385	1.233	.8107	.9454	10'
36° 00'	.6283	.5878	1.701	.7265	1.376	1.236	.8090	.9425	54° 00'
		$\cos \theta$	$\sec \theta$	$\cot \theta$	$\tan \theta$	$\csc \theta$	$\sin \theta$	Radians	Degrees $m(\theta)$

Table 2 Trigonometric Functions of θ (θ in degrees and minutes)

Degrees	Radians	sin θ	csc θ	tan θ	cot θ	sec θ	cos θ		
36° 00′	.6283	.5878	1.701	.7265	1.376	1.236	.8090	.9425	54° 00′
10′	.6312	.5901	1.695	.7310	1.368	1.239	.8073	.9396	50′
20′	.6341	.5925	1.688	.7355	1.360	1.241	.8056	.9367	40′
30′	.6370	.5948	1.681	.7400	1.351	1.244	.8039	.9338	30′
40′	.6400	.5972	1.675	.7445	1.343	1.247	.8021	.9308	20′
50′	.6429	.5995	1.668	.7490	1.335	1.249	.8004	.9279	10′
37° 00′	.6458	.6018	1.662	.7536	1.327	1.252	.7986	.9250	53° 00′
10′	.6487	.6041	1.655	.7581	1.319	1.255	.7969	.9221	50′
20′	.6516	.6065	1.649	.7627	1.311	1.258	.7951	.9192	40′
30′	.6545	.6088	1.643	.7673	1.303	1.260	.7934	.9163	30′
40′	.6574	.6111	1.636	.7720	1.295	1.263	.7916	.9134	20′
50′	.6603	.6134	1.630	.7766	1.288	1.266	.7898	.9105	10′
38° 00′	.6632	.6157	1.624	.7813	1.280	1.269	.7880	.9076	52° 00′
10′	.6661	.6180	1.618	.7860	1.272	1.272	.7862	.9047	50′
20′	.6690	.6202	1.612	.7907	1.265	1.275	.7844	.9018	40′
30′	.6720	.6225	1.606	.7954	1.257	1.278	.7826	.8988	30′
40′	.6749	.6248	1.601	.8002	1.250	1.281	.7808	.8959	20′
50′	.6778	.6271	1.595	.8050	1.242	1.284	.7790	.8930	10′
39° 00′	.6807	.6293	1.589	.8098	1.235	1.287	.7771	.8901	51° 00′
10′	.6836	.6316	1.583	.8146	1.228	1.290	.7753	.8872	50′
20′	.6865	.6338	1.578	.8195	1.220	1.293	.7735	.8843	40′
30′	.6894	.6361	1.572	.8243	1.213	1.296	.7716	.8814	30′
40′	.6923	.6383	1.567	.8292	1.206	1.299	.7698	.8785	20′
50′	.6952	.6406	1.561	.8342	1.199	1.302	.7679	.8756	10′
40° 00′	.6981	.6428	1.556	.8391	1.192	1.305	.7660	.8727	50° 00′
10′	.7010	.6450	1.550	.8441	1.185	1.309	.7642	.8698	50′
20′	.7039	.6472	1.545	.8491	1.178	1.312	.7623	.8668	40′
30′	.7069	.6494	1.540	.8541	1.171	1.315	.7604	.8639	30′
40′	.7098	.6517	1.535	.8591	1.164	1.318	.7585	.8610	20′
50′	.7127	.6539	1.529	.8642	1.157	1.322	.7566	.8581	10′
41° 00′	.7156	.6561	1.524	.8693	1.150	1.325	.7547	.8552	49° 00′
10′	.7185	.6583	1.519	.8744	1.144	1.328	.7528	.8523	50′
20′	.7214	.6604	1.514	.8796	1.137	1.332	.7509	.8494	40′
30′	.7243	.6626	1.509	.8847	1.130	1.335	.7490	.8465	30′
40′	.7272	.6648	1.504	.8899	1.124	1.339	.7470	.8436	20′
50′	.7301	.6670	1.499	.8952	1.117	1.342	.7451	.8407	10′
42° 00′	.7330	.6691	1.494	.9004	1.111	1.346	.7431	.8378	48° 00′
10′	.7359	.6713	1.490	.9057	1.104	1.349	.7412	.8348	50′
20′	.7389	.6734	1.485	.9110	1.098	1.353	.7392	.8319	40′
30′	.7418	.6756	1.480	.9163	1.091	1.356	.7373	.8290	30′
40′	.7447	.6777	1.476	.9217	1.085	1.360	.7353	.8261	20′
50′	.7476	.6799	1.471	.9271	1.079	1.364	.7333	.8232	10′
43° 00′	.7505	.6820	1.466	.9325	1.072	1.367	.7314	.8203	47° 00′
10′	.7534	.6841	1.462	.9380	1.066	1.371	.7294	.8174	50′
20′	.7563	.6862	1.457	.9435	1.060	1.375	.7274	.8145	40′
30′	.7592	.6884	1.453	.9490	1.054	1.379	.7254	.8116	30′
40′	.7621	.6905	1.448	.9545	1.048	1.382	.7234	.8087	20′
50′	.7650	.6926	1.444	.9601	1.042	1.386	.7214	.8058	10′
44° 00′	.7679	.6947	1.440	.9657	1.036	1.390	.7193	.8029	46° 00′
10′	.7709	.6967	1.435	.9713	1.030	1.394	.7173	.7999	50′
20′	.7738	.6988	1.431	.9770	1.024	1.398	.7153	.7970	40′
30′	.7767	.7009	1.427	.9827	1.018	1.402	.7133	.7941	30′
40′	.7796	.7030	1.423	.9884	1.012	1.406	.7112	.7912	20′
50′	.7825	.7050	1.418	.9942	1.006	1.410	.7092	.7883	10′
45° 00′	.7854	.7071	1.414	1.000	1.000	1.414	.7071	.7854	45° 00′
		cos θ	sec θ	cot θ	tan θ	csc θ	sin θ	Radians	Degrees
									$m(\theta)$

Table 3 Trigonometric Functions of θ (θ in radians)

$m^R(\theta)$	$m°(\theta)$	$\sin\theta$	$\csc\theta$	$\tan\theta$	$\cot\theta$	$\sec\theta$	$\cos\theta$
0.00	0° 00′	0.0000	Undefined	0.0000	Undefined	1.000	1.000
.01	0° 34′	.0100	100.0	.0100	100.0	1.000	1.000
.02	1° 09′	.0200	50.00	.0200	49.99	1.000	0.9998
.03	1° 43′	.0300	33.34	.0300	33.32	1.000	0.9996
.04	2° 18′	.0400	25.01	.0400	24.99	1.001	0.9992
0.05	2° 52′	0.0500	20.01	0.0500	19.98	1.001	0.9988
.06	3° 26′	.0600	16.68	.0601	16.65	1.002	.9982
.07	4° 01′	.0699	14.30	.0701	14.26	1.002	.9976
.08	4° 35′	.0799	12.51	.0802	12.47	1.003	.9968
.09	5° 09′	.0899	11.13	.0902	11.08	1.004	.9960
0.10	5° 44′	0.0998	10.02	0.1003	9.967	1.005	0.9950
.11	6° 18′	.1098	9.109	.1104	9.054	1.006	.9940
.12	6° 53′	.1197	8.353	.1206	8.293	1.007	.9928
.13	7° 27′	.1296	7.714	.1307	7.649	1.009	.9916
.14	8° 01′	.1395	7.166	.1409	7.096	1.010	.9902
0.15	8° 36′	0.1494	6.692	0.1511	6.617	1.011	0.9888
.16	9° 10′	.1593	6.277	.1614	6.197	1.013	.9872
.17	9° 44′	.1692	5.911	.1717	5.826	1.015	.9856
.18	10° 19′	.1790	5.586	.1820	5.495	1.016	.9838
.19	10° 53′	.1889	5.295	.1923	5.200	1.018	.9820
0.20	11° 28′	0.1987	5.033	0.2027	4.933	1.020	0.9801
.21	12° 02′	.2085	4.797	.2131	4.692	1.022	.9780
.22	12° 36′	.2182	4.582	.2236	4.472	1.025	.9759
.23	13° 11′	.2280	4.386	.2341	4.271	1.027	.9737
.24	13° 45′	.2377	4.207	.2447	4.086	1.030	.9713
0.25	14° 19′	0.2474	4.042	0.2553	3.916	1.032	0.9689
.26	14° 54′	.2571	3.890	.2660	3.759	1.035	.9664
.27	15° 28′	.2667	3.749	.2768	3.613	1.038	.9638
.28	16° 03′	.2764	3.619	.2876	3.478	1.041	.9611
.29	16° 37′	.2860	3.497	.2984	3.351	1.044	.9582
0.30	17° 11′	0.2955	3.384	0.3093	3.233	1.047	0.9553
.31	17° 46′	.3051	3.278	.3203	3.122	1.050	.9523
.32	18° 20′	.3146	3.179	.3314	3.018	1.053	.9492
.33	18° 55′	.3240	3.086	.3425	2.920	1.057	.9460
.34	19° 29′	.3335	2.999	.3537	2.827	1.061	.9428
0.35	20° 03′	0.3429	2.916	0.3650	2.740	1.065	0.9394
.36	20° 38′	.3523	2.839	.3764	2.657	1.068	.9359
.37	21° 12′	.3616	2.765	.3879	2.578	1.073	.9323
.38	21° 46′	.3709	2.696	.3994	2.504	1.077	.9287
.39	22° 21′	.3802	2.630	.4111	2.433	1.081	.9249
0.40	22° 55′	0.3894	2.568	0.4228	2.365	1.086	0.9211
.41	23° 30′	.3986	2.509	.4346	2.301	1.090	.9171
.42	24° 04′	.4078	2.452	.4466	2.239	1.095	.9131
.43	24° 38′	.4169	2.399	.4586	2.180	1.100	.9090
.44	25° 13′	.4259	2.348	.4708	2.124	1.105	.9048
0.45	25° 47′	0.4350	2.299	0.4831	2.070	1.111	0.9004
.46	26° 21′	.4439	2.253	.4954	2.018	1.116	.8961
.47	26° 56′	.4529	2.208	.5080	1.969	1.122	.8916
.48	27° 30′	.4618	2.166	.5206	1.921	1.127	.8870
.49	28° 05′	.4706	2.125	.5334	1.875	1.133	.8823

Table 3 Trigonometric Functions of θ (θ in radians)

$m^R(\theta)$	$m°(\theta)$	$\sin \theta$	$\csc \theta$	$\tan \theta$	$\cot \theta$	$\sec \theta$	$\cos \theta$
0.50	28° 39′	0.4794	2.086	0.5463	1.830	1.139	0.8776
.51	29° 13′	.4882	2.048	.5594	1.788	1.146	.8727
.52	29° 48′	.4969	2.013	.5726	1.747	1.152	.8678
.53	30° 22′	.5055	1.978	.5859	1.707	1.159	.8628
.54	30° 56′	.5141	1.945	.5994	1.668	1.166	.8577
0.55	31° 31′	0.5227	1.913	0.6131	1.631	1.173	0.8525
.56	32° 05′	.5312	1.883	.6269	1.595	1.180	.8473
.57	32° 40′	.5396	1.853	.6410	1.560	1.188	.8419
.58	33° 14′	.5480	1.825	.6552	1.526	1.196	.8365
.59	33° 48′	.5564	1.797	.6696	1.494	1.203	.8309
0.60	34° 23′	0.5646	1.771	0.6841	1.462	1.212	0.8253
.61	34° 57′	.5729	1.746	.6989	1.431	1.220	.8196
.62	35° 31′	.5810	1.721	.7139	1.401	1.229	.8139
.63	36° 06′	.5891	1.697	.7291	1.372	1.238	.8080
.64	36° 40′	.5972	1.674	.7445	1.343	1.247	.8021
0.65	37° 15′	0.6052	1.652	0.7602	1.315	1.256	0.7961
.66	37° 49′	.6131	1.631	.7761	1.288	1.266	.7900
.67	38° 23′	.6210	1.610	.7923	1.262	1.276	.7838
.68	38° 58′	.6288	1.590	.8087	1.237	1.286	.7776
.69	39° 32′	.6365	1.571	.8253	1.212	1.297	.7712
0.70	40° 06′	0.6442	1.552	0.8423	1.187	1.307	0.7648
.71	40° 41′	.6518	1.534	.8595	1.163	1.319	.7584
.72	41° 15′	.6594	1.517	.8771	1.140	1.330	.7518
.73	41° 50′	.6669	1.500	.8949	1.117	1.342	.7452
.74	42° 24′	.6743	1.483	.9131	1.095	1.354	.7385
0.75	42° 58′	0.6816	1.467	0.9316	1.073	1.367	0.7317
.76	43° 33′	.6889	1.452	.9505	1.052	1.380	.7248
.77	44° 07′	.6961	1.437	.9697	1.031	1.393	.7179
.78	44° 41′	.7033	1.422	.9893	1.011	1.407	.7109
.79	45° 16′	.7104	1.408	1.009	.9908	1.421	.7038
0.80	45° 50′	0.7174	1.394	1.030	0.9712	1.435	0.6967
.81	46° 25′	.7243	1.381	1.050	.9520	1.450	.6895
.82	46° 59′	.7311	1.368	1.072	.9331	1.466	.6822
.83	47° 33′	.7379	1.355	1.093	.9146	1.482	.6749
.84	48° 08′	.7446	1.343	1.116	.8964	1.498	.6675
0.85	48° 42′	0.7513	1.331	1.138	0.8785	1.515	0.6600
.86	49° 17′	.7578	1.320	1.162	.8609	1.533	.6524
.87	49° 51′	.7643	1.308	1.185	.8437	1.551	.6448
.88	50° 25′	.7707	1.297	1.210	.8267	1.569	.6372
.89	51° 00′	.7771	1.287	1.235	.8100	1.589	.6294
0.90	51° 34′	0.7833	1.277	1.260	0.7936	1.609	0.6216
.91	52° 08′	.7895	1.267	1.286	.7774	1.629	.6137
.92	52° 43′	.7956	1.257	1.313	.7615	1.651	.6058
.93	53° 17′	.8016	1.247	1.341	.7458	1.673	.5978
.94	53° 52′	.8076	1.238	1.369	.7303	1.696	.5898
0.95	54° 26′	0.8134	1.229	1.398	0.7151	1.719	0.5817
.96	55° 00′	.8192	1.221	1.428	.7001	1.744	.5735
.97	55° 35′	.8249	1.212	1.459	.6853	1.769	.5653
.98	56° 09′	.8305	1.204	1.491	.6707	1.795	.5570
.99	56° 43′	.8360	1.196	1.524	.6563	1.823	.5487

Table 3 Trigonometric Functions of θ (θ in radians)

$m^R(\theta)$	$m°(\theta)$	$\sin\theta$	$\csc\theta$	$\tan\theta$	$\cot\theta$	$\sec\theta$	$\cos\theta$
1.00	57° 18′	0.8415	1.188	1.557	0.6421	1.851	0.5403
1.01	57° 52′	.8468	1.181	1.592	.6281	1.880	.5319
1.02	58° 27′	.8521	1.174	1.628	.6142	1.911	.5234
1.03	59° 01′	.8573	1.166	1.665	.6005	1.942	.5148
1.04	59° 35′	.8624	1.160	1.704	.5870	1.975	.5062
1.05	60° 10′	0.8674	1.153	1.743	0.5736	2.010	0.4976
1.06	60° 44′	.8724	1.146	1.784	.5604	2.046	.4889
1.07	61° 18′	.8772	1.140	1.827	.5473	2.083	.4801
1.08	61° 53′	.8820	1.134	1.871	.5344	2.122	.4713
1.09	62° 27′	.8866	1.128	1.917	.5216	2.162	.4625
1.10	63° 02′	0.8912	1.122	1.965	0.5090	2.205	0.4536
1.11	63° 36′	.8957	1.116	2.014	.4964	2.249	.4447
1.12	64° 10′	.9001	1.111	2.066	.4840	2.295	.4357
1.13	64° 45′	.9044	1.106	2.120	.4718	2.344	.4267
1.14	65° 19′	.9086	1.101	2.176	.4596	2.395	.4176
1.15	65° 53′	0.9128	1.096	2.234	0.4475	2.448	0.4085
1.16	66° 28′	.9168	1.091	2.296	.4356	2.504	.3993
1.17	67° 02′	.9208	1.086	2.360	.4237	2.563	.3902
1.18	67° 37′	.9246	1.082	2.428	.4120	2.625	.3809
1.19	68° 11′	.9284	1.077	2.498	.4003	2.691	3717
1.20	68° 45′	0.9320	1.073	2.572	0.3888	2.760	0.3624
1.21	69° 20′	.9356	1.069	2.650	.3773	2.833	.3530
1.22	69° 54′	.9391	1.065	2.733	.3659	2.910	.3436
1.23	70° 28′	.9425	1.061	2.820	.3546	2.992	.3342
1.24	71° 03′	.9458	1.057	2.912	.3434	3.079	.3248
1.25	71° 37′	0.9490	1.054	3.010	0.3323	3.171	0.3153
1.26	72° 12′	.9521	1.050	3.113	.3212	3.270	.3058
1.27	72° 46′	.9551	1.047	3.224	.3102	3.375	.2963
1.28	73° 20′	.9580	1.044	3.341	.2993	3.488	.2867
1.29	73° 55′	.9608	1.041	3.467	.2884	3.609	.2771
1.30	74° 29′	0.9636	1.038	3.602	0.2776	3.738	0.2675
1.31	75° 03′	.9662	1.035	3.747	.2669	3.878	.2579
1.32	75° 38′	.9687	1.032	3.903	.2562	4.029	.2482
1.33	76° 12′	.9711	1.030	4.072	.2456	4.193	.2385
1.34	76° 47′	.9735	1.027	4.256	.2350	4.372	.2288
1.35	77° 21′	0.9757	1.025	4.455	0.2245	4.566	0.2190
1.36	77° 55′	.9779	1.023	4.673	.2140	4.779	.2092
1.37	78° 30′	.9799	1.021	4.913	.2035	5.014	.1994
1.38	79° 04′	.9819	1.018	5.177	.1931	5.273	.1896
1.39	79° 39′	.9837	1.017	5.471	.1828	5.561	.1798
1.40	80° 13′	0.9854	1.015	5.798	0.1725	5.883	0.1700
1.41	80° 47′	.9871	1.013	6.165	.1622	6.246	.1601
1.42	81° 22′	.9887	1.011	6.581	.1519	6.657	.1502
1.43	81° 56′	.9901	1.010	7.055	.1417	7.126	.1403
1.44	82° 30′	.9915	1.009	7.602	.1315	7.667	.1304
1.45	83° 05′	0.9927	1.007	8.238	0.1214	8.299	0.1205
1.46	83° 39′	.9939	1.006	8.989	.1113	9.044	.1106
1.47	84° 14′	.9949	1.005	9.887	.1011	9.938	.1006
1.48	84° 48′	.9959	1.004	10.98	.0911	11.03	.0907
1.49	85° 22′	.9967	1.003	12.35	.0810	12.39	.0807

Table 3 Trigonometric Functions of θ (θ in radians)

$m^R(\theta)$	$m°(\theta)$	$\sin\theta$	$\csc\theta$	$\tan\theta$	$\cot\theta$	$\sec\theta$	$\cos\theta$
1.50	85° 57′	0.9975	1.003	14.10	0.0709	14.14	0.0707
1.51	86° 31′	.9982	1.002	16.43	.0609	16.46	.0608
1.52	87° 05′	.9987	1.001	19.67	.0508	19.70	.0508
1.53	87° 40′	.9992	1.001	24.50	.0408	24.52	.0408
1.54	88° 14′	.9995	1.000	32.46	.0308	32.48	.0308
1.55	88° 49′	0.9998	1.000	48.08	0.0208	48.09	0.0208
1.56	89° 23′	.9999	1.000	92.62	.0108	92.63	.0108
1.57	89° 57′	1.000	1.000	1256	.0008	1256	.0008

Table 4 Squares and Square Roots

N	N^2	$\sqrt{N}$	$\sqrt{10N}$	N	N^2	$\sqrt{N}$	$\sqrt{10N}$
1.0	1.00	1.000	3.162	**5.5**	30.25	2.345	7.416
1.1	1.21	1.049	3.317	**5.6**	31.36	2.366	7.483
1.2	1.44	1.095	3.464	**5.7**	32.49	2.387	7.550
1.3	1.69	1.140	3.606	**5.8**	33.64	2.408	7.616
1.4	1.96	1.183	3.742	**5.9**	34.81	2.429	7.681
1.5	2.25	1.225	3.873	**6.0**	36.00	2.449	7.746
1.6	2.56	1.265	4.000	**6.1**	37.21	2.470	7.810
1.7	2.89	1.304	4.123	**6.2**	38.44	2.490	7.874
1.8	3.24	1.342	4.243	**6.3**	39.69	2.510	7.937
1.9	3.61	1.378	4.359	**6.4**	40.96	2.530	8.000
2.0	4.00	1.414	4.472	**6.5**	42.25	2.550	8.062
2.1	4.41	1.449	4.583	**6.6**	43.56	2.569	8.124
2.2	4.84	1.483	4.690	**6.7**	44.89	2.588	8.185
2.3	5.29	1.517	4.796	**6.8**	46.24	2.608	8.246
2.4	5.76	1.549	4.899	**6.9**	47.61	2.627	8.307
2.5	6.25	1.581	5.000	**7.0**	49.00	2.646	8.367
2.6	6.76	1.612	5.099	**7.1**	50.41	2.665	8.426
2.7	7.29	1.643	5.196	**7.2**	51.84	2.683	8.485
2.8	7.84	1.673	5.292	**7.3**	53.29	2.702	8.544
2.9	8.41	1.703	5.385	**7.4**	54.76	2.720	8.602
3.0	9.00	1.732	5.477	**7.5**	56.25	2.739	8.660
3.1	9.61	1.761	5.568	**7.6**	57.76	2.757	8.718
3.2	10.24	1.789	5.657	**7.7**	59.29	2.775	8.775
3.3	10.89	1.817	5.745	**7.8**	60.84	2.793	8.832
3.4	11.56	1.844	5.831	**7.9**	62.41	2.811	8.888
3.5	12.25	1.871	5.916	**8.0**	64.00	2.828	8.944
3.6	12.96	1.897	6.000	**8.1**	65.61	2.846	9.000
3.7	13.69	1.924	6.083	**8.2**	67.24	2.864	9.055
3.8	14.44	1.949	6.164	**8.3**	68.89	2.881	9.110
3.9	15.21	1.975	6.245	**8.4**	70.56	2.898	9.165
4.0	16.00	2.000	6.325	**8.5**	72.25	2.915	9.220
4.1	16.81	2.025	6.403	**8.6**	73.96	2.933	9.274
4.2	17.64	2.049	6.481	**8.7**	75.69	2.950	9.327
4.3	18.49	2.074	6.557	**8.8**	77.44	2.966	9.381
4.4	19.36	2.098	6.633	**8.9**	79.21	2.983	9.434
4.5	20.25	2.121	6.708	**9.0**	81.00	3.000	9.487
4.6	21.16	2.145	6.782	**9.1**	82.81	3.017	9.539
4.7	22.09	2.168	6.856	**9.2**	84.64	3.033	9.592
4.8	23.04	2.191	6.928	**9.3**	86.49	3.050	9.644
4.9	24.01	2.214	7.000	**9.4**	88.36	3.066	9.695
5.0	25.00	2.236	7.071	**9.5**	90.25	3.082	9.747
5.1	26.01	2.258	7.141	**9.6**	92.16	3.098	9.798
5.2	27.04	2.280	7.211	**9.7**	94.09	3.114	9.849
5.3	28.09	2.302	7.280	**9.8**	96.04	3.130	9.899
5.4	29.16	2.324	7.348	**9.9**	98.01	3.146	9.950
5.5	30.25	2.345	7.416	**10**	100.00	3.162	10.000

Table 5 Common Logarithms of Numbers *

N	0	1	2	3	4	5	6	7	8	9
10	0000	0043	0086	0128	0170	0212	0253	0294	0334	0374
11	0414	0453	0492	0531	0569	0607	0645	0682	0719	0755
12	0792	0828	0864	0899	0934	0969	1004	1038	1072	1106
13	1139	1173	1206	1239	1271	1303	1335	1367	1399	1430
14	1461	1492	1523	1553	1584	1614	1644	1673	1703	1732
15	1761	1790	1818	1847	1875	1903	1931	1959	1987	2014
16	2041	2068	2095	2122	2148	2175	2201	2227	2253	2279
17	2304	2330	2355	2380	2405	2430	2455	2480	2504	2529
18	2553	2577	2601	2625	2648	2672	2695	2718	2742	2765
19	2788	2810	2833	2856	2878	2900	2923	2945	2967	2989
20	3010	3032	3054	3075	3096	3118	3139	3160	3181	3201
21	3222	3243	3263	3284	3304	3324	3345	3365	3385	3404
22	3424	3444	3464	3483	3502	3522	3541	3560	3579	3598
23	3617	3636	3655	3674	3692	3711	3729	3747	3766	3784
24	3802	3820	3838	3856	3874	3892	3909	3927	3945	3962
25	3979	3997	4014	4031	4048	4065	4082	4099	4116	4133
26	4150	4166	4183	4200	4216	4232	4249	4265	4281	4298
27	4314	4330	4346	4362	4378	4393	4409	4425	4440	4456
28	4472	4487	4502	4518	4533	4548	4564	4579	4594	4609
29	4624	4639	4654	4669	4683	4698	4713	4728	4742	4757
30	4771	4786	4800	4814	4829	4843	4857	4871	4886	4900
31	4914	4928	4942	4955	4969	4983	4997	5011	5024	5038
32	5051	5065	5079	5092	5105	5119	5132	5145	5159	5172
33	5185	5198	5211	5224	5237	5250	5263	5276	5289	5302
34	5315	5328	5340	5353	5366	5378	5391	5403	5416	5428
35	5441	5453	5465	5478	5490	5502	5514	5527	5539	5551
36	5563	5575	5587	5599	5611	5623	5635	5647	5658	5670
37	5682	5694	5705	5717	5729	5740	5752	5763	5775	5786
38	5798	5809	5821	5832	5843	5855	5866	5877	5888	5899
39	5911	5922	5933	5944	5955	5966	5977	5988	5999	6010
40	6021	6031	6042	6053	6064	6075	6085	6096	6107	6117
41	6128	6138	6149	6160	6170	6180	6191	6201	6212	6222
42	6232	6243	6253	6263	6274	6284	6294	6304	6314	6325
43	6335	6345	6355	6365	6375	6385	6395	6405	6415	6425
44	6435	6444	6454	6464	6474	6484	6493	6503	6513	6522
45	6532	6542	6551	6561	6571	6580	6590	6599	6609	6618
46	6628	6637	6646	6656	6665	6675	6684	6693	6702	6712
47	6721	6730	6739	6749	6758	6767	6776	6785	6794	6803
48	6812	6821	6830	6839	6848	6857	6866	6875	6884	6893
49	6902	6911	6920	6928	6937	6946	6955	6964	6972	6981
50	6990	6998	7007	7016	7024	7033	7042	7050	7059	7067
51	7076	7084	7093	7101	7110	7118	7126	7135	7143	7152
52	7160	7168	7177	7185	7193	7202	7210	7218	7226	7235
53	7243	7251	7259	7267	7275	7284	7292	7300	7308	7316
54	7324	7332	7340	7348	7356	7364	7372	7380	7388	7396

*Mantissas; decimal points omitted. Characteristics are found by inspection.

Table 5 Common Logarithms of Numbers

N	0	1	2	3	4	5	6	7	8	9
55	7404	7412	7419	7427	7435	7443	7451	7459	7466	7474
56	7482	7490	7497	7505	7513	7520	7528	7536	7543	7551
57	7559	7566	7574	7582	7589	7597	7604	7612	7619	7627
58	7634	7642	7649	7657	7664	7672	7679	7686	7694	7701
59	7709	7716	7723	7731	7738	7745	7752	7760	7767	7774
60	7782	7789	7796	7803	7810	7818	7825	7832	7839	7846
61	7853	7860	7868	7875	7882	7889	7896	7903	7910	7917
62	7924	7931	7938	7945	7952	7959	7966	7973	7980	7987
63	7993	8000	8007	8014	8021	8028	8035	8041	8048	8055
64	8062	8069	8075	8082	8089	8096	8102	8109	8116	8122
65	8129	8136	8142	8149	8156	8162	8169	8176	8182	8189
66	8195	8202	8209	8215	8222	8228	8235	8241	8248	8254
67	8261	8267	8274	8280	8287	8293	8299	8306	8312	8319
68	8325	8331	8338	8344	8351	8357	8363	8370	8376	8382
69	8388	8395	8401	8407	8414	8420	8426	8432	8439	8445
70	8451	8457	8463	8470	8476	8482	8488	8494	8500	8506
71	8513	8519	8525	8531	8537	8543	8549	8555	8561	8567
72	8573	8579	8585	8591	8597	8603	8609	8615	8621	8627
73	8633	8639	8645	8651	8657	8663	8669	8675	8681	8686
74	8692	8698	8704	8710	8716	8722	8727	8733	8739	8745
75	8751	8756	8762	8768	8774	8779	8785	8791	8797	8802
76	8808	8814	8820	8825	8831	8837	8842	8848	8854	8859
77	8865	8871	8876	8882	8887	8893	8899	8904	8910	8915
78	8921	8927	8932	8938	8943	8949	8954	8960	8965	8971
79	8976	8982	8987	8993	8998	9004	9009	9015	9020	9025
80	9031	9036	9042	9047	9053	9058	9063	9069	9074	9079
81	9085	9090	9096	9101	9106	9112	9117	9122	9128	9133
82	9138	9143	9149	9154	9159	9165	9170	9175	9180	9186
83	9191	9196	9201	9206	9212	9217	9222	9227	9232	9238
84	9243	9248	9253	9258	9263	9269	9274	9279	9284	9289
85	9294	9299	9304	9309	9315	9320	9325	9330	9335	9340
86	9345	9350	9355	9360	9365	9370	9375	9380	9385	9390
87	9395	9400	9405	9410	9415	9420	9425	9430	9435	9440
88	9445	9450	9455	9460	9465	9469	9474	9479	9484	9489
89	9494	9499	9504	9509	9513	9518	9523	9528	9533	9538
90	9542	9547	9552	9557	9562	9566	9571	9576	9581	9586
91	9590	9595	9600	9605	9609	9614	9619	9624	9628	9633
92	9638	9643	9647	9652	9657	9661	9666	9671	9675	9680
93	9685	9689	9694	9699	9703	9708	9713	9717	9722	9727
94	9731	9736	9741	9745	9750	9754	9759	9763	9768	9773
95	9777	9782	9786	9791	9795	9800	9805	9809	9814	9818
96	9823	9827	9832	9836	9841	9845	9850	9854	9859	9863
97	9868	9872	9877	9881	9886	9890	9894	9899	9903	9908
98	9912	9917	9921	9926	9930	9934	9939	9943	9948	9952
99	9956	9961	9965	9969	9974	9978	9983	9987	9991	9996

Table 6　Four-place Logarithms of Trigonometric Functions*

Angle	L Sin	d	L Tan	cd	L Cot	d	L Cos	Angle
0° 0′	——		——		——	0	10.0000	90° 0′
10′	7.4637	3011	7.4637	3011	12.5363	0	10.0000	50′
20′	7.7648	1760	7.7648	1761	12.2352	0	10.0000	40′
30′	7.9408	1250	7.9409	1249	12.0591	0	10.0000	30′
40′	8.0658	969	8.0658	969	11.9342	0	10.0000	20′
50′	8.1627	792	8.1627	792	11.8373	1	10.0000	10′
1° 0′	8.2419	669	8.2419	670	11.7581	0	9.9999	89° 0′
10′	8.3088	580	8.3089	580	11.6911	0	9.9999	50′
20′	8.3668	511	8.3669	512	11.6331	0	9.9999	40′
30′	8.4179	458	8.4181	457	11.5819	1	9.9999	30′
40′	8.4637	413	8.4638	415	11.5362	0	9.9998	20′
50′	8.5050	378	8.5053	378	11.4947	1	9.9998	10′
2° 0′	8.5428	348	8.5431	348	11.4569	0	9.9997	88° 0′
10′	8.5776	321	8.5779	322	11.4221	1	9.9997	50′
20′	8.6097	300	8.6101	300	11.3899	0	9.9996	40′
30′	8.6397	280	8.6401	281	11.3599	1	9.9996	30′
40′	8.6677	263	8.6682	263	11.3318	0	9.9995	20′
50′	8.6940	248	8.6945	249	11.3055	1	9.9995	10′
3° 0′	8.7188	235	8.7194	235	11.2806	1	9.9994	87° 0′
10′	8.7423	222	8.7429	223	11.2571	0	9.9993	50′
20′	8.7645	212	8.7652	213	11.2348	1	9.9993	40′
30′	8.7857	202	8.7865	202	11.2135	1	9.9992	30′
40′	8.8059	192	8.8067	194	11.1933	1	9.9991	20′
50′	8.8251	185	8.8261	185	11.1739	1	9.9990	10′
4° 0′	8.8436	177	8.8446	178	11.1554	1	9.9989	86° 0′
10′	8.8613	170	8.8624	171	11.1376	0	9.9989	50′
20′	8.8783	163	8.8795	165	11.1205	1	9.9988	40′
30′	8.8946	158	8.8960	158	11.1040	1	9.9987	30′
40′	8.9104	152	8.9118	154	11.0882	1	9.9986	20′
50′	8.9256	147	8.9272	148	11.0728	2	9.9985	10′
5° 0′	8.9403	142	8.9420	143	11.0580	1	9.9983	85° 0′
10′	8.9545	137	8.9563	138	11.0437	1	9.9982	50′
20′	8.9682	134	8.9701	135	11.0299	1	9.9981	40′
30′	8.9816	129	8.9836	130	11.0164	1	9.9980	30′
40′	8.9945	125	8.9966	127	11.0034	2	9.9979	20′
50′	9.0070	122	9.0093	123	10.9907	1	9.9977	10′
6° 0′	9.0192	119	9.0216	120	10.9784	1	9.9976	84° 0′
10′	9.0311	115	9.0336	117	10.9664	1	9.9975	50′
20′	9.0426	113	9.0453	114	10.9547	2	9.9973	40′
30′	9.0539	109	9.0567	111	10.9433	1	9.9972	30′
40′	9.0648	107	9.0678	108	10.9322	1	9.9971	20′
50′	9.0755	104	9.0786	105	10.9214	2	9.9969	10′
7° 0′	9.0859	102	9.0891	104	10.9109	1	9.9968	83° 0′
10′	9.0961	99	9.0995	101	10.9005	2	9.9966	50′
20′	9.1060	97	9.1096	98	10.8904	2	9.9964	40′
30′	9.1157	95	9.1194	97	10.8806	1	9.9963	30′
40′	9.1252	93	9.1291	94	10.8709	2	9.9961	20′
50′	9.1345	91	9.1385	93	10.8615	2	9.9959	10′
8° 0′	9.1436	89	9.1478	91	10.8522	1	9.9958	82° 0′
10′	9.1525	87	9.1569	89	10.8431	2	9.9956	50′
20′	9.1612	85	9.1658	87	10.8342	2	9.9954	40′
30′	9.1697	84	9.1745	86	10.8255	2	9.9952	30′
40′	9.1781	82	9.1831	84	10.8169	2	9.9950	20′
50′	9.1863	80	9.1915	82	10.8085	2	9.9948	10′
9° 0′	9.1943		9.1997		10.8003		9.9946	81° 0′
	L Cos	d	L Cot	cd	L Tan	d	L Sin	Angle

*These tables give the logarithms increased by 10. Hence in each case 10 should be subtracted.

Table 6 Four-place Logarithms of Trigonometric Functions

Angle	L Sin	d	L Tan	cd	L Cot	d	L Cos	
9° 0′	9.1943	79	9.1997	81	10.8003	2	9.9946	81° 0′
10′	9.2022	78	9.2078	80	10.7922	2	9.9944	50′
20′	9.2100	76	9.2158	78	10.7842	2	9.9942	40′
30′	9.2176	75	9.2236	77	10.7764	2	9.9940	30′
40′	9.2251	73	9.2313	76	10.7687	2	9.9938	20′
50′	9.2324	73	9.2389	74	10.7611	2	9.9936	10′
10° 0′	9.2397	71	9.2463	73	10.7537	3	9.9934	80° 0′
10′	9.2468	70	9.2536	73	10.7464	2	9.9931	50′
20′	9.2538	68	9.2609	71	10.7391	2	9.9929	40′
30′	9.2606	68	9.2680	70	10.7320	3	9.9927	30′
40′	9.2674	66	9.2750	69	10.7250	2	9.9924	20′
50′	9.2740	66	9.2819	68	10.7181	3	9.9922	10′
11° 0′	9.2806	64	9.2887	66	10.7113	2	9.9919	79° 0′
10′	9.2870	64	9.2953	67	10.7047	3	9.9917	50′
20′	9.2934	63	9.3020	65	10.6980	2	9.9914	40′
30′	9.2997	61	9.3085	64	10.6915	3	9.9912	30′
40′	9.3058	61	9.3149	63	10.6851	2	9.9909	20′
50′	9.3119	60	9.3212	63	10.6788	3	9.9907	10′
12° 0′	9.3179	59	9.3275	61	10.6725	3	9.9904	78° 0′
10′	9.3238	58	9.3336	61	10.6664	2	9.9901	50′
20′	9.3296	57	9.3397	61	10.6603	3	9.9899	40′
30′	9.3353	57	9.3458	59	10.6542	3	9.9896	30′
40′	9.3410	56	9.3517	59	10.6483	3	9.9893	20′
50′	9.3466	55	9.3576	58	10.6424	3	9.9890	10′
13° 0′	9.3521	54	9.3634	57	10.6366	3	9.9887	77° 0′
10′	9.3575	54	9.3691	57	10.6309	3	9.9884	50′
20′	9.3629	53	9.3748	56	10.6252	3	9.9881	40′
30′	9.3682	52	9.3804	55	10.6196	3	9.9878	30′
40′	9.3734	52	9.3859	55	10.6141	3	9.9875	20′
50′	9.3786	51	9.3914	54	10.6086	3	9.9872	10′
14° 0′	9.3837	50	9.3968	53	10.6032	3	9.9869	76° 0′
10′	9.3887	50	9.4021	53	10.5979	3	9.9866	50′
20′	9.3937	49	9.4074	53	10.5926	4	9.9863	40′
30′	9.3986	49	9.4127	51	10.5873	3	9.9859	30′
40′	9.4035	48	9.4178	52	10.5822	3	9.9856	20′
50′	9.4083	47	9.4230	51	10.5770	4	9.9853	10′
15° 0′	9.4130	47	9.4281	50	10.5719	3	9.9849	75° 0′
10′	9.4177	46	9.4331	50	10.5669	3	9.9846	50′
20′	9.4223	46	9.4381	49	10.5619	4	9.9843	40′
30′	9.4269	45	9.4430	49	10.5570	3	9.9839	30′
40′	9.4314	45	9.4479	48	10.5521	4	9.9836	20′
50′	9.4359	44	9.4527	48	10.5473	4	9.9832	10′
16° 0′	9.4403	44	9.4575	47	10.5425	3	9.9828	74° 0′
10′	9.4447	44	9.4622	47	10.5378	4	9.9825	50′
20′	9.4491	42	9.4669	47	10.5331	4	9.9821	40′
30′	9.4533	43	9.4716	46	10.5284	3	9.9817	30′
40′	9.4576	42	9.4762	46	10.5238	4	9.9814	20′
50′	9.4618	41	9.4808	45	10.5192	4	9.9810	10′
17° 0′	9.4659	41	9.4853	45	10.5147	4	9.9806	73° 0′
10′	9.4700	41	9.4898	45	10.5102	4	9.9802	50′
20′	9.4741	40	9.4943	44	10.5057	4	9.9798	40′
30′	9.4781	40	9.4987	44	10.5013	4	9.9794	30′
40′	9.4821	40	9.5031	44	10.4969	4	9.9790	20′
50′	9.4861	39	9.5075	43	10.4925	4	9.9786	10′
18° 0′	9.4900		9.5118		10.4882		9.9782	72° 0′
	L Cos	d	L Cot	cd	L Tan	d	L Sin	Angle

Table 6   Four-place Logarithms of Trigonometric Functions

Angle	L Sin	d	L Tan	cd	L Cot	d	L Cos	Angle
18° 0'	9.4900		9.5118		10.4882	4	9.9782	72° 0'
10'	9.4939	39	9.5161	43	10.4839	4	9.9778	50'
20'	9.4977	38	9.5203	42	10.4797	4	9.9774	40'
30'	9.5015	38	9.5245	42	10.4755	4	9.9770	30'
40'	9.5052	37	9.5287	42	10.4713	5	9.9765	20'
50'	9.5090	38	9.5329	42	10.4671	4	9.9761	10'
		36		41		4		
19° 0'	9.5126		9.5370		10.4630		9.9757	71° 0'
10'	9.5163	37	9.5411	41	10.4589	5	9.9752	50'
20'	9.5199	36	9.5451	40	10.4549	4	9.9748	40'
30'	9.5235	36	9.5491	40	10.4509	5	9.9743	30'
40'	9.5270	35	9.5531	40	10.4469	4	9.9739	20'
50'	9.5306	36	9.5571	40	10.4429	5	9.9734	10'
		35		40		4		
20° 0'	9.5341		9.5611		10.4389		9.9730	70° 0'
10'	9.5375	34	9.5650	39	10.4350	5	9.9725	50'
20'	9.5409	34	9.5689	39	10.4311	4	9.9721	40'
30'	9.5443	34	9.5727	38	10.4273	5	9.9716	30'
40'	9.5477	34	9.5766	39	10.4234	5	9.9711	20'
50'	9.5510	33	9.5804	38	10.4196	5	9.9706	10'
		33		38		4		
21° 0'	9.5543		9.5842		10.4158		9.9702	69° 0'
10'	9.5576	33	9.5879	37	10.4121	5	9.9697	50'
20'	9.5609	33	9.5917	38	10.4083	5	9.9692	40'
30'	9.5641	32	9.5954	37	10.4046	5	9.9687	30'
40'	9.5673	32	9.5991	37	10.4009	5	9.9682	20'
50'	9.5704	31	9.6028	37	10.3972	5	9.9677	10'
		32		36		5		
22° 0'	9.5736		9.6064		10.3936		9.9672	68° 0'
10'	9.5767	31	9.6100	36	10.3900	5	9.9667	50'
20'	9.5798	31	9.6136	36	10.3864	6	9.9661	40'
30'	9.5828	30	9.6172	36	10.3828	5	9.9656	30'
40'	9.5859	31	9.6208	36	10.3792	5	9.9651	20'
50'	9.5889	30	9.6243	35	10.3757	5	9.9646	10'
		30		36		6		
23° 0'	9.5919		9.6279		10.3721		9.9640	67° 0'
10'	9.5948	29	9.6314	35	10.3686	5	9.9635	50'
20'	9.5978	30	9.6348	34	10.3652	6	9.9629	40'
30'	9.6007	29	9.6383	35	10.3617	5	9.9624	30'
40'	9.6036	29	9.6417	34	10.3583	6	9.9618	20'
50'	9.6065	29	9.6452	35	10.3548	5	9.9613	10'
		28		34		6		
24° 0'	9.6093		9.6486		10.3514		9.9607	66° 0'
10'	9.6121	28	9.6520	34	10.3480	5	9.9602	50'
20'	9.6149	28	9.6553	33	10.3447	6	9.9596	40'
30'	9.6177	28	9.6587	34	10.3413	6	9.9590	30'
40'	9.6205	28	9.6620	33	10.3380	6	9.9584	20'
50'	9.6232	27	9.6654	34	10.3346	5	9.9579	10'
		27		33		6		
25° 0'	9.6259		9.6687		10.3313		9.9573	65° 0'
10'	9.6286	27	9.6720	33	10.3280	6	9.9567	50'
20'	9.6313	27	9.6752	32	10.3248	6	9.9561	40'
30'	9.6340	27	9.6785	33	10.3215	6	9.9555	30'
40'	9.6366	26	9.6817	32	10.3183	6	9.9549	20'
50'	9.6392	26	9.6850	33	10.3150	6	9.9543	10'
		26		32		6		
26° 0'	9.6418		9.6882		10.3118		9.9537	64° 0'
10'	9.6444	26	9.6914	32	10.3086	7	9.9530	50'
20'	9.6470	26	9.6946	32	10.3054	6	9.9524	40'
30'	9.6495	25	9.6977	31	10.3023	6	9.9518	30'
40'	9.6521	26	9.7009	32	10.2991	6	9.9512	20'
50'	9.6546	25	9.7040	31	10.2960	7	9.9505	10'
		24		32		6		
27° 0'	9.6570		9.7072		10.2928		9.9499	63° 0'
	L Cos	d	L Cot	cd	L Tan	d	L Sin	Angle

Table 6 Four-place Logarithms of Trigonometric Functions

Angle	L Sin	d	L Tan	cd	L Cot	d	L Cos	
27° 0′	9.6570		9.7072		10.2928		9.9499	63° 0′
10′	9.6595	25	9.7103	31	10.2897	7	9.9492	50′
20′	9.6620	25	9.7134	31	10.2866	6	9.9486	40′
30′	9.6644	24	9.7165	31	10.2835	7	9.9479	30′
40′	9.6668	24	9.7196	31	10.2804	6	9.9473	20′
50′	9.6692	24	9.7226	30	10.2774	7	9.9466	10′
28° 0′	9.6716	24	9.7257	31	10.2743	7	9.9459	62° 0′
10′	9.6740	24	9.7287	30	10.2713	6	9.9453	50′
20′	9.6763	23	9.7317	30	10.2683	7	9.9446	40′
30′	9.6787	24	9.7348	31	10.2652	7	9.9439	30′
40′	9.6810	23	9.7378	30	10.2622	7	9.9432	20′
50′	9.6833	23	9.7408	30	10.2592	7	9.9425	10′
29° 0′	9.6856	23	9.7438	30	10.2562	7	9.9418	61° 0′
10′	9.6878	22	9.7467	29	10.2533	7	9.9411	50′
20′	9.6901	23	9.7497	30	10.2503	7	9.9404	40′
30′	9.6923	22	9.7526	29	10.2474	7	9.9397	30′
40′	9.6946	23	9.7556	30	10.2444	7	9.9390	20′
50′	9.6968	22	9.7585	29	10.2415	7	9.9383	10′
30° 0′	9.6990	22	9.7614	29	10.2386	8	9.9375	60° 0′
10′	9.7012	22	9.7644	30	10.2356	7	9.9368	50′
20′	9.7033	21	9.7673	29	10.2327	7	9.9361	40′
30′	9.7055	22	9.7701	28	10.2299	8	9.9353	30′
40′	9.7076	21	9.7730	29	10.2270	7	9.9346	20′
50′	9.7097	21	9.7759	29	10.2241	8	9.9338	10′
31° 0′	9.7118	21	9.7788	29	10.2212	7	9.9331	59° 0′
10′	9.7139	21	9.7816	28	10.2184	8	9.9323	50′
20′	9.7160	21	9.7845	29	10.2155	8	9.9315	40′
30′	9.7181	21	9.7873	28	10.2127	8	9.9308	30′
40′	9.7201	20	9.7902	29	10.2098	7	9.9300	20′
50′	9.7222	21	9.7930	28	10.2070	8	9.9292	10′
32° 0′	9.7242	20	9.7958	28	10.2042	8	9.9284	58° 0′
10′	9.7262	20	9.7986	28	10.2014	8	9.9276	50′
20′	9.7282	20	9.8014	28	10.1986	8	9.9268	40′
30′	9.7302	20	9.8042	28	10.1958	8	9.9260	30′
40′	9.7322	20	9.8070	28	10.1930	8	9.9252	20′
50′	9.7342	20	9.8097	27	10.1903	8	9.9244	10′
33° 0′	9.7361	19	9.8125	28	10.1875	8	9.9236	57° 0′
10′	9.7380	19	9.8153	28	10.1847	8	9.9228	50′
20′	9.7400	20	9.8180	27	10.1820	9	9.9219	40′
30′	9.7419	19	9.8208	28	10.1792	8	9.9211	30′
40′	9.7438	19	9.8235	27	10.1765	8	9.9203	20′
50′	9.7457	19	9.8263	28	10.1737	9	9.9194	10′
34° 0′	9.7476	19	9.8290	27	10.1710	8	9.9186	56° 0′
10′	9.7494	18	9.8317	27	10.1683	9	9.9177	50′
20′	9.7513	19	9.8344	27	10.1656	8	9.9169	40′
30′	9.7531	18	9.8371	27	10.1629	9	9.9160	30′
40′	9.7550	19	9.8398	27	10.1602	9	9.9151	20′
50′	9.7568	18	9.8425	27	10.1575	9	9.9142	10′
35° 0′	9.7586	18	9.8452	27	10.1548	8	9.9134	55° 0′
10′	9.7604	18	9.8479	27	10.1521	9	9.9125	50′
20′	9.7622	18	9.8506	27	10.1494	9	9.9116	40′
30′	9.7640	18	9.8533	27	10.1467	9	9.9107	30′
40′	9.7657	17	9.8559	26	10.1441	9	9.9098	20′
50′	9.7675	18	9.8586	27	10.1414	9	9.9089	10′
36° 0′	9.7692	17	9.8613	27	10.1387	9	9.9080	54° 0′
	L Cos	d	L Cot	cd	L Tan	d	L Sin	Angle

Table 6 Four-place Logarithms of Trigonometric Functions

Angle	L Sin	d	L Tan	cd	L Cot	d	L Cos	
36° 0'	9.7692		9.8613		10.1387		9.9080	**54° 0'**
10'	9.7710	18	9.8639	26	10.1361	10	9.9070	50'
20'	9.7727	17	9.8666	27	10.1334	9	9.9061	40'
30'	9.7744	17	9.8692	26	10.1308	9	9.9052	30'
40'	9.7761	17	9.8718	26	10.1282	10	9.9042	20'
50'	9.7778	17	9.8745	27	10.1255	9	9.9033	10'
37° 0'	9.7795	17	9.8771	26	10.1229	10	9.9023	**53° 0'**
10'	9.7811	16	9.8797	26	10.1203	9	9.9014	50'
20'	9.7828	17	9.8824	27	10.1176	10	9.9004	40'
30'	9.7844	16	9.8850	26	10.1150	9	9.8995	30'
40'	9.7861	17	9.8876	26	10.1124	10	9.8985	20'
50'	9.7877	16	9.8902	26	10.1098	10	9.8975	10'
38° 0'	9.7893	16	9.8928	26	10.1072	10	9.8965	**52° 0'**
10'	9.7910	17	9.8954	26	10.1046	10	9.8955	50'
20'	9.7926	16	9.8980	26	10.1020	10	9.8945	40'
30'	9.7941	15	9.9006	26	10.0994	10	9.8935	30'
40'	9.7957	16	9.9032	26	10.0968	10	9.8925	20'
50'	9.7973	16	9.9058	26	10.0942	10	9.8915	10'
39° 0'	9.7989	16	9.9084	26	10.0916	10	9.8905	**51° 0'**
10'	9.8004	15	9.9110	25	10.0890	10	9.8895	50'
20'	9.8020	16	9.9135	26	10.0865	11	9.8884	40'
30'	9.8035	15	9.9161	26	10.0839	10	9.8874	30'
40'	9.8050	15	9.9187	25	10.0813	10	9.8864	20'
50'	9.8066	16	9.9212	26	10.0788	11	9.8853	10'
40° 0'	9.8081	15	9.9238	26	10.0762	10	9.8843	**50° 0'**
10'	9.8096	15	9.9264	26	10.0736	11	9.8832	50'
20'	9.8111	15	9.9289	25	10.0711	11	9.8821	40'
30'	9.8125	14	9.9315	26	10.0685	11	9.8810	30'
40'	9.8140	15	9.9341	26	10.0659	10	9.8800	20'
50'	9.8155	15	9.9366	25	10.0634	11	9.8789	10'
41° 0'	9.8169	14	9.9392	26	10.0608	11	9.8778	**49° 0'**
10'	9.8184	15	9.9417	25	10.0583	11	9.8767	50'
20'	9.8198	14	9.9443	26	10.0557	11	9.8756	40'
30'	9.8213	15	9.9468	25	10.0532	11	9.8745	30'
40'	9.8227	14	9.9494	26	10.0506	12	9.8733	20'
50'	9.8241	14	9.9519	25	10.0481	11	9.8722	10'
42° 0'	9.8255	14	9.9544	25	10.0456	11	9.8711	**48° 0'**
10'	9.8269	14	9.9570	26	10.0430	12	9.8699	50'
20'	9.8283	14	9.9595	25	10.0405	11	9.8688	40'
30'	9.8297	14	9.9621	26	10.0379	12	9.8676	30'
40'	9.8311	14	9.9646	25	10.0354	11	9.8665	20'
50'	9.8324	13	9.9671	25	10.0329	12	9.8653	10'
43° 0'	9.8338	14	9.9697	26	10.0303	12	9.8641	**47° 0'**
10'	9.8351	13	9.9722	25	10.0278	12	9.8629	50'
20'	9.8365	14	9.9747	25	10.0253	11	9.8618	40'
30'	9.8378	13	9.9772	25	10.0228	12	9.8606	30'
40'	9.8391	13	9.9798	26	10.0202	12	9.8594	20'
50'	9.8405	14	9.9823	25	10.0177	12	9.8582	10'
44° 0'	9.8418	13	9.9848	25	10.0152	13	9.8569	**46° 0'**
10'	9.8431	13	9.9874	26	10.0126	12	9.8557	50'
20'	9.8444	13	9.9899	25	10.0101	12	9.8545	40'
30'	9.8457	13	9.9924	25	10.0076	13	9.8532	30'
40'	9.8469	12	9.9949	25	10.0051	12	9.8520	20'
50'	9.8482	13	9.9975	26	10.0025	13	9.8507	10'
45° 0'	9.8495	13	10.0000	25	10.0000	12	9.8495	**45° 0'**
	L Cos	d	L Cot	cd	L Tan	d	L Sin	Angle

SUMMARY OF FORMULAS

Uniform Circular Motion

Let P be a point moving counterclockwise in uniform circular motion around a point O. The **linear speed** v of P as P moves along an arc s in one unit of time t is given by

$$v = \frac{s}{t} \quad \text{or} \quad s = vt.$$ **45**

The **angular speed** ω of P with respect to time t at which the angle θ generated by $\overline{OP}$ changes is given by

$$\omega = \frac{\theta}{t} \quad \text{or} \quad \theta = \omega t.$$ **45**

If ω represents the angular speed of P in radians per unit of time, the linear speed v of P is given by

$$v = r\omega,$$

where r is the radius of P's circular path. **45**

Basic Trigonometric Identities **94**

Solution of Oblique Triangles

Law of Cosines: $c^2 = a^2 + b^2 - 2ab \cos C$
$$b^2 = a^2 + c^2 - 2ac \cos B$$
$$a^2 = b^2 + c^2 - 2bc \cos A$$ **101**

Law of Sines: $\dfrac{\sin A}{a} = \dfrac{\sin B}{b} = \dfrac{\sin C}{c}$ **107**

Area Formulas for Triangle ABC

$\text{Area} = \dfrac{1}{2} ab \sin C = \dfrac{1}{2} bc \sin A = \dfrac{1}{2} ac \sin B$ **112–113**

$\text{Area} = \sqrt{s(s-a)(s-b)(s-c)} \text{ where } s = \dfrac{a+b+c}{2}$

(Hero's Formula) **114**

$\text{Area} = \dfrac{1}{2} a^2 \dfrac{\sin B \sin C}{\sin A} = \dfrac{1}{2} b^2 \dfrac{\sin A \sin C}{\sin B} = \dfrac{1}{2} c^2 \dfrac{\sin A \sin B}{\sin C}$ **114**

Sinusoidal Variation

The equation of any **sinusoidal variation** can be written in the form

$$y = a \sin b(x - c) + d \quad \text{or} \quad y = a \cos b(x - c) + d,$$

where a, b, c, and d are constants.   **124**

The position y of an object in **simple harmonic motion** at the time t can be given by an equation of the form

$$y = a \sin (\omega t + \beta) \quad \text{or} \quad y = a \cos (\omega t + \beta),$$

where $\omega > 0$ and a and β are constants.   **133**

Vectors

If $\mathbf{a} = a_1 \mathbf{i} + a_2 \mathbf{j}$ then $\|\mathbf{a}\| = \sqrt{a_1^2 + a_2^2}$. (**norm** of **a**)   **151**

$\mathbf{a} \cdot \mathbf{b} = \|\mathbf{a}\| \|\mathbf{b}\| \cos \theta$, where θ is the angle between $\mathbf{a}$ and $\mathbf{b}$.
  (**dot product** or **inner product**)   **161**

If $\mathbf{a} = a_1 \mathbf{i} + a_2 \mathbf{j}$ and $\mathbf{b} = b_1 \mathbf{i} + b_2 \mathbf{j}$ then $\mathbf{a} \cdot \mathbf{b} = a_1 b_1 + a_2 b_2$.   **162**

If θ is the angle between two nonzero vectors $\mathbf{a}$ and $\mathbf{b}$ then

$$\cos \theta = \frac{\mathbf{a} \cdot \mathbf{b}}{\|\mathbf{a}\| \|\mathbf{b}\|} \quad \text{and} \quad \theta = \text{Cos}^{-1} \frac{\mathbf{a} \cdot \mathbf{b}}{\|\mathbf{a}\| \|\mathbf{b}\|}.$$ **162**

$\mathbf{v_a} = \dfrac{\mathbf{a} \cdot \mathbf{v}}{\mathbf{a} \cdot \mathbf{a}} \mathbf{a}$ (**vector projection** of $\mathbf{v}$ onto $\mathbf{a}$)   **164**

$\|\mathbf{v_a}\| = \dfrac{\mathbf{a} \cdot \mathbf{v}}{\|\mathbf{a}\|}$ (**scalar projection** of $\mathbf{v}$ onto $\mathbf{a}$)   **164**

The work W done by a force $\mathbf{F}$ moving an object from A to B is given by

$$W = \|\mathbf{F}\| \|\mathbf{d}\|$$

when $\mathbf{F}$ has the same direction as the displacement vector $\mathbf{d} = \overrightarrow{AB}$ and by

$$W = \mathbf{F} \cdot \mathbf{d}$$

when $\mathbf{F}$ does not have the same direction as $\mathbf{d} = \overrightarrow{AB}$.   **166–167**

Polar Coordinates

Conversion from polar to rectangular coordinates:

$$x = r \cos \theta \qquad y = r \sin \theta$$ **176**

Conversion from rectangular to polar coordinates:

$$r = \pm\sqrt{x^2 + y^2} \qquad \cos\theta = \frac{x}{r}, \; \sin\theta = \frac{y}{r}$$

176

Complex Numbers

For $w = u + vi$ and $z = x + yi$:

$w + z = (u + x) + (v + y)i$ (**sum** of complex numbers)   **189**

$wz = (ux - vy) + (uy + vx)i$ (**product** of complex numbers)   **189**

$\bar{z} = x - yi$ (**conjugate** of a complex number)   **189**

$z\bar{z} = x^2 + y^2$ **189**

$|z| = \sqrt{x^2 + y^2}$ (**modulus,** or **absolute value**)   **189**

$|w + z| \le |w| + |z|$ (**triangle inequality**)   **192**

If $z_1 = r_1(\cos\theta_1 + i\sin\theta_1)$ and $z_2 = r_2(\cos\theta_2 + i\sin\theta_2)$ then

$z_1 z_2 = r_1 r_2 [\cos(\theta_1 + \theta_2) + i\sin(\theta_1 + \theta_2)]$ and   **195**

$$\frac{z_1}{z_2} = \frac{r_1}{r_2}[\cos(\theta_1 - \theta_2) + i\sin(\theta_1 - \theta_2)].$$

196

$[r(\cos\theta + i\sin\theta)]^n = r^n(\cos n\theta + i\sin n\theta)$, for every integer n.
    (**De Moivre's theorem**)   **198**

The n nth roots of $r(\cos\theta + i\sin\theta)$ are given by

$$r^{\frac{1}{n}}\left(\cos\frac{\theta + k \cdot 360°}{n} + i\sin\frac{\theta + k \cdot 360°}{n}\right),$$

where $k = 0, 1, 2, \ldots, n - 1$.   **202**

Transformations

Under the **translation** that moves the origin to the point having old coordinates (h, k), the equations that relate the new coordinates (x', y') of a point and its old coordinates (x, y) are:

$$\begin{array}{lll} x = x' + h & y = y' + k & \text{or} \\ x' = x - h & y' = y - k. \end{array}$$

214

Under the **rotation** that moves the coordinate axes through an angle ϕ about the origin, the equations that relate the old (x, y)-coordinates of a point to the new (x', y')-coordinates are:

$$\begin{array}{lll} x = x'\cos\phi - y'\sin\phi & y = x'\sin\phi + y'\cos\phi & \text{or} \\ x' = x\cos\phi - y\sin\phi & y' = -x\sin\phi + y\cos\phi \end{array}$$

217

Vectors in Space

The distance between the points $P_1(x_1, y_1, z_1)$ and $P_2(x_2, y_2, z_2)$ is given by:

$$P_1P_2 = \sqrt{(x_2 - x_1)^2 + (y_2 - y_1)^2 + (z_2 - z_1)^2}$$ **234**

If $\mathbf{a} = a_1\mathbf{i} + a_2\mathbf{j} + a_3\mathbf{k}$ then $\|\mathbf{a}\| = \sqrt{a_1^2 + a_2^2 + a_3^2}$. (**norm** of **a**)   **237**

$\mathbf{a} \cdot \mathbf{b} = \|\mathbf{a}\|\|\mathbf{b}\| \cos \theta$, where θ is the angle between **a** and **b**
  (**dot product** or **inner product**)   **238**

If $\mathbf{a} = a_1\mathbf{i} + a_2\mathbf{j} + a_3\mathbf{k}$ and $\mathbf{b} = b_1\mathbf{i} + b_2\mathbf{j} + b_3\mathbf{k}$ then:

$$\mathbf{a} \cdot \mathbf{b} = a_1b_1 + a_2b_2 + a_3b_3$$ **238**

A vector equation of the line that passes through a point P_0 and is parallel to a given vector **m** is

$$\mathbf{r} = \mathbf{r}_0 + t\mathbf{m} \qquad (\mathbf{r} = \overrightarrow{OP}, \mathbf{r}_0 = \overrightarrow{OP_0}).$$ **242**

A vector equation of the line determined by two points P_0 and P_1 with position vectors $\mathbf{r}_0$ and $\mathbf{r}_1$, respectively, is

$$\mathbf{r} = (1 - t)\mathbf{r}_0 + t\mathbf{r}_1.$$ **243**

A vector equation of the plane Q through P with normal vector **n** is

$$\mathbf{n} \cdot (\mathbf{r} - \mathbf{r}_0) = 0,$$

where $\mathbf{r}_0 = \overrightarrow{OP_0}$ and $\mathbf{r} = \overrightarrow{OP}$. Letting $\mathbf{n} = a\mathbf{i} + b\mathbf{j} + c\mathbf{k}$, $\mathbf{r}_0 = x_0\mathbf{i} + y_0\mathbf{j} + z_0\mathbf{k}$, and $\mathbf{r} = x\mathbf{i} + y\mathbf{j} + z\mathbf{k}$, a scalar equation of Q is

$$a(x - x_0) + b(y - y_0) + c(z - z_0) = 0.$$ **247**

The distance from the point $P_1(x_1, y_1, z_1)$ to the plane $ax + by + cz + d = 0$ is given by

$$D = \frac{|ax_1 + by_1 + cz_1 + d|}{\sqrt{a^2 + b^2 + c^2}}.$$ **249**

Spherical Trigonometry

Formulas for solving right spherical triangles:

(1) $\cos c = \cos a \cos b$

(2) $\sin a = \sin c \sin A$

(3) $\sin b = \sin c \sin B$

(4) $\tan a = \sin b \tan A$

(5) $\tan b = \sin a \tan B$

(6) $\tan b = \tan c \cos A$

(7) $\tan a = \tan c \cos B$

(8) $\cos A = \cos a \sin B$

(9) $\cos B = \cos b \sin A$

(10) $\cos c = \cot A \cot B$ **264**

Solution of Oblique Spherical Triangles

Law of Sines: $\dfrac{\sin a}{\sin A} = \dfrac{\sin b}{\sin B} = \dfrac{\sin c}{\sin C}$ **271**

Law of Cosines for Sides: $\cos a = \cos b \cos c + \sin b \sin c \cos A$ with analogous formulas for $\cos b$ and $\cos c$ **271**

Law of Cosines for Angles: $\cos A = -\cos B \cos C + \sin B \sin C \cos a$ with analogous formulas for $\cos B$ and $\cos C$ **271**

Infinite Series

$\sin x = x - \dfrac{x^3}{3!} + \dfrac{x^5}{5!} - \dfrac{x^7}{7!} + \cdots$ **(Taylor series** for $\sin x$)   **286**

$\cos x = 1 - \dfrac{x^2}{2!} + \dfrac{x^4}{4!} - \dfrac{x^6}{6!} + \cdots$ **(Taylor series** for $\cos x$)   **286**

$e^x = 1 + x + \dfrac{x^2}{2!} + \dfrac{x^3}{3!} + \dfrac{x^4}{4!} + \cdots$ **(exponential function)**   **287**

$\cosh x = \dfrac{1}{2}(e^x + e^{-x}) = 1 + \dfrac{x^2}{2!} + \dfrac{x^4}{4!} + \dfrac{x^6}{6!} + \cdots$

(hyperbolic cosine)   **289**

$\sinh x = \dfrac{1}{2}(e^x - e^{-x}) = x + \dfrac{x^3}{3!} + \dfrac{x^5}{5!} + \dfrac{x^7}{7!} + \cdots$

(hyperbolic sine)   **289**

$e^x = \cosh x + \sinh x$ **290**

$e^{ix} = \cos x + i \sin x$ **290**

Mixed Review: Chapters 1–3

1. Find the exact values of the six trigonometric functions of 60°.

2. Draw a detailed graph of $y = 2 \sin \left(x + \frac{\pi}{3} \right)$, $0 \le x \le 2\pi$.

3. Use Table 3, pages 365–368, to find $\cos 6.03$ to two significant digits.

4. If $\csc \theta = -\frac{13}{12}$ and $90° < \theta < 270°$, find $\tan \theta$.

5. Prove: $\csc 2x + \cot 2x = \cot x$.

6. Rewrite $-8 \cos \theta + 6 \sin \theta$ in the form $C \cos (\theta - \phi)$ for $-180° \le \phi < 180°$. Find ϕ to the nearest degree.

7. Give the degree measure of an angle between 0° and 360° that is coterminal with $-412°39'$.

8. Find the radius of a circle in which a central angle of 144° intercepts an arc of length 8π cm.

9. Give the exact values of $\sin 285°$ and $\tan 285°$.

10. Give the general solution: $2 \cos 2x = \sin 4x$.

11. Find the coordinates of the final position of a point P moving counter-clockwise in uniform circular motion around a circle of radius 4 centered at the origin at $\omega = \frac{3\pi}{4}$ radians/s if P starts at the point $(4, 0)$ and moves for 0.75 min.

12. The base of an isosceles triangle is 20.8 cm long and each base angle measures 41°. Find the length of the altitude to the base.

13. Is $\text{Tan}^{-1}(-x) = -\text{Tan}^{-1} x$ for all x in the domain of Tan^{-1}? Explain.

14. If $\sin \theta = \frac{2}{5}$, find the value of θ between 0° and 90°. Round your answer to the nearest minute.

15. Express in terms of a single trigonometric function:
$$\frac{\sec \theta + 1}{\tan \theta} + \frac{\tan \theta}{\sec \theta + 1}.$$

16. Evaluate: **(a)** $\text{Sin}^{-1} \left(\tan \frac{7\pi}{4} \right)$ **(b)** $\cos \left(\text{Tan}^{-1} \frac{3}{4} \right)$

17. Solve $\triangle ABC$ in which $\angle C = 90°$, $a = 12.0$, and $c = 37.0$.

18. Find the exact values of the six trigonometric functions for an angle θ, $0 < \theta \le 360°$, in standard position whose terminal side passes through $(-6, -9)$. Give the measure of θ to the nearest tenth of a degree.

19. Prove the identity $\dfrac{\cos^3 x}{(\cot x - \cos x)(1 + \sin x)} = \sin x.$

20. Express $5°20'42''$ in decimal degrees.

21. Express $\sin \theta(\sin \theta - \csc \theta)$ in terms of $\cos \theta$.

22. A point P travels uniformly at an angular speed of $\omega = \dfrac{5\pi}{3}$ radians/s in a counterclockwise direction around a circular path of radius 2 with center at the origin. Find the coordinates of P and the distance that P has traveled 6 s after P moves from $(\sqrt{3}, -1)$.

23. Use the formula for $\sin(\alpha + \beta)$ to find the exact value of $\sin 75°$. (Use $\alpha = 45°$ and $\beta = 30°$.)

24. Solve for $0 \le x < 2\pi$: $\cos 3x \cos x + \sin 3x \sin x = -\dfrac{\sqrt{2}}{2}.$

25. Convert $\dfrac{5\pi^R}{9}$ to degree measure.

26. Prove: $\dfrac{\cot \alpha - \tan \alpha}{\sin \alpha \cos \alpha} = \csc^2 \alpha - \sec^2 \alpha.$

If $\cos \theta = -\dfrac{3}{4}$ and $180° \le \theta < 360°$, evaluate each expression.

27. $\cos 2\theta$ 28. $\tan 2\theta$ 29. $\sin \dfrac{\theta}{2}$ 30. $\tan \dfrac{\theta}{2}$

31. Find x to three significant digits.

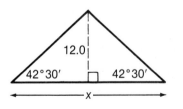

32. Find θ to the nearest $10'$.

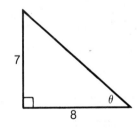

33. Draw the graph of $y = \tan x$, $0 \le x \le 2\pi$ where $x \ne \dfrac{\pi}{2}$ or $\dfrac{3\pi}{2}$.

34. Solve $5 \cos \theta + 12 \sin \theta = 1$ for $0° \le \theta < 360°$ to the nearest degree.

35. Prove: $\left(\sin \dfrac{1}{2}\alpha + \cos \dfrac{1}{2}\alpha\right)^2 = \sin \alpha + 1.$

36. Prove: $\cot\left(\dfrac{\pi}{2} - x\right) = \tan x.$

Mixed Review: Chapters 1–7

1. Evaluate: (a) $\text{Cos}^{-1}\left(\dfrac{\sqrt{3}}{2}\right)$ (b) $\text{Tan}^{-1}\left(\sin\left(-\dfrac{3\pi}{4}\right)\right)$

2. Give the domain and range of $y = \text{Cos}^{-1} 2x$ and graph the function.

3. An object suspended from a spring is pulled 4 cm below its equilibrium point at time $t = 0$ and released. If the object completes an oscillation every 3 s, give an equation in terms of the cosine function for the position of the object. Also give the frequency of the oscillation.

4. How much energy in kilowatt-hours is needed to lift a 1500 kg girder 50 m?

5. Express $(-\sqrt{2} - i\sqrt{2})^8$ in the form $x + iy$.

6. $\triangle XYZ$ has area 21, $\angle Y = 150°$, and $x = 7$. Find z.

7. Find the degree-minute-second measure of the supplement of an angle of $12°5'14''$.

8. Two forces act on a particle P: $\mathbf{F}_1$ of 8 N with a bearing of 210° and $\mathbf{F}_2$ of 6 N with a bearing of 140°. Find the magnitude to the nearest tenth and the bearing to the nearest tenth of a degree of the addition force $\mathbf{F}$ that will keep P stationary.

9. A triangle has sides of lengths 4, 7, and 10. Find the length of the median to the shortest side.

10. Find to the nearest cm/s the linear speed of a point on the rim of a record of radius 15 cm turning at an angular speed of $33\dfrac{1}{3}$ rpm.

11. Prove a distributive property for dot products: $(\mathbf{b} + \mathbf{c}) \cdot \mathbf{a} = \mathbf{b} \cdot \mathbf{a} + \mathbf{c} \cdot \mathbf{a}$.

12. Show that for all real numbers if $f(x) = \sin x - \cos x$, then $f(x + \pi) = -f(x)$.

13. Convert 184.5° to radian measure in terms of π.

14. Solve $\triangle RST$ if $\angle R = 57°40'$, $\angle S = 38°30'$, and $t = 50$. Give lengths to the nearest tenth.

15. Find the value of x in the diagram to the nearest tenth.

16. State the domain of the function $\text{Tan}^{-1}(\tan)$. Give a counterexample to show that $\text{Tan}^{-1}(\tan x) = x$ is not true for all x in the domain.

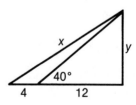

17. Use a half-angle formula to evaluate $\cos 165°$.

18. (Optional) Determine what type of conic has the polar equation $r = \dfrac{2}{1 + \sin \theta}$. Sketch the graph of the conic.

19. Sketch the graph of $r = 6 \sin \theta - 8 \cos \theta$.

20. Find unit vectors (a) parallel to and (b) perpendicular to the vector $-3\mathbf{i} + \mathbf{j}$.

21. Find (a) the conjugate, (b) the modulus, and (c) the reciprocal of $2\sqrt{2} + i$.

22. Each leg of an isosceles triangle is 250 cm long and the vertex angle measures 24°. Find the lengths of the base and the altitude to the base.

23. Write $-6 - 3i$ in polar form. Express the argument θ to the nearest tenth of a degree.

24. Graph $y = 2 \sin x + \sin 2x$ for $0 \le x \le 2\pi$.

25. Rectangle $ABCD$ is inscribed in a circle. If $DC = 10$ and $\angle DAC = 27°$, find the perimeter of the rectangle to the nearest unit.

26. Prove: $\dfrac{1 - \tan \theta}{1 + \tan \theta} = \dfrac{\cot \theta - 1}{\cot \theta + 1}$.

27. Solve for the scalars s and t: $(-2s\mathbf{i} + 3t\mathbf{j}) + (4\mathbf{i} - 2\mathbf{j}) = 3\mathbf{j}$.

28. Find a polar equation for $y = -3x$.

29. If $z = \cos(-100°) + i \sin(-100°)$, find the polar forms of (a) iz and (b) $\dfrac{z}{i}$. Then graph z, iz, and $\dfrac{z}{i}$ in the same complex plane.

30. At what heading and air speed should a plane fly if a wind of 30 km/h is blowing from 300° and the pilot wants to maintain a true course of due south and a ground speed of 400 km/h?

31. Suppose that a point P moves counterclockwise a distance $\dfrac{7\pi}{4}$ on a circle of radius $\dfrac{1}{2}$ with center at origin. If P starts at the point $\left(\dfrac{1}{2}, 0\right)$, what are the coordinates of its final position $P(x, y)$?

32. Rewrite $y = -2 \cos x - 2\sqrt{3} \sin x$ in the form $y = C \cos(x - \phi)$ and graph the function.

33. Find the sine, cosine, and tangent of 479° 30′ to four significant digits.

34. Use the quadratic formula and De Moivre's theorem to find the roots of $z^2 - 4z + (4 - i) = 0$.

35. Express $\dfrac{\tan^2 \theta + 1}{\tan^2 \theta}$ in terms of a single trigonometric function.

36. Find the general solution: $\cos\theta + \sec\theta = -\dfrac{5}{2}$.

37. If $\csc\theta = \dfrac{29}{20}$ and $90° < \theta < 180°$, find the exact values of the other five trigonometric functions of θ.

38. Solve $\triangle ABC$ if $\angle C = 42°$, $c = 100$, and $a = 120$. If no triangle can be formed, so state. If there is more than one solution, find each one.

39. Prove that $\csc\left(\dfrac{3\pi}{2} - x\right) = -\sec x$.

40. Solve for $0° \le \theta < 360°$ to the nearest degree: $-\sqrt{3}\cos\theta - \sin\theta = \sqrt{2}$.

41. Use the Law of Sines to show that in $\triangle ABC$, $\dfrac{a + b}{b} = \dfrac{\sin A + \sin B}{\sin B}$.

42. Use the vectors **a** and **b** at the right. Draw a diagram to show that $\mathbf{a} + (\mathbf{b} - \mathbf{a}) = \mathbf{b}$.

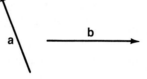

43. Let $\mathbf{a} = a_1\mathbf{i} + a_2\mathbf{j}$ and $\mathbf{b} = b_1\mathbf{i} + b_2\mathbf{j}$. Verify that $\mathbf{a} + (\mathbf{b} - \mathbf{a}) = \mathbf{b}$.

44. If $z_1 = 4(\cos 180° + i\sin 180°)$ and $z_2 = \sqrt{2}(\cos 225° + i\sin 225°)$, find (a) z_1z_2 and (b) $\dfrac{z_1}{z_2}$ in polar form and in the form $x + iy$.

45. Graph $y = -2\sin\dfrac{\pi}{3}x$ for $0 \le x \le 2\pi$.

46. Express $\dfrac{\tan^2 x}{1 - \cos x}$ in terms of $\sec x$.

47. Find the cube roots of $-i$ in the form $x + iy$.

48. Find the lengths of the diagonals of a parallelogram given with sides of lengths 4 and 6 and an angle of measure 128°.

49. Use angle-sum formulas to find the exact values of $\sin 255°$, $\cos 255°$, and $\tan 255°$ in simplest radical form.

50. Find the work done by the force $\mathbf{F} = 7\mathbf{i} - \mathbf{j}$ in moving an object from $A(-7, -3)$ to $B(-4, -5)$.

51. Find all possible values for $\angle R$ in $\triangle RST$ if $\angle S = 120°$, $r = 8$, and $s = 2\sqrt{3}$.

52. Name two different pairs of polar coordinates for point $(-4, 4)$.

53. Describe how to obtain the graph of $y = 1 - \cos(x + \pi)$ from the graph of $y = \cos x$.

54. Solve for $0° \le \theta < 360°$: $3\sin^2\theta + \cos\theta = 1$.

55. Graph the polar equation $r = 3 \sin 3\theta$ (three-leaved rose).

56. Find the area to the nearest tenth of $\triangle ABC$ with $a = 20$, $\angle B = 119°20'$, and $\angle C = 9°10'$.

57. Simplify: (a) $\dfrac{1}{i^{203}}$ (b) $\dfrac{1}{i^{4n+2}}$ for a negative integer n.

58. Refer to Exercise 3. Express the equation in terms of the sine function.

59. Express in the form $x + iy$: $\dfrac{[2(\cos 28° + i \sin 28°)]^8}{[4(\cos 11° + i \sin 11°)]}$

60. A boat sails due east for 80 km and then due northeast for 80 km more. Find the distance and bearing of the boat from its starting point.

61. Prove: $2 \csc 2\theta = \sec \theta \csc \theta$.

62. Graph: $y = \cos x - \cos x$ for $0 \le x \le 2\pi$.

63. Use a trigonometric identity to find the general solution for $\dfrac{\sin 2x}{1 + \cos 2x} = -1$.

64. If $w = -7 + 2i$ and $z = 3 - 2i$, find (a) $\overline{w + z}$, (b) $\overline{w - z}$, and (c) $\overline{w} - \overline{z}$.

GLOSSARY

Absolute value of a complex number (p. 189): For $z = x + iy$, $|z| = \sqrt{x^2 + y^2}$.

Addition of vectors (p. 148): For vectors **a** and **b,** place the initial point of **b** at the terminal point of **a.** Then **a** + **b** is the vector extending from the initial point of **a** to the terminal point of **b.**

Air speed (p. 155): Speed in still air.

Alias (p. 220): The interpretation of transformations that considers the points of a curve as fixed but permits the coordinate axes to move.

Alibi (p. 221): The interpretation of transformations that regards the points of a curve as moving into new positions while the coordinate axes remain fixed.

Ambiguous case (p. 110): The case of an oblique triangle in which two sides and a nonincluded angle are known.

Amplitude of a periodic function (p. 125): One half the difference between the maximum and minimum values of the function.

Angular speed (p. 45): The rate with respect to time t at which the angle generated by $\overrightarrow{OP}$ changes when P is a point moving about point O in uniform circular motion.

Arcsin x (p. 57): The inverse of Sin x.

Argand plane (p. 191): The Cartesian plane when used to represent points in the complex plane. The x-axis is called the real axis, and the y-axis is called the imaginary axis.

Argument of a complex number (p. 193): The angle θ is called the argument of a complex number when the number is written as $r(\cos \theta + i \sin \theta)$.

Argument of a function (p. 88): The expression or variable upon which a function operates.

Asymptote (p. 50): A line that a curve approaches more and more closely.

Bearing of a vector (p. 154): The angle θ ($0° \leq \theta < 360°$), measured clockwise, that a vector makes with another vector pointing due north.

Catenary (p. 97): The curve defined by an equation of the form $y = a \cosh \dfrac{x}{a}$.

Circular functions (p. 40): The functions sine, cosine, tangent, etc. defined for real numbers through the use of the unit circle.

Complex number (p. 188): A number of the form $x + iy$, where x and y are real numbers and i is the imaginary unit. x is called the real part and y is called the imaginary part.

Complex plane (p. 191): *See* Argand plane.

Conic sections (p. 183): The curves formed when a plane intersects a cone. (p. 184): The set of all points whose distances from a fixed point, the focus, and a fixed line, the directrix, have a constant ratio, e, the eccentricity.

Conjugate of a complex number (p. 189): The conjugate of $z = x + yi$ is $\overline{z} = x - yi$.

Coordinate system in space (p. 233): Use of three axes to describe the position of points in space.

Cosecant (p. 17): $\csc \theta = \dfrac{1}{\sin \theta}$, θ an acute angle of a right triangle.

(p. 25): $\csc \theta = \dfrac{r}{y}$, where θ is an angle in standard position, $P(x, y)$ a point on the terminal side of θ, r the distance OP, and $y \neq 0$.

Cosine (p. 7): $\cos \theta = \dfrac{\text{length of side adjacent to } \theta}{\text{length of hypotenuse}}$, where θ is an acute angle of a right triangle.

(p. 25): $\cos \theta = \dfrac{x}{r}$, where θ is an angle in standard position, $P(x, y)$ a point on the terminal side of θ, and r the distance OP.

Cos x (p. 58): $\cos x$ with domain restricted to $0 \leq x \leq \pi$.

Cos^{-1}x (p. 58): The inverse of Cos x.

Cotangent (p. 17): $\cot \theta = \dfrac{1}{\tan \theta}$, θ an acute angle of a right triangle.

(p. 25): $\cot \theta = \dfrac{x}{y}$, where θ is an angle in standard position, $P(x, y)$ a point on the terminal side of θ, and $y \neq 0$.

Coterminal angles (p. 4): Two angles of different measure that have coincident terminal sides when in standard position.

Cycloid (p. 171): The path traced by a fixed point of a circle rolling along a line.

De Moivre's theorem (p. 198): For every integer n, $[r(\cos \theta + i \sin \theta)]^n = r^n(\cos n\theta + i \sin n\theta)$.

Dilation (p. 229): A linear transformation that "expands" or "contracts" the coordinate system radially along the lines through the origin. Also known as "magnification" or "shrinking."

Directrix (p. 184): *See* Conic sections.

Displacement (p. 154): Movement of an object from A to B indicated by the vector $\overrightarrow{AB}$.

Dot product (p. 161): When θ is the angle between $\mathbf{a}$ and $\mathbf{b}$, $\mathbf{a} \cdot \mathbf{b} = \|\mathbf{a}\|\|\mathbf{b}\|\cos \theta$.

Eccentricity (e) (p. 184): *See* Conic sections.

Ellipse (p. 184): A conic section in which the eccentricity is between 0 and 1. (p. 186): A set of points in a plane such that, for each point of the set, the sum of its distances to two fixed points is a constant.

Equal vectors (p. 147): Vectors having the same length and the same direction.

Even function (p. 49): A function f such that whenever x is in its domain, $-x$ is also in its domain, and $f(-x) = f(x)$ for all x in the domain.

Field (p. 190): An algebraic system that consists of a set F together with two binary operations that satisfy certain axioms.

Focus (p. 184): *See* Conic Sections.

Fourier series (p. 291): An infinite series of the form $a_0 = a_1 \cos x + b_1 \sin x + a_2 \cos 2x + b_2 \sin 2x + a_3 \cos 3x + b_3 \sin 3x + \ldots$ where the a's and b's are constants.

Frequency of a simple harmonic motion (p. 133): Denotes oscillations per unit of time $(f = \dfrac{1}{T} = \dfrac{\omega}{2\pi})$.

Fundamental (p. 136): The note of lowest frequency when a string (as on a violin) is vibrated.

Fundamental period (p. 50): The least positive period, if there is one, of a periodic function.

Great circle (p. 259): The circle determined by a plane cutting through the surface of a sphere and its center.

Ground speed (p. 155): Speed relative to the ground.

Heading (p. 154, 155): Direction in which a craft is headed, or pointed.

Horizontal asymptote (p. 50): The line $y = a$ which $f(x)$ approaches for progressively larger positive or negative values of x.

Hyperbola (p. 184): A conic section in which the eccentricity is greater than 1.

Hyperbolic functions (p. 96): The pair of functions related to the "unit hyperbola" $x^2 - y^2 = 1$ and called the hyperbolic sine (sinh) and the hyperbolic cosine (cosh).

Identity (p. 69): An equation that is true for all values of the variable for which both sides of the equation are defined.

Imaginary unit (p. 188): i, or $\sqrt{-1}$.

Initial point of a vector (p. 147): The tail of the arrow representing a vector.

Inverse function (p. 56): The function g such that for functions f and g, $g(y) = x$ if and only if $f(x) = y$. The inverse of f is denoted by f^{-1}. In general, a function f has an inverse if and only if the function is one-to-one; that is, if and only if $f(x_1) = f(x_2)$ implies $x_1 = x_2$.

Joule (p. 166): The basic unit of work and energy in the metric system; the work done by a force of 1 newton in moving an object 1 m.

Kilowatt-hour (p. 166): $1 \text{kW} \cdot \text{h} = 3.6 \times 10^6 \text{ J}$

Latitude (p. 276): The distance of a point from the equator, measured in degrees.

Law of cosines (p. 101): In any triangle, the square of the length of any side equals the sum of the squares of the lengths of the other two sides decreased by twice the product of the lengths of these two sides and the cosine of the included angle.

Law of sines (p. 107): The sines of the angles of any triangle are proportional to the lengths of the corresponding sides.

Linear combinations of a_1, a_2, ..., a_n (p. 149): Expressions of the form $t_1 a_1 + t_2 a_2 + \ldots + t_n a_n$, where the **a**'s are vectors and the t's are scalars.

Linear interpolation (p. 13): A process for approximating the value of a function $f(c)$, where $a < c < b$ and $f(a)$ and $f(b)$ are known, which uses the segment with endpoints $(a, f(a))$ and $(b, f(b))$ to approximate the graph of f over the interval with endpoints a and b.

Linear speed (p. 45): The distance a point P moving in uniform circular motion travels along an arc s in one unit of time t.

Longitude (p. 276): The angle between the meridian of a point and the prime meridian.

Matrix of a linear transformation (p. 229): A rectangular array of constants that uniquely determine the linear transformation.

Meridians (p. 260): Great circles containing the geographic North and South Poles.

Modulus of a complex number (p. 189): For $z = x + iy$, the modulus is $|z| = \sqrt{x^2 + y^2}$. (p. 193): The number r when a complex number is written as $r(\cos \theta + i \sin \theta)$.

Multiplication of a vector by a scalar (p. 149): For vector **a** and scalar t, $t\mathbf{a}$ is the product of the length of **a** by $|t|$. If $t < 0$, the direction is reversed.

Nautical mile (p. 276): The length of a one-minute great-circle arc; about 1.15 land miles or 1.85 kilometers.

Newton (p. 93) (Ex. 27), (p. 158): Basic unit of force in the metric system; the force necessary to accelerate a 1 kg mass 1 m/s^2.

Norm of a vector (p. 151): The magnitude of a vector, **a**, denoted by $\|\mathbf{a}\|$.

Oblique triangle (p. 103): A triangle having no right angle.

Octant (p. 234): One of the eight regions into which space is divided by coordinate planes.

Odd function (p. 49): A function f such that whenever x is in its domain, $-x$ is also in its domain, $f(-x) = -f(x)$ for all x in the domain.

Orthogonal vectors (p. 163): Perpendicular vectors.

Overtones (p. 136): Tones weaker than the fundamental when a string is vibrated.

Parabola (p. 184): A conic section in which the eccentricity is 1.

Parallelogram rule (p. 148): *See* Addition of vectors.

Parallels (p. 260): Small circles cut by planes parallel to the equator.

Period of simple harmonic motion (p. 133): The time necessary for an object moving in simple harmonic motion to complete one "round trip" in its path.

Periodic function (p. 50): A function f such that there is some fixed positive number p having the property that whenever x is in the domain of f, both $x - p$ and $x + p$ are also in the domain and $f(x - p) = f(x) = f(x + p)$ for all x in the domain. If there is a least such positive number p, then p is called the *fundamental period* of f.

Phase shift (p. 128–129): In the case of a sinusoid, the number of units a graph is translated left or right.

Pi (π) (p. 64): Ratio of the circumference of a circle to its diameter.

Polar axis (p. 175): The ray used as a reference in the system of polar coordinates.

Polar coordinate system (p. 175): A system for locating points in the plane by their distance and direction from a given point and given ray respectively.

Polar coordinates (p. 175): The ordered pair (r, θ) used to locate a point in the system of polar coordinates.

Polar form of a complex number (p. 192) The polar form of a complex number $z = x + iy$ is $z = r(\cos \theta + i \sin \theta)$ for $r \geq 0$.

Polar triangle (p. 262): $\triangle A'B'C'$ is the polar triangle of $\triangle ABC$ provided A' is the pole of $\overparen{BC}$ on the same side of $\overparen{BC}$ as A, B' is the pole of $\overparen{AC}$ on the same side of $\overparen{AC}$ as B, and C' is the pole of $\overparen{AB}$ on the same side of $\overparen{AB}$ as C.

Pole (p. 175): The origin in the reference system for polar coordinates.

Poles (p. 262): The two points that are 90° away from a great-circle arc.

Power series (p. 285–286): A series of the form $a_0 + a_1 x + a_2 x^2 + a_3 x^3 + \ldots$, where the a's are constants.

Primitive root (p. 204): An nth root of unity, w, is primitive if every nth root of unity is an integral power of n.

Pythagorean identities (p. 70): $\sin^2 x + \cos^2 x = 1$; $1 + \cot^2 x = \csc^2 x$; $1 + \tan^2 x = \sec^2 x$.

Quadrantal angles (p. 26): Angles whose terminal sides lie along an axis.
Quadrantal triangle (p. 269): A spherical triangle having a 90° side. ﹒
Quadrants (p. 26): The four regions into which the coordinate axes divide the plane.

Radian (p. 37): A unit of angle measure. The measure of a central angle of a circle whose intercepted arc is equal in length to the radius of the circle.
Reference angle (p. 27): The positive acute angle x between the x-axis and the terminal side of any angle θ.
Rotation (p. 217): A method for simplifying the equation of a given curve by rotating the coordinate axes through an angle θ about the origin.

Scalar (p. 147): A quantity having magnitude but without direction.
Scalar projection of v onto a (p. 164): The norm of $\mathbf{v_a}$.
Secant (p. 17): $\sec \theta = \dfrac{1}{\cos \theta}$, θ an acute angle.

(p. 25): $\sec \theta = \dfrac{r}{x}$, where θ is an angle in standard position, $P(x, y)$ a point on the terminal side of θ, r the distance OP, and $x \neq 0$.
Significant digit (p. 8–9): Any nonzero digit or any zero that has a purpose other than indicating the position of the decimal point.
Simple harmonic motion (p. 132): The motion that occurs when an object is displaced from an equilibrium (at rest) position.
Sin x (p. 56): $\sin x$ with domain restricted to $-\dfrac{\pi}{2} \leq x \leq \dfrac{\pi}{2}$.

Sin^{-1} x (p. 57): The inverse of Sin x.
Sine (p. 7): $\sin \theta = \dfrac{\text{length of side opposite } \theta}{\text{length of hypotenuse}}$, where θ is an acute angle of a right triangle.

(p. 25): $\sin \theta = \dfrac{y}{r}$, where θ is an angle in standard position, $P(x, y)$ a point on the terminal side of θ, and r the distance OP.
Sinusoid (p. 124): Graph of an oscillating function.
Solving a triangle (p. 20): Finding values for all the sides and angles of a triangle.
Spherical angle (p. 260): The angle formed by two great-circle arcs having a common endpoint.
Spherical triangle (p. 261): The triangle formed by any three great-circle arcs.
Standard position of a vector (p. 150): A vector with its endpoint at the origin.

Tan *x* (p. 59): tan *x* with domain restricted to $-\frac{\pi}{2} \leq x \leq \frac{\pi}{2}$.

Tan^{-1} *x* (p. 59): The inverse of Tan *x*.

Tangent (p. 16–17): $\tan \theta = \dfrac{\text{length of side opposite } \theta}{\text{length of side adjacent to } \theta}$, θ an acute angle
of a right triangle.

(p. 25): $\tan \theta = \dfrac{y}{x}$ where θ is an angle in standard position, $P(x, y)$ a point
on the terminal side of θ, and $x \neq 0$.

Taylor series (p. 286): A series which approximates a function to any desired degree of accuracy. Also called a power series expansion.

Tension (p. 158): In a rope or cable, the magnitude of the force it exerts.

Terminal point of a vector (p. 147): The tip of the arrow representing a vector.

Translation (p. 128): A shift of the graph of a function in which the final position of the graph is parallel to or coincident with its original position.

Triangle inequality (p. 192): The property $|w + z| < |w| + |z|$, where w and z are complex numbers.

Trigonometric equation (p. 86): An equation involving one or more circular or trigonometric functions.

True course (p. 155): Bearing of a plane's path relative to the ground.

Uniform circular motion (p. 45): The motion of a point P moving with a constant speed in a circular path.

Unit circle (p. 41): Circle with radius 1 and center at the origin.

Unit vector (p. 151): A vector having norm 1.

Vector quantities (p. 147): Quantities that have both magnitude and direction.

Vector space (p. 152): Any system of objects {**a, b, c,** . . .} satisfying the vector-space axioms (Exs. 21–28).

Vertical asymptote (p. 50): The line $x = a$, if as x approaches a, $|f(x)|$ gets larger and larger without bound.

Vertical shift (p. 128–129): In the case of a sinusoid, the number of units a graph is translated upward or downward.

Zero vector (p. 149): Vector with magnitude 0; direction not defined.

INDEX

SELECTED ANSWERS

CHAPTER 1 TRIGONOMETRIC FUNCTIONS

Exercises 1-1, pages 4–5

1.

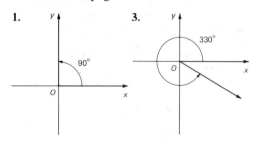

3.

5.

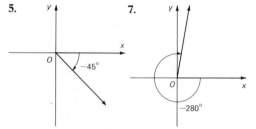

7.

9.

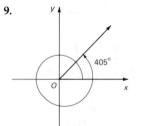

11.

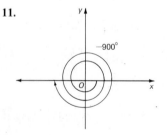

13. 53°24′ **15.** −22°7′48″ **17.** −61.15°
19. 12.075° **21.** 80° **23.** 45° **25.** 180°
27. 310°30′ or 310.5° **29.** 310°6′ or 310.1°
31. 218°18′ or 218.3° **33.** 37°45′10″
35. $(65 + 360k)°$ **37.** $(\theta + 360k)°$
41. 17 revolutions

Exercises 1-2, pages 9–11
1. (a) $\sin \theta = 0.8$, $\cos \theta = 0.6$ (b) 55°
3. (a) $\sin \theta = 0.6$, $\cos \theta = 0.8$ (b) 35°
5. (a) $\sin \theta = 0.96$, $\cos \theta = 0.28$ (b) $\theta = 75°$
7. (a) $\sin \theta = 0.6614$, $\cos \theta = 0.75$
(b) $\theta = 40°$ **9.** (a) $\sin \theta = 0.6667$,
$\cos \theta = 0.7454$ (b) $\theta = 40°$
11. (a) $\sin \theta = 0.1429$, $\cos \theta = 0.9897$
(b) $\theta = 10°$ **13.** $x = 171$ **15.** $\theta = 40°$
19. 426 cm **21.** 4.34 cm **23.** 17.7 m
29. $2 \sin \dfrac{\theta}{2}$

Exercises 1-3, pages 13–16 **1.** 45° **3.** 45°
5. 30° **7.** 30° **9.** $7\sqrt{2}$ **11.** $\dfrac{5}{2}\sqrt{3}$

13. $\sin 12°50′ = 0.2221$, $\cos 12°50′ = 0.9750$
15. $\sin 56°10′ = 0.8307$, $\cos 56°10′ = 0.5568$
17. 41.50° or 41°30′ **19.** 79.33° or 79°20′
21. 48.19° or 48°11′ **23.** $\dfrac{\sqrt{6} - \sqrt{2}}{4}$

25. 116 **27.** The distance between Earth and the sun is about 300 times the distance between Earth and the moon.
29. 1.4 m **31.** (b) $\dfrac{\sqrt{5} - 1}{4}$
33. 2.260×10^8 m/sec

Exercises 1-4, pages 19–20 **1.** $\sin \theta = \dfrac{12}{13}$;
$\cos \theta = \dfrac{5}{13}$; $\tan \theta = \dfrac{12}{5}$; $\cot \theta = \dfrac{5}{12}$;
$\sec \theta = \dfrac{13}{5}$; $\csc \theta = \dfrac{13}{12}$ **3.** $\sin \theta = \dfrac{\sqrt{7}}{4}$;
$\cos \theta = \dfrac{3}{4}$; $\tan \theta = \dfrac{\sqrt{7}}{3}$; $\cot \theta = \dfrac{3\sqrt{7}}{7}$;
$\sec \theta = \dfrac{4}{3}$; $\csc \theta = \dfrac{4\sqrt{7}}{7}$ **5.** $\sin \theta = \dfrac{3\sqrt{13}}{13}$;
$\cos \theta = \dfrac{2\sqrt{13}}{13}$; $\tan \theta = \dfrac{3}{2}$; $\cot \theta = \dfrac{2}{3}$;
$\sec \theta = \dfrac{\sqrt{13}}{2}$; $\csc \theta = \dfrac{\sqrt{13}}{3}$ **7.** $\sin \theta = \dfrac{2}{3}$;
$\cos \theta = \dfrac{\sqrt{5}}{3}$; $\tan \theta = \dfrac{2\sqrt{5}}{5}$; $\cot \theta = \dfrac{\sqrt{5}}{2}$;
$\sec \theta = \dfrac{3\sqrt{5}}{5}$; $\csc \theta = \dfrac{3}{2}$ **9.** 62.1 **11.** 110.7

13. 67.7° or 67°40′ **15.** 1 **17.** $\dfrac{2\sqrt{3}}{3}$

19. $\sqrt{2}$ **21.** $\sqrt{2}$ **23.** 65° **25.** No **27.** 59 cm

Exercises 1-5, pages 22–24
1. $\angle B = 67°20'$; $a = 92.5$; $b = 221$
3. $\angle A = 33°20'$; $b = 98.8$; $c = 118$
5. $\angle B = 26°$; $b = 39.0$; $c = 89.0$
7. $c = 97.0$; $\angle A = 42.1°$ or $42°10'$;
$\angle B = 47.9°$ or $47°50'$ **9.** $a = 45.0$;
$\angle B = 31.9°$ or $31°50'$; $\angle A = 58.1°$ or
$58°10'$ **11.** $b = 5\sqrt{3}$; $\angle A = 45°$;
$\angle B = 45°$ **13.** $\angle B = 49.7°$ or $49°40'$
15. $c = 16.5$ cm; $a = 7.48$ cm; $b = 14.7$ cm
17. 36° **19.** 24.8° or 24°50' **21.** 146 m

Exercises 1-6, pages 28–29
1. $\sin 210° = -\dfrac{1}{2}$; $\cos 210° = -\dfrac{\sqrt{3}}{2}$;
$\tan 210° = \dfrac{\sqrt{3}}{3}$; $\cot 210° = \sqrt{3}$;
$\sec 210° = -\dfrac{2\sqrt{3}}{3}$; $\csc 210° = -2$
3. $\sin 315° = -\dfrac{\sqrt{2}}{2}$; $\cos 315° = \dfrac{\sqrt{2}}{2}$;
$\tan 315° = -1$; $\cot 315° = -1$;
$\sec 315° = \sqrt{2}$; $\csc 315° = -\sqrt{2}$
5. $\sin 330° = -\dfrac{1}{2}$; $\cos 330° = \dfrac{\sqrt{3}}{2}$;
$\tan 330° = -\dfrac{\sqrt{3}}{3}$; $\cot 330° = -\sqrt{3}$;
$\sec 330° = \dfrac{2\sqrt{3}}{3}$; $\csc 330° = -2$
7. $\sin 90° = 1$; $\cos 90° = 0$; $\tan 90°$ is
undefined; $\cot 90° = 0$; $\sec 90°$ is
undefined; $\csc 90° = 1$ **9.** -0.9886
11. -0.9325 **13.** 1.213 **15.** -0.9063
17. (a) $\sin \theta = \dfrac{3}{5}$; $\cos \theta = -\dfrac{4}{5}$ (b) $143.1°$ or
$143°10'$ **19.** (a) $\sin \theta = -\dfrac{8}{17}$; $\cos \theta = \dfrac{15}{17}$
(b) $331.9°$ or $332°0'$ **21.** (a) $\sin \theta = -\dfrac{\sqrt{5}}{3}$;
$\cos \theta = -\dfrac{2}{3}$ (b) $228.2°$ or $228°10'$
23. (a) $\sin \theta = \dfrac{3}{4}$; $\cos \theta = -\dfrac{\sqrt{7}}{4}$ (b) $131.4°$
or $131°20'$ **29.** $\tan \theta = \dfrac{15}{8}$ **31.** $\tan \theta = -\dfrac{24}{7}$
33. $\tan \theta = -\dfrac{5\sqrt{11}}{11}$

CHAPTER 2 CIRCULAR FUNCTIONS AND THEIR GRAPHS

Exercises 2-1, pages 39–40   1. $-\dfrac{\pi}{2}$ **3.** $\dfrac{7\pi}{4}$
5. $\dfrac{2\pi}{3}$ **7.** $\dfrac{2\pi}{5}$ **9.** $\dfrac{\pi}{8}$ **11.** 4π **13.** 60°
15. $-120°$ **17.** $-330°$ **19.** 105°
21. 112.5° **23.** 900° **25.** $\theta = 210°$;
$s = 110.0$ **27.** $\theta = \dfrac{3\pi}{4}$; $s = 66.0$ **29.** $\theta = \dfrac{5\pi}{6}$;
$r = 7.2$ **31.** $s = r\theta$ **33.** $\dfrac{35\pi}{3}$ **35.** 36 cm
37. $\dfrac{3\pi}{2}$ **39.** 4.00×10^4 km
41. $(10\pi + 20)$ cm

Exercises 2-2, pages 43–44   1. $\sin \dfrac{7\pi}{6} = -\dfrac{1}{2}$;
$\cos \dfrac{7\pi}{6} = -\dfrac{\sqrt{3}}{2}$; $\tan \dfrac{7\pi}{6} = \dfrac{\sqrt{3}}{3}$
3. $\sin \dfrac{3\pi}{4} = \dfrac{\sqrt{2}}{2}$; $\cos \dfrac{3\pi}{4} = -\dfrac{\sqrt{2}}{2}$;
$\tan \dfrac{3\pi}{4} = -1$ **5.** $\sin\left(-\dfrac{\pi}{2}\right) = -1$;
$\cos\left(-\dfrac{\pi}{2}\right) = 0$; $\tan\left(-\dfrac{\pi}{2}\right)$ is undefined
7. $\sin \dfrac{8\pi}{3} = \dfrac{\sqrt{3}}{2}$; $\cos \dfrac{8\pi}{3} = -\dfrac{1}{2}$;
$\tan \dfrac{8\pi}{3} = -\sqrt{3}$ **9.** $\sec\left(-\dfrac{7\pi}{4}\right) = \sqrt{2}$;
$\csc\left(-\dfrac{7\pi}{4}\right) = \sqrt{2}$; $\cot\left(-\dfrac{7\pi}{4}\right) = 1$
11. $\sec \dfrac{5\pi}{3} = 2$; $\csc \dfrac{5\pi}{3} = -\dfrac{2\sqrt{3}}{3}$;
$\cot \dfrac{5\pi}{3} = -\dfrac{\sqrt{3}}{3}$ **13.** $\sec\left(-\dfrac{4\pi}{3}\right) = -2$;
$\csc\left(-\dfrac{4\pi}{3}\right) = \dfrac{2\sqrt{3}}{3}$; $\cot\left(-\dfrac{4\pi}{3}\right) = -\dfrac{\sqrt{3}}{3}$
15. $\sec\left(-\dfrac{11\pi}{6}\right) = \dfrac{2\sqrt{3}}{3}$; $\csc\left(-\dfrac{11\pi}{3}\right) = 2$;
$\cot\left(-\dfrac{11\pi}{3}\right) = \sqrt{3}$ **17.** -0.91 **19.** -0.43
21. -0.65 **23.** 1.0 **25.** -0.54 **27.** -0.64
29. $P\left(\dfrac{\sqrt{2}}{2}, -\dfrac{\sqrt{2}}{2}\right)$ **31.** $P(-1, -\sqrt{3})$
33. $P\left(\dfrac{3\sqrt{3}}{4}, -\dfrac{3}{4}\right)$ **37.** $(-t + \cos t, \sin t)$

Exercises 2-3, pages 46–48 **1.** 198 cm/s
3. 3.49 radians/s **5.** 204 cm/s **7.** (2, 0)
9. (6, 0) **11.** (−10, 0) **13.** (3, 3 $\sqrt{3}$)
15. (4, 0) **17.** (0, −1); $s = \dfrac{20\pi}{3}$ cm
19. (−2 $\sqrt{3}$, −2); $s = 18\pi$ cm
21. (−1.553, 5.795); $s = 8\pi$ cm
23. $\dfrac{\pi}{12}$ radians/h; 1662 km/h
27. 831.2 km/h **29.** The linear speed will
be multiplied by $\sqrt{2}$.

Exercises 2-4, page 55

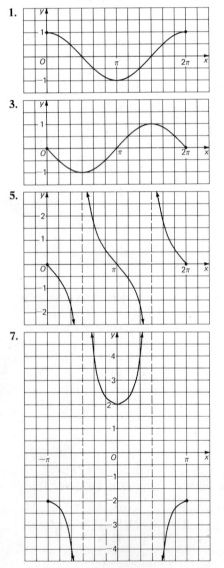

9.

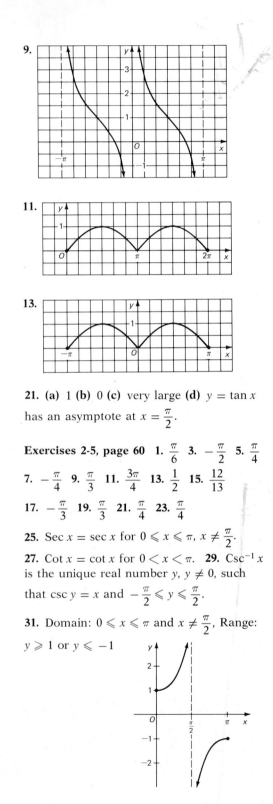

11.

13.

21. **(a)** 1 **(b)** 0 **(c)** very large **(d)** $y = \tan x$
has an asymptote at $x = \dfrac{\pi}{2}$.

Exercises 2-5, page 60 **1.** $\dfrac{\pi}{6}$ **3.** $-\dfrac{\pi}{2}$ **5.** $\dfrac{\pi}{4}$
7. $-\dfrac{\pi}{4}$ **9.** $\dfrac{\pi}{3}$ **11.** $\dfrac{3\pi}{4}$ **13.** $\dfrac{1}{2}$ **15.** $\dfrac{12}{13}$
17. $-\dfrac{\pi}{3}$ **19.** $\dfrac{\pi}{3}$ **21.** $\dfrac{\pi}{4}$ **23.** $\dfrac{\pi}{4}$
25. Sec x = sec x for $0 \leqslant x \leqslant \pi$, $x \neq \dfrac{\pi}{2}$.
27. Cot x = cot x for $0 < x < \pi$. **29.** Csc^{-1} x
is the unique real number y, $y \neq 0$, such
that csc $y = x$ and $-\dfrac{\pi}{2} \leqslant y \leqslant \dfrac{\pi}{2}$.
31. Domain: $0 \leqslant x \leqslant \pi$ and $x \neq \dfrac{\pi}{2}$, Range:
$y \geqslant 1$ or $y \leqslant -1$

33. Domain: x is a real number; No
35. Yes

Cumulative Review • Chapters 1 and 2,
pages 62–63 Chapter 1 1. $247.5°$ or

$247°30'$ **3.** $60°$ **5.** $\sqrt{2} + \sqrt{3}$ **7.** $\sin \theta = \dfrac{\sqrt{7}}{4}$,

$\cos \theta = -\dfrac{3}{4}$, $\tan \theta = -\dfrac{\sqrt{7}}{3}$, $\cot \theta = -\dfrac{3\sqrt{7}}{7}$,

$\sec \theta = -\dfrac{4}{3}$, $\csc \theta = \dfrac{4\sqrt{7}}{7}$ **9.** $\tan 90°$,

$\cot 0°$, $\cot 180°$, $\sec 90°$, $\csc 0°$, $\csc 180°$

Chapter 2 1. (a) 0 **(b)** $\dfrac{\pi}{4}$ **(c)** $\dfrac{14\pi}{9}$ **3.** $\dfrac{175\pi}{12}$

5. $\dfrac{615}{7}$ rad/s

7.

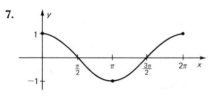

9. (a) $\dfrac{\pi}{6}$ **(b)** $\dfrac{4}{3}$ **(c)** undefined **(d)** $\dfrac{\pi}{6}$

CHAPTER 3 PROPERTIES OF TRIGONOMETRIC FUNCTIONS

Exercises 3-1, pages 71–72 1. $\dfrac{\sin^2 x}{1 - \sin^2 x}$

3. $\dfrac{\sin x}{1 - \sin^2 x}$ **5.** $\sin x$ **7.** $\dfrac{1}{\sin x}$ **9.** $\dfrac{1}{\cos^2 x}$

11. $\dfrac{1}{\cos^2 x}$ **13.** $1 - \cos x$ **15.** $\dfrac{1}{1 - \cos x}$

17. $\cot x$ **19.** $\cot t$ **21.** $\tan x$ **23.** $\sin x$
25. $\sec \theta$ **27.** $\cot y$ **29.** $\cos \theta$ **31.** $\csc x$
35. 0 **37.** 0 **39.** 0

Exercises 3-3, pages 79–81 1. $\dfrac{\sqrt{6} + \sqrt{2}}{4}$

3. $\dfrac{\sqrt{6} - \sqrt{2}}{4}$ **5.** $2 + \sqrt{3}$ **7.** $\dfrac{\sqrt{6} - \sqrt{2}}{4}$

9. $\dfrac{\sqrt{2} + \sqrt{6}}{4}$ **11.** $-2 - \sqrt{3}$ **13.** $\dfrac{63}{65}$ **15.** $\dfrac{56}{65}$

17. $-\dfrac{84}{85}$ **19.** $\dfrac{77}{85}$ **21.** $-\dfrac{36}{77}$ **41.** $\dfrac{84}{85}$

43. $-\dfrac{204}{325}$ **45.** $\dfrac{\sqrt{6} - \sqrt{2}}{4}$

Exercises 3-4, pages 84–86 1. (a) $\dfrac{24}{25}$,

(b) $-\dfrac{7}{25}$ **3. (a)** $-\dfrac{120}{169}$, **(b)** $-\dfrac{119}{169}$

5. (a) $\dfrac{3\sqrt{7}}{8}$, **(b)** $-\dfrac{1}{8}$ **7.** $-\dfrac{24}{7}$ **9.** $\dfrac{120}{119}$

11. $-3\sqrt{7}$ **13.** $\dfrac{336}{527}$ **15.** $-\dfrac{28,560}{239}$

17. $\dfrac{3\sqrt{7}}{31}$ **19.** $\dfrac{\sqrt{2 - \sqrt{3}}}{2}$

21. $\dfrac{\sqrt{2 - \sqrt{2}}}{2}$ **23.** $2 + \sqrt{3}$

25. $\dfrac{\sqrt{34}}{10}$ **27.** $\dfrac{3}{4}$

Exercises 3-5, pages 89–90 1. $135°$, $315°$
3. $116.6°$ or $116°30'$, $296.6°$ or $296°30'$
5. $60°$, $120°$, $240°$, $300°$ **7.** $90°$, $270°$
9. $90°$, $30°$, $150°$ **11.** $0°$, $180°$, $60°$, $300°$

13. $\dfrac{\pi}{6}$, $\dfrac{5\pi}{6}$, $\dfrac{3\pi}{2}$ **15.** 0, π, $\dfrac{2\pi}{3}$, $\dfrac{4\pi}{3}$

17. $\dfrac{3\pi}{8}$, $\dfrac{7\pi}{8}$, $\dfrac{11\pi}{8}$, $\dfrac{15\pi}{8}$ **19.** $\dfrac{\pi}{12} + k\pi$, $\dfrac{5\pi}{12} + k\pi$,

$\dfrac{3\pi}{4} + k\pi$ **21.** $\dfrac{\pi}{6} + 2k\pi$, $\dfrac{5\pi}{6} + 2k\pi$

23. $0.72 + 2k\pi$, $5.56 + 2k\pi$, $(2k + 1)\pi$ **25.** $\dfrac{\pi}{6}$,

$\dfrac{\pi}{3}$, $\dfrac{7\pi}{6}$, $\dfrac{4\pi}{3}$ **27.** $\dfrac{2\pi}{3}$, $\dfrac{4\pi}{3}$ **29.** $\dfrac{\pi}{12}$, $\dfrac{5\pi}{12}$

31. $15°$ or $75°$ **33.** $15°$, $75°$, $90°$ **35.** $30°$,
$150°$, $210°$, $330°$ **37.** $0°$, $180°$, $45°$, $135°$,
$225°$, $315°$

Exercises 3-6, page 92–93 1. $\sqrt{2} \cos 335°$
3. $2 \cos 310°$ **5.** $2\sqrt{2} \cos(\theta + 45°)$
7. $4 \cos(\theta - 60°)$ **9.** $5 \cos(\theta + 37°)$
11. $13 \cos(\theta - 113°)$ **13.** $3 \cos(\theta + 109°)$
15. $105°$, $345°$ **17.** $46°$, $254°$ **199.** $97°$,

$337°$ **21.** $C = \sqrt{A^2 + B^2}$, $\cos \phi = \dfrac{B}{C}$,

$\sin \phi = \dfrac{A}{C}$ **23.** $105°$, $345°$ **25.** $46°$, $254°$

27. $7°$ **29.** $15°$ or $75°$

CHAPTER 4 OBLIQUE TRIANGLES

Exercises 4-1, pages 103–106 1. 7 **3.** 19
5. $60°$ **7.** $132°10'$ or $132.2°$ **9.** $2\sqrt{13}$ and
$2\sqrt{37}$ **15.** 95 cm **17.** 15 m **19.** $3\sqrt{2}$
21. 13 **23.** $2\sqrt{6}$ **25.** $a = 3$ or $a = 4$

27. 4.4 m   **29.** 8   **33.** 13.47 m

Exercises 4-2, pages 107–109
1. $\angle C = 79.5°$, $b = 17.0$, $c = 17.7$
3. $\angle A = 139°30'$ or $139.5°$, $a = 130.0$,
$c = 54.0$ **5.** $\angle C = 32°20'$ or $32.3°$, $b = 29.7$,
$c = 17.8$ **7.** $\angle A = 20°$, $b = 64.0$, $c = 44.3$
9. $\angle B = 98°$, $a = 10.1$, $c = 15.9$ **11.** 6
17. 0.793 km

Exercises 4-3, pages 111–112   **1.** 29.5° or
29°30'   **3.** 30°, 150°   **5.** 22.9° or 22°50',
157.1° or 157°10'   **7.** No triangle can be
formed.   **9.** 72.2° or 72°10', 107.8° or
107°50'   **11.** No triangle can be formed.
13. 58°, 122°   **17.** 71.3° or 71°20'

Exercises 4-4, pages 115–116   **1.** 30.0
3. 117.0   **5.** 288.0   **7.** 24.0   **9.** 29.9   **11.** 20.7
13. 48.0   **15.** 195.2   **21.** 4 or $2\sqrt{13}$
23. 114,000 cm^2

**CHAPTER 5   SINUSOIDAL
VARIATION**

Exercises 5-1, pages 127–128

1.

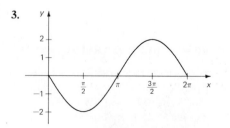

3.

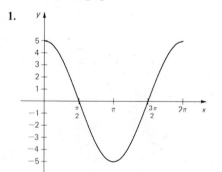

5.

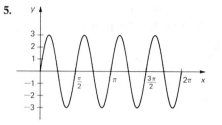

7.

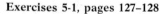
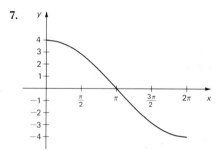

9.

11.

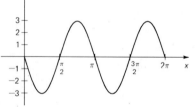

13.

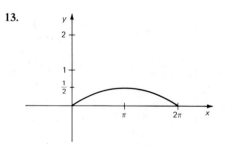

15. (portion of graph shown)

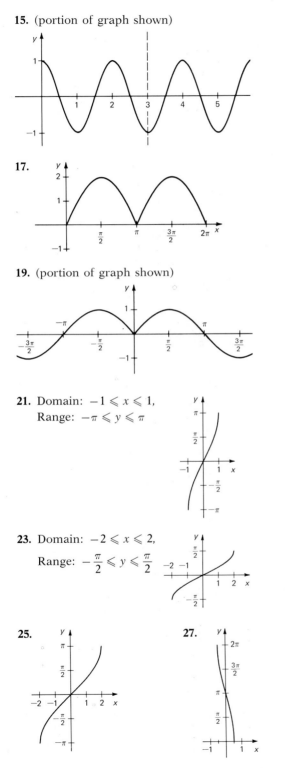

17.

19. (portion of graph shown)

21. Domain: $-1 \leqslant x \leqslant 1$,
Range: $-\pi \leqslant y \leqslant \pi$

23. Domain: $-2 \leqslant x \leqslant 2$,
Range: $-\dfrac{\pi}{2} \leqslant y \leqslant \dfrac{\pi}{2}$

25.

27.

29.

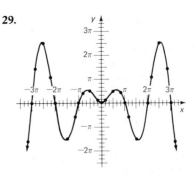

Exercises 5-2, pages 130–131

1.

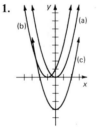

3.

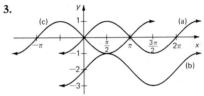

5.

7. (a), (b), (c)

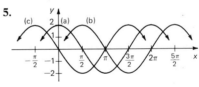

9.

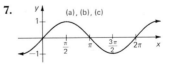

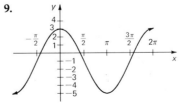

11.

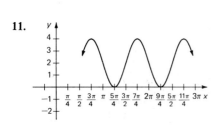

13.

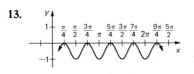

15.

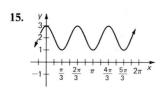

17.

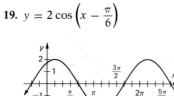

19. $y = 2 \cos \left(x - \dfrac{\pi}{6} \right)$

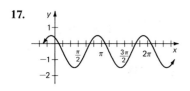

21. $y = 4 \cos \left(x - \dfrac{7\pi}{4} \right)$

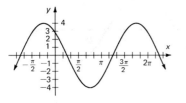

23. $y = 6 \cos \left(x - \dfrac{2\pi}{3} \right)$

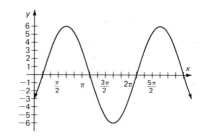

25. $y = \cos \left(2x - \dfrac{5\pi}{6} \right)$ **27.** $y = \sin \left(3x - \dfrac{\pi}{2} \right)$

29. Domain: $0 \leqslant x \leqslant 2$, Range: $0 \leqslant y \leqslant \pi$

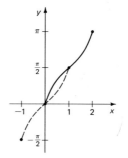

Exercises 5-3, pages 134–136

1. (a) $y = 4 \sin \left(\dfrac{\pi}{2} t + \dfrac{3\pi}{2} \right)$

(b)

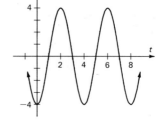

(c) $\dfrac{1}{4}$ cycle/s

3. (a) $y = 0.01 \sin \left(400\pi\, t + \dfrac{3\pi}{2} \right)$

(b) See next column.

(c) 200 cycles/s

(b)

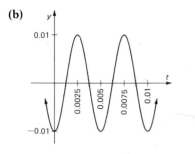

(b)

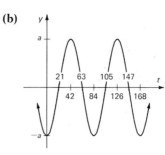

5. (a) $y = 30 \sin \left(\pi t + \dfrac{3\pi}{2} \right)$

(b)

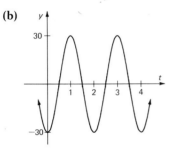

(c) $\dfrac{1}{2}$ cycle/s

7. (a) $y = \dfrac{\pi}{4} \sin \left(\dfrac{\pi}{20} t + \dfrac{\pi}{2} \right)$

(b)

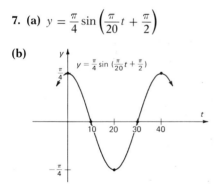

$y = \dfrac{\pi}{4} \sin \left(\dfrac{\pi}{20} t + \dfrac{\pi}{2} \right)$

(c) $\dfrac{1}{40}$ cycle/s

9. (a) $y = a \sin \left(\dfrac{\pi}{42} t + \dfrac{3\pi}{2} \right)$

(b) See next column.

(c) $\dfrac{1}{84}$ cycle/min

11. $y = 4 \cos \left(\dfrac{\pi}{2} t + \pi \right)$

13. $y = 0.01 \cos(400\pi t + \pi)$

15. $y = 30 \cos(\pi t + \pi)$ **17.** $y = \dfrac{\pi}{4} \cos \dfrac{\pi}{20} t$

19. $y = a \cos \left(\dfrac{\pi}{42} t + \pi \right)$

21. (a) $y = 120 \sin(120\pi t + \beta)$

(b) $y = 120 \sin \left(120\pi t + \dfrac{5\pi}{3} \right)$, $\beta = \dfrac{5\pi}{3}$

23. $x = a \cos \omega t$

27. $t = -\dfrac{\beta}{\omega} + \dfrac{2n\pi}{\omega}$, if $a > 0$

$t = -\dfrac{\beta}{\omega} + \dfrac{(2n+1)\pi}{\omega}$, if $a < 0$

Exercises 5-4, page 138

1.

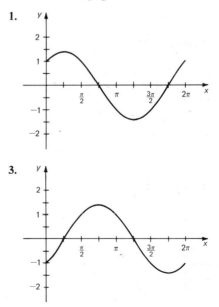

3.

5.

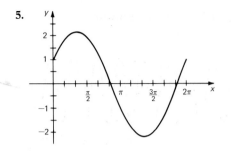

7.

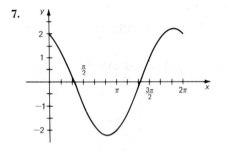

9.

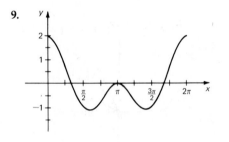

11.

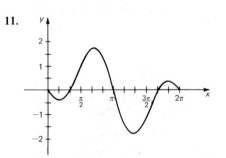

13.

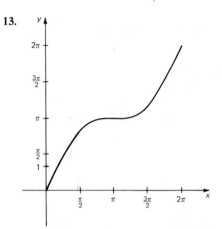

15.

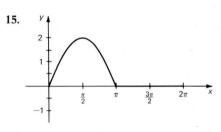

17. $y = \sin 3x + \cos x$,

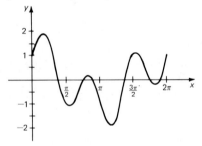

19. $y = \dfrac{1}{2}\sin 4x + \dfrac{1}{2}\cos 2x$,

(portion of graph shown)

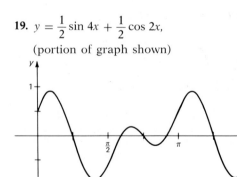

21. $y = \dfrac{1}{2}\cos 2x + \dfrac{1}{2}\cos x$,

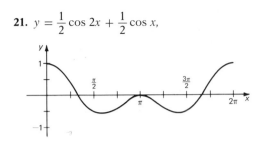

Cumulative Review • Chapters 3–5, pages 140–141

Chapter 3   1. $-\cos x$ **5. (a)** $-\dfrac{3\sqrt{3}+4}{10}$

(b) $-\dfrac{4\sqrt{3}+3}{10}$ **(c)** $\dfrac{48+25\sqrt{3}}{39}$

7. $\sqrt{2}-1$ **9.** $0, \dfrac{7\pi}{6}, \dfrac{11\pi}{6}$

Chapter 4   1. $3\sqrt{6}$ **3.** $70°$ **5.** 468 cm^2

Chapter 5

1. (a) Period $= \pi$, Amplitude $= 3$

(b)

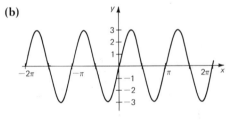

3. (a) Period $= \dfrac{2}{3}\pi$, Amplitude $= 2$

(b) (portion of graph shown)

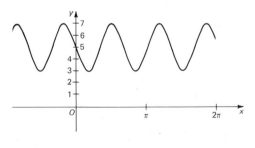

5. (a) $H = \sin\left(2\pi t + \dfrac{3\pi}{2}\right) + 1.5$

(b) (portion of graph shown)

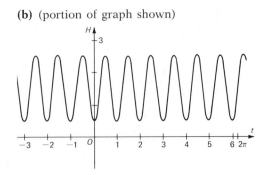

(c) 1 cycle/s

(d) $H = \cos(2\pi t + \pi) + 1.5$

CHAPTER 6   VECTORS IN THE PLANE

Exercises 6-1, page 151–153   1. $\|\mathbf{a}\| = 1$; $\|\mathbf{b}\| = 1$; $\|\mathbf{a}+\mathbf{b}\| = \sqrt{2}$; $\|\mathbf{a}-\mathbf{b}\| = \sqrt{2}$

3. $\|\mathbf{a}\| = \sqrt{5}$; $\|\mathbf{b}\| = 2\sqrt{5}$; $\|\mathbf{a}+\mathbf{b}\| = \sqrt{41}$; $\|\mathbf{a}-\mathbf{b}\| = 3$

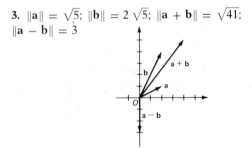

5. $\|\mathbf{a}\| = 5$; $\|\mathbf{b}\| = \sqrt{5}$; $\|\mathbf{a}+\mathbf{b}\| = \sqrt{10}$; $\|\mathbf{a}-\mathbf{b}\| = 5\sqrt{2}$

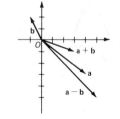

7. $\hat{a} = \dfrac{2\sqrt{5}}{5}i + \dfrac{\sqrt{5}}{5}j$; $\hat{b} = \dfrac{\sqrt{5}}{5}i + \dfrac{2\sqrt{5}}{5}j$;

$\widehat{a+b} = \dfrac{4\sqrt{41}}{41}i + \dfrac{5\sqrt{41}}{41}j$; $\widehat{a-b} = -j$

9. $11i$ **11.** $\dfrac{5}{2}i - j$ **13.** $s = -1$ and $t = 1$

15. $s = 6$ and $t = 7$ **17. (a)** $u = a + b$;
$v = 2a + b$ **(b)** $a = -u + v$; $b = 2u - v$

19. (a) $u = \dfrac{1}{2}a + \dfrac{1}{2}b$; $v = \dfrac{1}{2}a - \dfrac{1}{2}b$

(b) $a = u + v$; $b = u - v$ **31.** $r = -1$;
$s = 2$; $t = 1$ **33.** $r = 1$; $s = 1$; $t = 1$

Exercises 6-2, pages 156–157 **1.** Distance is 634 km; bearing is 96.8° or 96°50′
3. Ground speed is 325 km/h; true course is 96.9° or 96°50′ **5.** Distance is 65 km; heading is 330° **7. (a)** 24 m **(b)** 287°
9. 363 km **11.** Compass heading is 316.6° or 316°40′; the ETA is 1:54 P.M.

Exercises 6-3, pages 160–161 **1.** The magnitude of the parallel component is 110 N; the magnitude of the perpendicular component is 273 N. **3.** 1140 N **5.** 2490 N and 3810 N **7.** $\|F\| = 8.3$ N; the bearing is 338.7° or 338°40′ **9.** $\|F\| = 55$ N; the bearing is 153.5° or 153°30′ **11.** 91 kg.

Exercises 6-4, pages 164–166 **1.** 0
3. $p^2 - q^2$ **5. (a)** $b \parallel d$ **(b)** $b \perp c$ and $c \perp d$
7. 38 **9.** 51 **11.** 60° **13.** 26.6° or 26°30′

15. 121.3° or 121°20′ **17. (a)** $\dfrac{\sqrt{2}}{2}i - \dfrac{\sqrt{2}}{2}j$

or $-\dfrac{\sqrt{2}}{2}i + \dfrac{\sqrt{2}}{2}j$ **(b)** $\dfrac{\sqrt{2}}{2}i + \dfrac{\sqrt{2}}{2}j$ or

$-\dfrac{\sqrt{2}}{2}i - \dfrac{\sqrt{2}}{2}j$ **19. (a)** $\dfrac{4}{5}i - \dfrac{3}{5}j$ or

$-\dfrac{4}{5}i + \dfrac{3}{5}j$ **(b)** $\dfrac{3}{5}i + \dfrac{4}{5}j$ or $-\dfrac{3}{5}i - \dfrac{4}{5}j$

21. (a) $\dfrac{x}{\sqrt{x^2 + y^2}}i + \dfrac{y}{\sqrt{x^2 + y^2}}j$ or

$-\dfrac{x}{\sqrt{x^2 + y^2}}i - \dfrac{y}{\sqrt{x^2 + y^2}}j$

(b) $\dfrac{y}{\sqrt{x^2 + y^2}}i - \dfrac{x}{\sqrt{x^2 + y^2}}j$ or

$-\dfrac{y}{\sqrt{x^2 + y^2}}i + \dfrac{x}{\sqrt{x^2 + y^2}}j$ **23.** $\dfrac{\sqrt{2}}{2}$

25. $\dfrac{1}{5}$ **27.** $v = \dfrac{1}{2}a - \dfrac{1}{2}b$

29. $v = \dfrac{5}{17}a + \dfrac{14}{17}b$

Exercises 6-5, page 168 **1.** 10 J **3. (a)** 70 J
(b) 70 J **5.** 612,500 J or 0.1701 Kwh
7. 7638 J **9.** 3.995×10^7 J **11.** 1.135×10^7 J

CHAPTER 7 POLAR COORDINATES

Exercises 7-1, pages 178–179
1. $(\sqrt{3}, 1)$

3. $(-\sqrt{3}, -1)$

5. $(\sqrt{3}, -1)$

7. $(-\sqrt{3}, 1)$

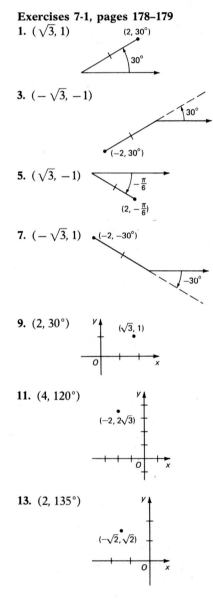

9. $(2, 30°)$

11. $(4, 120°)$

13. $(2, 135°)$

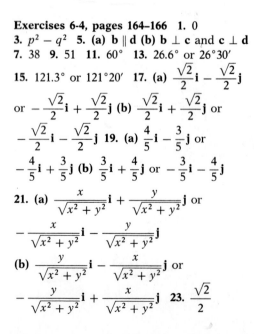

15. $(2\sqrt{2}, 30°)$

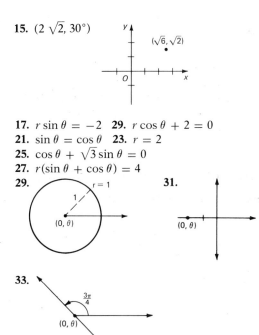

17. $r \sin \theta = -2$ **29.** $r \cos \theta + 2 = 0$
21. $\sin \theta = \cos \theta$ **23.** $r = 2$
25. $\cos \theta + \sqrt{3} \sin \theta = 0$
27. $r(\sin \theta + \cos \theta) = 4$
29.

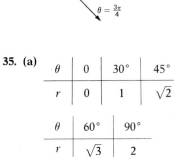

31.

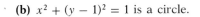

33.

35. (a)

θ	0	30°	45°
r	0	1	$\sqrt{2}$

θ	60°	90°
r	$\sqrt{3}$	2

(b) $x^2 + (y - 1)^2 = 1$ is a circle.

37. (a)

θ	0	30°	45°
r	1	$\dfrac{1 + \sqrt{3}}{2}$	$\sqrt{2}$

θ	60°	90°
r	$\dfrac{1 + \sqrt{3}}{2}$	1

(b) $\left(x - \dfrac{1}{2}\right)^2 + \left(y - \dfrac{1}{2}\right)^2 = \dfrac{1}{2}$ is a circle.
41. $(r \cos \theta - r_0 \cos \theta_0)^2 + (r \sin \theta - r_0 \sin \theta_0)^2$
$= a^2$

Exercises 7-2, pages 182–183

1.

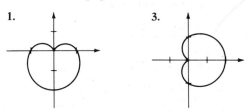

3.

5.

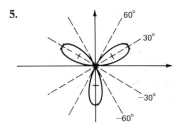

7.

9.

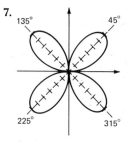

11.

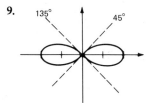

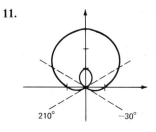

13.

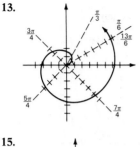

15.

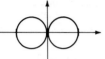

17.

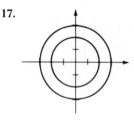

19.

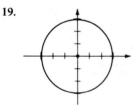

21.

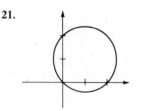

23.

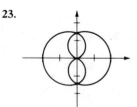

Exercises 7-3, page 187   1. parabola
3. ellipse   **5.** hyperbola   **7.** ellipse
9. parabola   **11.**

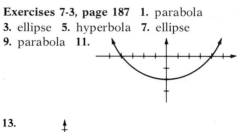

13.

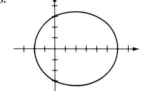

15.   **17.**

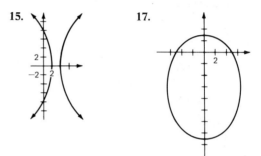

19.

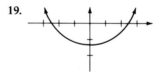

21. hyperbola

Exercises 7-4, pages 190–191   1. (a) 4
(b) $4 - 2i$ **(c)** $1 + 2i$ **3. (a)** $3 + 9i$
(b) $-20 + 15i$ **(c)** $\dfrac{4}{5} + \dfrac{3}{5}i$ **5. (a)** -2 **(b)** 4
(c) $-\dfrac{1}{2} - \dfrac{i}{2}\sqrt{3}$ **7. (a)** $2 - i$ **(b)** $\sqrt{5}$
(c) $\dfrac{2}{5} - \dfrac{1}{5}i$ **9. (a)** $\dfrac{\sqrt{3}}{2} + \dfrac{1}{2}i$ **(b)** 1
(c) $\dfrac{\sqrt{3}}{2} + \dfrac{1}{2}i$ **11.** $z^2 = -2i$; $z^3 = 2 + 2i$
13. $z^2 = -2 - 2i\sqrt{3}$; $z^3 = 8$ **15. (a)** i
(b) -1 **(c)** 1   **17.** i **19. (a)** $-i$ **(b)** -1 **(c)** i
(d) 1   **21.** -1 **23.** 1

Exercises 7-5, pages 193–194

1.

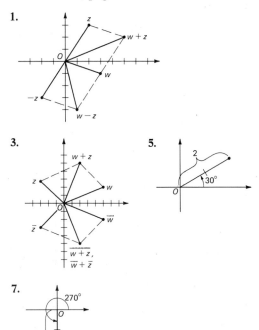

3.

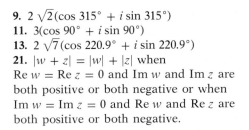

5.

7.

9. $2\sqrt{2}(\cos 315° + i \sin 315°)$
11. $3(\cos 90° + i \sin 90°)$
13. $2\sqrt{7}(\cos 220.9° + i \sin 220.9°)$
21. $|w + z| = |w| + |z|$ when
Re w = Re z = 0 and Im w and Im z are
both positive or both negative or when
Im w = Im z = 0 and Re w and Re z are
both positive or both negative.

Exercises 7-6, pages 196–197
1. $2\sqrt{2}(\cos 225° + i \sin 225°)$
3. $8(\cos 180° + i \sin 180°)$
5. a. $\dfrac{1}{2}(\cos 240° + i \sin 240°)$

b. $-\dfrac{1}{4} - \dfrac{i}{4}\sqrt{3}$

7. a. $\dfrac{\sqrt{2}}{2}(\cos 225° + i \sin 225°)$

b. $-\dfrac{1}{2} - \dfrac{1}{2}i$ **9.** $\dfrac{\sqrt{2}}{2}(\cos 45° + i \sin 45°)$

11. $\dfrac{1}{2}(\cos 300° + i \sin 300°)$

13. a. $2(\cos 190° + i \sin 190°)$
b. $2(\cos 10° + i \sin 10°)$

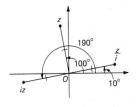

15. a. $2\sqrt{2}(\cos 315° + i \sin 315°)$
b. $2\sqrt{2}(\cos 135° + i \sin 135°)$

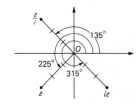

19.

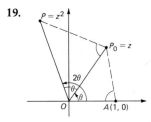

21.

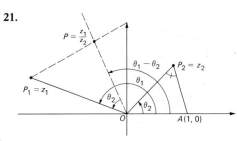

Exercises 7-7, pages 199–200   **1.** 64   **3.** $-\dfrac{1}{16}$

5. -1 **7.** -8 **9.** $32i$ **11.** $9i$ **13.** $\dfrac{1}{4}i$

15. -1 **17.** $\dfrac{\sqrt{3}}{4} + \dfrac{1}{4}i$

21. $\cos 5\theta = \cos^5 \theta - 10 \cos^3 \theta \sin^2 \theta + 5 \cos \theta \sin^4 \theta$ or $\cos 5\theta = 16 \cos^5 \theta - 20 \cos^3 \theta + 5 \cos \theta$; $\sin 5\theta = \sin^5 \theta - 10 \cos^2 \theta \sin^3 \theta + 5 \cos^4 \theta \sin \theta$ or $\sin 5\theta = 16 \sin^5 \theta - 20 \sin^3 \theta + 5 \sin \theta$

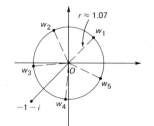

Exercises 7-8, pages 203–205 **1.** $1, -\dfrac{1}{2} + \dfrac{\sqrt{3}}{2}i, -\dfrac{1}{2} - \dfrac{\sqrt{3}}{2}i$

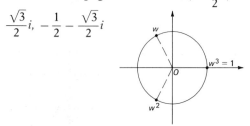

3. $1, 0.3090 + 0.9511i, -0.8090 + 0.5878i, -0.8090 - 0.5878i, 0.3090 - 0.9511i$

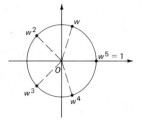

5. $\dfrac{\sqrt{3}}{2} + \dfrac{1}{2}i, -\dfrac{\sqrt{3}}{2} + \dfrac{1}{2}i, -i$

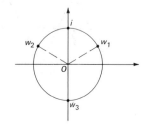

7. $2^{\frac{1}{10}}(\cos 45° + i \sin 45°),$
$2^{\frac{1}{10}}(\cos 117° + i \sin 117°),$
$2^{\frac{1}{10}}(\cos 189° + i \sin 189°),$
$2^{\frac{1}{10}}(\cos 261° + i \sin 261°),$
$2^{\frac{1}{10}}(\cos 333° + i \sin 333°)$

9. $2^{\frac{1}{6}}(\cos 25° + i \sin 25°),$
$2^{\frac{1}{6}}(\cos 85° + i \sin 85°),$
$2^{\frac{1}{6}}(\cos 145° + i \sin 145°),$
$2^{\frac{1}{6}}(\cos 205° + i \sin 205°),$
$2^{\frac{1}{6}}(\cos 265° + i \sin 265°),$
$2^{\frac{1}{6}}(\cos 325° + i \sin 325°),$

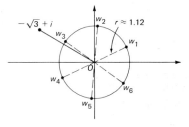

11. Since $z^2 + z + 1 = \dfrac{z^3 - 1}{z - 1}$ any cube root of unity will make $z^3 - 1$ equal to zero, hence $z^2 + z + 1$ will equal zero if $z \neq 1$. **13.** Any nth root of unity other than 1 satisfies $z^{n-1} + z^{n-2} + z^{n-3} + \cdots + z^2 + z + 1 = 0$ **15.** $z = -i, z = -2 + i$ **19.** w, w^5, w^7, w^{11}

CUMULATIVE REVIEW • Chapters 6 and 7, pages 207–208

Chapter 6 **1.**

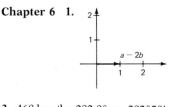

3. 468 km/h; 282.8° or 282°50′

5. $y = -\dfrac{5}{17}a + \dfrac{14}{17}b$ **7.** (2) is true.

Chapter 7 **1.** $\left(-\dfrac{7}{2}, \dfrac{7}{2}\sqrt{3}\right)$

3.

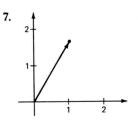

5. (a) $6 - 2i$ **(b)** $24\sqrt{2}(\cos 45° + i \sin 45°)$

7.

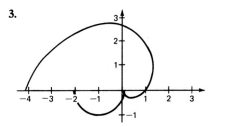

9. $-117.2 + 43.6i$

11. $\cos 22.5° + i \sin 22.5°$,
$\cos 112.5° + i \sin 112.5°$,
$\cos 202.5° + i \sin 202.5°$,
$\cos 292.5° + i \sin 292.5°$

CHAPTER 8   TRANSFORMATIONS

Exercises 8-1, pages 215–216

1. (a) $x = x' + 2$, $y = y' - 4$ **(b)** $y' = (x')^2$
(c)

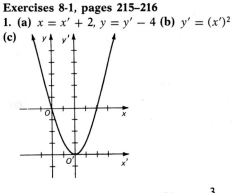

3. (a) $x = x' + 3$, $y = y' + 1$ **(b)** $y' = \dfrac{3}{4}x'$

(c)

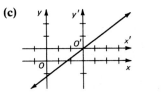

5. (a) $x = x' + 2$, $y = y' - 1$ **(b)** $x'y' = 1$

(c)

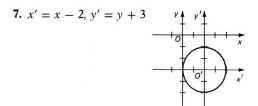

7. $x' = x - 2$, $y' = y + 3$

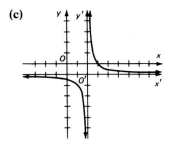

9. $x' = x$, $y' = y - 2$

11. $y' = \cos 2x'$

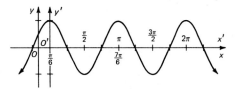

13. $y' = \cos 3x'$

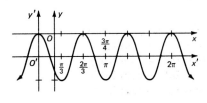

418   *Selected Answers*

15. $y' = y - 3, x' = x + 1$

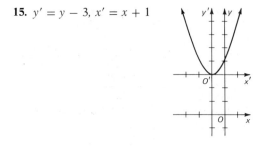

17. $x' = x - 1, y' = y - 2$

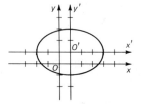

19. $x' = x + 3, y' = y$

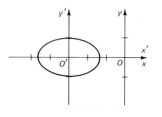

21. $x' = x - 2, y' = y + 2$

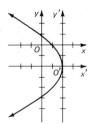

23. $x' = x + 2, y' = y - 4$

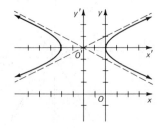

Exercises 8-2, pages 219–220

1. $y' = -3$

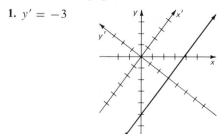

3. $x'y' = 1$

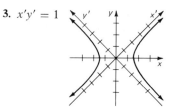

5. $x' = 0$ or $x' = 1$

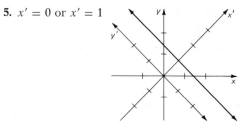

7. $y' = (x')^2$

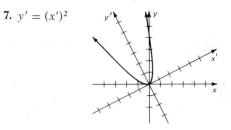

9. $(x')^2 - 3(y')^2 = 36$

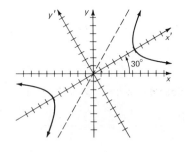

11. $-2(x')^2 + 3(y')^2 = 10$

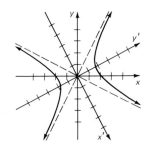

13. $x' = x \cos \theta + y \sin \theta,$
$y' = -x \sin \theta + y \cos \theta$ **15.** $y' = -2\sqrt{2}$
17. $y' = \dfrac{3\sqrt{10}}{5}$

Exercises 8-3, pages 222–223

1.

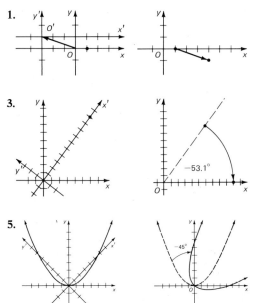

3.

5.

7. $y' = 2(x')^2$ **9.** $y' = 3(x')^2$

Exercises 8-4, pages 226–227 **1.** $x = \dfrac{x' - y'}{\sqrt{2}},$
$y = \dfrac{x' + y'}{\sqrt{2}}$ **3.** $x = \dfrac{\sqrt{3}x' - y'}{2},$

$y = \dfrac{x' + \sqrt{3}y'}{2}$ **5.** $x = \dfrac{x' - 2y'}{\sqrt{5}},$
$y = \dfrac{2x' + y'}{\sqrt{5}}$ **7.** $x = \dfrac{x' - 2y'}{\sqrt{5}},\ y = \dfrac{2x' + y'}{\sqrt{5}}$
9. $x = \dfrac{x' - y'}{\sqrt{2}},\ y = \dfrac{x' + y'}{\sqrt{2}}$
11. (a) $y' = -(x')^2$ **(b)** No translation is
needed. **(c)**

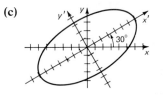

13. (a) $(x')^2 + 5(y')^2 = 50$ **(b)** No
translation is needed.

(c)

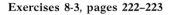

15. (a) $\dfrac{(x')^2}{80} + \dfrac{(y')^2}{30} = 1$
(b) No further transformation needed.

(c)

17. (a) $(y')^2 - 4(x')^2 = 4$ **(b)** No translation
is needed. **(c)**

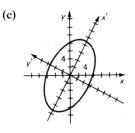

19. (a) $3(x' - 1)^2 + (y' - 3)^2 = 36$
(b) $x'' = x' - 1$, $y'' = y' - 3$,
$3(x'')^2 + (y'')^2 = 36$

(c)
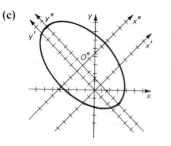

(a)
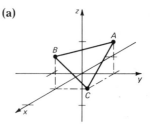

7. $x^2 + y^2 + z^2 - 25 = 0$
9. $x^2 + y^2 + z^2 - 4x + 2y + 4z = 0$
11. center is $(2, 0, 0)$, radius is 1   **13.** center
is $(2, -3, -1)$, radius is 3   **15.** $(5, 0, 3)$,
$(5, 0, -1)$, $(2, 0, -1)$, $(2, 2, -1)$, $(2, 2, 3)$,
$(5, 2, 3)$ **17.** $(0, 0, 4)$ **19.** $x - y = 0$
21. $x^2 + y^2 + z^2 - 6x + 8z = 0$
23. $x^2 + y^2 + z^2 - 4x \pm 4y - 6z + 13 = 0$
27. $36y^2 + 100z^2 - 250 = 0$

CHAPTER 9   VECTORS IN SPACE

Exercises 9-1, pages 235–236

1. (a)
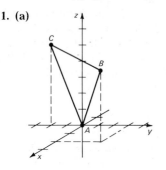

(b) $\triangle ABC$ is neither right nor isosceles.

Exercises 9-2, pages 239–241
1. (a) $3\mathbf{i} + 3\mathbf{k}$; $\mathbf{i} - 8\mathbf{j} + 5\mathbf{k}$ **(b)** $\|\mathbf{a}\| = 3\sqrt{2}$;
$\|\mathbf{b}\| = 3$; $\|\mathbf{a} + \mathbf{b}\| = 3$; $\|\mathbf{a} - \mathbf{b}\| = 3\sqrt{5}$
3. (a) $3\mathbf{i} + \mathbf{j} + 2\mathbf{k}$ **(b)** $3\mathbf{a} - 5\mathbf{b} + 6\mathbf{c}$
5. (a) $2\mathbf{i} + 6\mathbf{j} + \mathbf{k}$ **(b)** $-2\mathbf{a} - \mathbf{b} + 5\mathbf{c}$
7. (a) $\mathbf{a} \perp \mathbf{c}$; $\mathbf{c} \perp \mathbf{d}$ **(b)** $\mathbf{a} \parallel \mathbf{d}$ **9.** $15.8°$ or
$15°50'$ **11.** $107.3°$ or $107°20'$

13. $\dfrac{1}{2}\mathbf{a} - \dfrac{2}{3}\mathbf{b} + \dfrac{1}{6}\mathbf{c}$ **15.** $8\mathbf{i} - \mathbf{j} - 7\mathbf{k}$

3. (a)
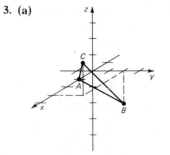

(b) $\triangle ABC$ is neither right nor isosceles.

5. (a) See next column.
(b) $\triangle ABC$ is isosceles and not right.

Exercises 9-3, pages 245–246
1. $\mathbf{r} = (1 + t)\mathbf{i} + (2 - t)\mathbf{j} + 3t\mathbf{k}$
3. $\mathbf{r} = 2t\mathbf{i} + 3t\mathbf{j} - t\mathbf{k}$
5. $\mathbf{r} = (1 + t)\mathbf{i} + (2 + t)\mathbf{j} + t\mathbf{k}$
7. $\mathbf{r} = t\mathbf{i} + (-1 + 2t)\mathbf{j} + (3 - t)\mathbf{k}$
9. $(5, 4, 0)$ **11.** $2\sqrt{5}$ **13.** $\dfrac{5\sqrt{3}}{3}$
17. $(4, 2, -2)$, $(-2, -1, 1)$ **19.** $(1, 2, 1)$

Exercises 9-4, pages 250–251
1. (a) $2x + y + 2z - 4 = 0$

(b)
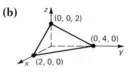

3. (a) $x + 2z - 10 = 0$

(b)

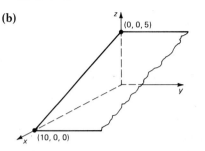

5. $\mathbf{r} = (3 + 2t)\mathbf{i} + (2 + t)\mathbf{j} + (1 - 2t)\mathbf{k}$
7. $\mathbf{r} = 3t\,\mathbf{i} + (-2 + 2t)\mathbf{j} + (2 - 6t)\mathbf{k}$ **9.** 4
11. $\dfrac{20}{7}$ **13.** $2x - y + 1 = 0$ **15.** $\dfrac{17}{6}$
17. $(0, -6, 4)$
21. $\mathbf{r} = (2 + 2t)\mathbf{i} + (1 - t)\mathbf{j} + t\,\mathbf{k}$
23. (a) one **(b)** many **(c)** many **(d)** one
(e) none **(f)** none

CUMULATIVE REVIEW • page 254
Chapter 8   1. (a) $x = x' + 1$, $y = y' + 2$
(b) $(x' - 2)^2 + (y' + 6)^2 = 13$

3. y'

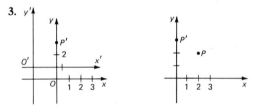

Chapter 9   1. $\triangle ABC$
is right and not isosceles.
3. (a) $3\mathbf{i} + 7\mathbf{j}$, $\mathbf{i} + 5\mathbf{j} - 4\mathbf{k}$ **(b)** $\sqrt{21}$
5. $\mathbf{r} = (1 + 4t)\mathbf{i} + (1 + 3t)\mathbf{j} + (1 - 12t)\mathbf{k}$

**CHAPTER 10   SPHERICAL
TRIGONOMETRY**

Exercises 10-1, page 263
1. 1.000×10^4 km   **3.** 1.750×10^3 km
5. 9.00° or 9°00′   **7.** 1.00° or 1°00′
9. $a = 30°$, $b = 90°$, $c = 90°$ **11.** $\angle A = 90°$,
$\angle B = 90°$, $c = 70°$ **13.** $\angle A = 150°$,
$\angle B = 90°$, $\angle C = 90°$ **15.** A triangle
formed by three points equally spaced
around the equator.   **17.** The poles lie on

the equator and are the points of
intersection of the equator and a meridian
circle that intersects the given meridian
circle at 90°.

Exercises 10-2, pages 266–267
1. $c = 100°40′$; $\angle A = 113°20′$;
$\angle B = 66°30′$ **3.** $\angle A = 37°20′$; $b = 70°0′$;
$c = 73°50′$ **5.** $a = 26°20′$; $b = 121°10′$;
$c = 117°40′$ **7.** $a = 145.7°$; $\angle A = 142.4°$;
$\angle B = 73.6°$ or 106.4°   **9.** $\angle A = 77.7°$;
$a = 70.7°$; $b = 38.4°$ or 141.6°

Exercises 10-3, pages 270–271
1. $\cos c = \cos a \cos b$; $\sin b = \tan a \cot A$;
$\sin a = \tan b \cot B$ **3.** $\sin b = \tan a \cot A$;
$\sin a = \sin c \sin A$; $\cos A = \cos a \sin B$
5. $\cos B = \tan a \cot c$; $\sin b = \sin c \sin B$;
$\cos c = \cot A \cot B$ **7.** $b_1 = 13°$,
$\angle B_1 = 21°$, $c_1 = 40°$; $b_2 = 167°$,
$\angle B_2 = 159°$, $c_2 = 140°$ **9.** There is no such
triangle   **11.** $\angle A = 76°10′$; $\angle B = 123°20′$;
$\angle C = 82°30′$ **13.** $\angle A = 36°40′$; $a = 42°30′$;
$\angle B = 125°50′$ **15.** $a = 25°20′$; $b = 84°30′$;
$\angle C = 101°40′$ **23.** If $a > A$ then the
triangle will have no solution.

Exercises 10-4, pages 275–276
1. $\angle A = 64°30′$; $\angle B = 113°20′$;
$\angle C = 87°10′$ **3.** $a = 41°0′$; $b = 143°20′$;
$\angle C = 66°10′$ **5.** $\angle A = 123.3°$; $b = 91.3°$;
$\angle C = 32.9°$ **7.** $a = 79.1°$; $b = 79.1°$;
$c = 52.7°$ **9.** $\angle B_1 = 60°10′$, $c_1 = 72°50′$,
$\angle C_1 = 100°40′$; $\angle B_2 = 119°50′$, $c_2 = 24°30′$,
$\angle C_2 = 25°10′$ **11.** There is no solution.
13. $a_1 = 129.8°$, $\angle A_1 = 136.2°$,
$\angle B_1 = 58.5°$ **15.** $b_1 = 39°$, $c_1 = 28°$,
$\angle C_1 = 61°$

Exercises 10-5, pages 279–280   1. (a) 1785
nautical miles **(b)** 3615 nautical miles
(c) 7185 nautical miles   **3.** 27°58′ N,
169°09′ W; 243°32′   **5.** 48°36′ N, 34°30′ W
7. 60° N   **9. (a)** 3544 nautical miles
(b) 59°36′ **(c)** 121°11′   **11. (a)** 6569 nautical
miles **(b)** 81°12′ **(c)** 132°39′   **13. (a)** 4361
nautical miles **(b)** 130°31′ **(c)** 134°11′
15. (a) 5831 nautical miles **(b)** 6235
nautical miles

CHAPTER 11   INFINITE SERIES

Exercises 11-1, pages 287–288   1. (a) 0.9998
(b) 0.4969   **3. (a)** 0.908 **(b)** 0.909   **5.** 0.367
7. (a) 2.594 **(b)** 2.692 **(c)** 2.704 **(d)** 2.716

Exercises 11-3, pages 293–294

1.

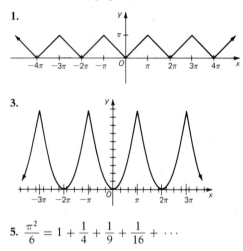

3.

5. $\dfrac{\pi^2}{6} = 1 + \dfrac{1}{4} + \dfrac{1}{9} + \dfrac{1}{16} + \cdots$

CUMULATIVE REVIEW • Chapters 10–11,

page 298   Chapter 10   1. $a = 50°$;
$\angle B = 90°$; $\angle C = 90°$ **3.** $\angle A = 62°50'$;
$\angle B = 112°40'$; $\angle C = 100°10'$ **5.** 33°10′ N,
119°40′ W; course is 249°20′   **Chapter 11**
1. 0.866   **3.** 1.350   **5.** 3.105

7. (a) $-\dfrac{1}{2} + i\dfrac{\sqrt{3}}{2}$

(b) $\dfrac{\sqrt{3}}{2} + \dfrac{1}{2}i$

EXTRA PRACTICE EXERCISES

Chapter 1, page 312

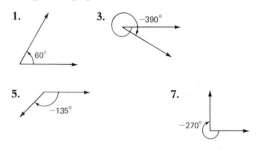

9. (a) 0.8000, 0.6000 **(b)** 50°
11. (a) 0.9428, 0.3333 **(b)** 70°
13. (a) 0.7454, 0.6667 **(b)** 50°
15. (a) 0.8944, 0.4472 **(b)** 65°   **17.** 0.4014,
0.9159   **19.** 0.9075, 0.4200   **21.** 76.17° or
76°10′   **23.** 57.17° or 57°10′   **25.** $\sin \theta = \dfrac{5}{13}$,

$\cos \theta = \dfrac{12}{13}$, $\tan \theta = \dfrac{5}{12}$, $\csc \theta = \dfrac{13}{5}$,

$\sec \theta = \dfrac{13}{12}$, $\cot \theta = \dfrac{12}{5}$ **27.** $\sin \theta = \dfrac{21}{29}$,

$\cos \theta = \dfrac{20}{29}$, $\tan \theta = \dfrac{21}{20}$, $\csc \theta = \dfrac{29}{21}$,

$\sec \theta = \dfrac{29}{20}$, $\cot \theta = \dfrac{20}{21}$ **29.** $\sin \theta = \dfrac{\sqrt{2}}{2}$,

$\cos \theta = \dfrac{\sqrt{2}}{2}$, $\tan \theta = 1$, $\csc \theta = \sqrt{2}$,

$\sec \theta = \sqrt{2}$, $\cot \theta = 1$ **31.** $\sin \theta = \dfrac{2\sqrt{5}}{5}$,

$\cos \theta = \dfrac{\sqrt{5}}{5}$, $\tan \theta = 2$, $\csc \theta = \dfrac{\sqrt{5}}{2}$,

$\sec \theta = \sqrt{5}$, $\cot \theta = \dfrac{1}{2}$ **33.** $a = 100.8$,
$\angle A = 57.2°$ or 57°10′, $\angle B = 32.8°$ or
32°50′   **35.** $c = 40.8$, $\angle A = 59°$, $\angle B = 31°$
37. $a = 99.1$, $c = 122.5$, $\angle A = 54°$
39. $c = 30.0$, $\angle A = 36.9°$ or 36°50′,

$\angle B = 53.1°$ or 53°10′   **41.** $\sin \theta = -\dfrac{\sqrt{2}}{2}$,

$\cos \theta = -\dfrac{\sqrt{2}}{2}$, $\tan \theta = 1$, $\csc \theta = -\sqrt{2}$,

$\sec \theta = -\sqrt{2}$, $\cot \theta = 1$ **43.** $\sin \theta = -1$,
$\cos \theta = 0$, $\tan \theta =$ undef., $\csc \theta = -1$,
$\sec \theta =$ undef., $\cot \theta = 0$ **45.** 0.3640
47. −0.9833   **49.** 10.39

Chapter 2, pages 313–314   1. $\dfrac{5\pi}{4}$ **3.** $-\dfrac{5\pi}{9}$

5. $-\dfrac{\pi}{3}$ **7.** 135°   **9.** 210°   **11.** $-\dfrac{\sqrt{3}}{2}$, $-\dfrac{1}{2}$,

$\sqrt{3}$, -2, $-\dfrac{2\sqrt{3}}{3}$, $\dfrac{\sqrt{3}}{3}$ **13.** $\dfrac{1}{2}$, $\dfrac{\sqrt{3}}{2}$, $\dfrac{\sqrt{3}}{3}$,

$\dfrac{2\sqrt{3}}{3}$, 2, $\sqrt{3}$ **15.** $-\dfrac{\sqrt{3}}{2}$, $\dfrac{1}{2}$, $-\sqrt{3}$, 2, $-\dfrac{2\sqrt{3}}{3}$,

$-\dfrac{\sqrt{3}}{3}$ **17.** $-\dfrac{\sqrt{3}}{2}$, $\dfrac{1}{2}$, $-\sqrt{3}$, 2, $-\dfrac{2\sqrt{3}}{3}$,

$-\dfrac{\sqrt{3}}{3}$ **19.** (3, 0)   **21.** (−2, 0)   **23.** (4, 0)

25. (4, 0)   **27.**

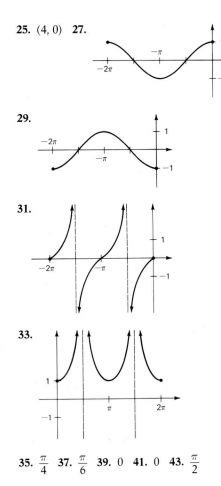

29.

31.

33.

35. $\dfrac{\pi}{4}$ **37.** $\dfrac{\pi}{6}$ **39.** 0   **41.** 0   **43.** $\dfrac{\pi}{2}$

Chapter 3, pages 314–315   1. $\dfrac{1 - 2\sin^2 x}{\sin^2 x}$

3. $\dfrac{\cos^2 x}{\sqrt{1 - \cos^2 x}}$ **5.** $\dfrac{-\cos^3 x}{1 - \cos^2 x}$ **7.** $2\tan x$

17. $-\dfrac{63}{65}$ **19.** $-\dfrac{56}{65}$ **21.** $\dfrac{16}{65}$ **23.** $-\dfrac{56}{65}$

25. (a) $\dfrac{120}{169}$, **(b)** $-\dfrac{119}{169}$ **27. (a)** $-\dfrac{24}{25}$,

(b) $-\dfrac{7}{25}$ **29.** $-\dfrac{\sqrt{34}}{12}$ **31.** $\dfrac{1}{7}$ **33.** $\dfrac{\pi}{6}, \dfrac{5\pi}{6}, \dfrac{3\pi}{2}$,

$\dfrac{\pi}{2}$ **35.** $0, \dfrac{\pi}{4}, \pi, \dfrac{5\pi}{4}, 2\pi$ **37.** $n\pi, \dfrac{\pi + n\pi}{2}$

39. $n\pi, \dfrac{\pi}{2} + n\pi$ **41.** $5\cos(\theta + 53°)$

43. $3\sqrt{3}\cos(\theta + 16°)$
45. $3\sqrt{2}\cos(\theta + 45°)$ **47.** $2\cos(\theta - 45°)$

Chapter 4, page 316   1. 48.2° or 48°10′
3. 4.58   **5.** 15.39   **7.** 49.34 and 34.87
9. $\angle B = 78°$, $a = 4.6$, $c = 6.3$
11. $\angle C = 31°40′$, $a = 23.2$, $b = 15.2$
13. $\angle A = 54°50′$, $b = 25.5$, $c = 21.8$
15. $\angle C = 74°20′$, $a = 12.0$, $b = 15.7$ **17.** no
such triangle   **19.** $\angle B = 23°10′$ or 156°50′
21. $\angle B = 90°$ **23.** no such triangle
25. 47.8   **27.** 236.9   **29.** 3.4   **31.** 27.5

Chapter 5, pages 317–318

1.

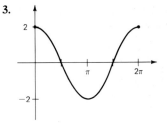

3.

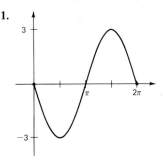

5.

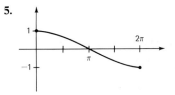

7.

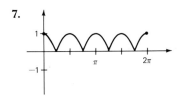

9.

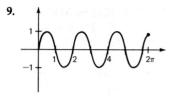

11.

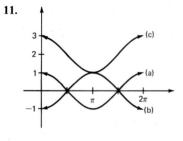

13.

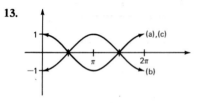

15.

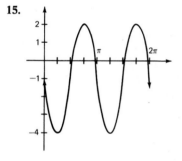

17.

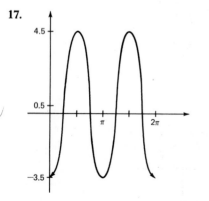

19. (a) $y = 7 \sin \left(\pi t - \dfrac{\pi}{2} \right)$

(b)

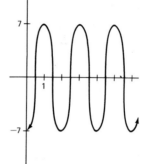

(c) $f = \dfrac{1}{2}$

21. (a) $y = 20 \sin \left(\pi t + \dfrac{11\pi}{6} \right)$

(b)

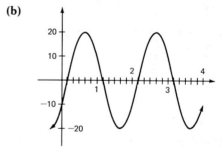

(c) $f = \dfrac{1}{2}$

23. $y = 7 \cos(\pi t - \pi)$

25. $y = 20 \cos \left(\pi t + \dfrac{4\pi}{3} \right)$

27.

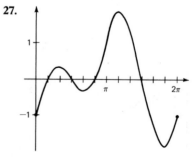

29.

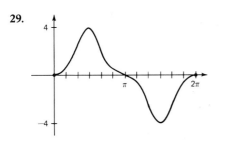

31.

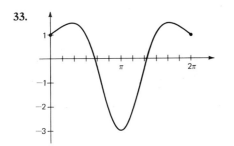

33.

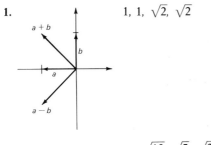

Chapter 6, pages 318–320

1.

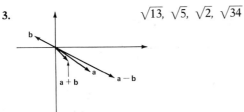

$1, 1, \sqrt{2}, \sqrt{2}$

3.

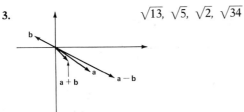

$\sqrt{13}, \sqrt{5}, \sqrt{2}, \sqrt{34}$

5. i + 13**j**   **7. i** + **j**   **9.** 206 km, 346°
11. 236°, 74.3 km   **13.** 174°50′, 329.4 km/h
15. 314.8 km, 278.6°, 98.6°   **17.** 670 N,
1842 N   **19.** 1347.5 N each   **21.** 1919 N,
2585.4 N   **23.** 14.3 N at 250.8°   **25.** −4, no,
no   **27.** −20, no, yes   **29.** $-\dfrac{1}{5}\mathbf{a} + \dfrac{3}{5}\mathbf{b}$

31. $\dfrac{11}{25}\mathbf{a} + \dfrac{2}{25}\mathbf{b}$ **33.** 1 J   **35. (a)** −16 J
(b) −16 J   **37.** 1.10×10^6 J   **39.** 2.57×10^6 J

Chapter 7, pages 320–322   **1.** $r = \dfrac{3}{\sin \theta}$

3. $r = 3$ **5.**

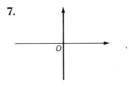

7.

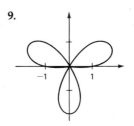

9.

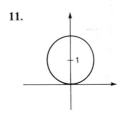

11.

13.

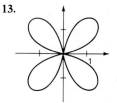

15.

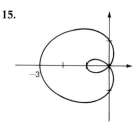

17. parabola **19.** hyperbola

21.

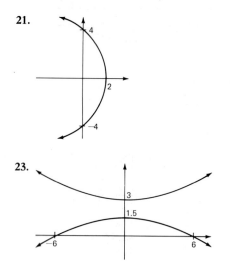

23.

25. $5 + 0i$, $7 + i$, $\dfrac{1}{2} + \dfrac{1}{2}i$ **27.** $5 + i$, $12 + 0i$,

$0 - \dfrac{2}{3}i$ **29.** $3 + i$, $\sqrt{10}$, $\dfrac{3}{10} + \dfrac{1}{10}i$, $8 - 6i$

31. $\dfrac{4}{5} - \dfrac{3}{5}i$, 1, $\dfrac{4}{5} - \dfrac{3}{5}i$, $\dfrac{7}{25} + \dfrac{24}{25}i$

33.

35.

37. $\sqrt{2}(\cos 135° + i \sin 135°)$

39. $2\sqrt{3}(\cos 330° + i \sin 330°)$

41. $4(\cos 240° + i \sin 240°)$

43. $2\sqrt{2}(\cos 345° + i \sin 345°)$

45. $3\sqrt{2}(\cos 90° + i \sin 90°)$

47. $\dfrac{\sqrt{2}}{2}(\cos 285° + i \sin 285°)$

49. $\dfrac{\sqrt{2}}{3}(\cos 30° + i \sin 30°)$

51. $\cos 60° + i \sin 60°$, $\dfrac{1}{2} + \dfrac{\sqrt{3}}{2}i$

53. $\dfrac{\sqrt{2}}{2}(\cos 45° + i \sin 45°)$, $\dfrac{1}{2} + \dfrac{1}{2}i$

55. $-\dfrac{27\sqrt{2}}{2} + \dfrac{27\sqrt{2}}{2}i$ **57.** $\dfrac{1}{32} + 0i$

59. $-\dfrac{1}{4} + 0i$ **61.** $-4 + 0i$

63.

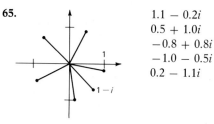

$1.1 + 0.3i$
$-0.8 + 0.8i$
$-0.3 - 1.1i$

65.

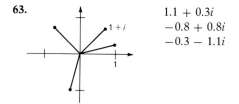

$1.1 - 0.2i$
$0.5 + 1.0i$
$-0.8 + 0.8i$
$-1.0 - 0.5i$
$0.2 - 1.1i$

67. $\cos 67.5° + i \sin 67.5°$
$\cos 157.5° + i \sin 157.5°$
$\cos 247.5° + i \sin 247.5°$
$\cos 337.5° + i \sin 337.5°$

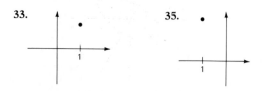

69.

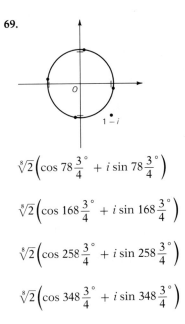

$$\sqrt[8]{2}\left(\cos 78\frac{3}{4}^{\circ} + i\sin 78\frac{3}{4}^{\circ}\right)$$

$$\sqrt[8]{2}\left(\cos 168\frac{3}{4}^{\circ} + i\sin 168\frac{3}{4}^{\circ}\right)$$

$$\sqrt[8]{2}\left(\cos 258\frac{3}{4}^{\circ} + i\sin 258\frac{3}{4}^{\circ}\right)$$

$$\sqrt[8]{2}\left(\cos 348\frac{3}{4}^{\circ} + i\sin 348\frac{3}{4}^{\circ}\right)$$

Chapter 8, pages 322–324

1. $x = x' + 3$
$y = y' + 4$

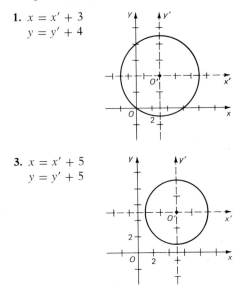

3. $x = x' + 5$
$y = y' + 5$

5. $y' = 2\cos 2x'$

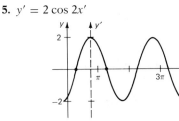

7. $y' = y - 2,\ x' = x - 3$

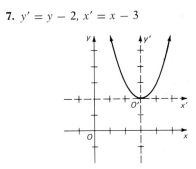

9. $x' = x - 1,\ y' = y - 1$

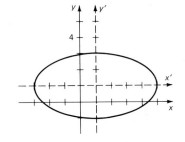

11. $\dfrac{(x')^2}{4} + \dfrac{(y')^2}{9} = 1$

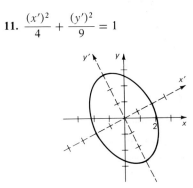

13. $\dfrac{(x')^2}{20} + \dfrac{(y')^2}{4} = 1$

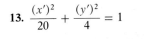
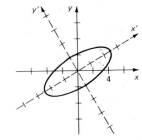

15. $\dfrac{(y')^2}{4} - \dfrac{(x')^2}{44} = 1$

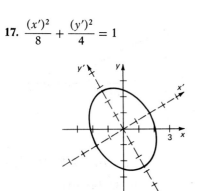

17. $\dfrac{(x')^2}{8} + \dfrac{(y')^2}{4} = 1$

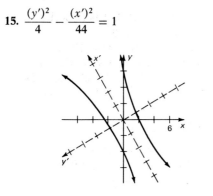

19.

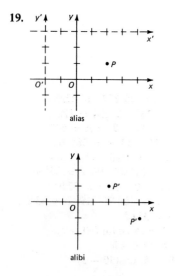

21.

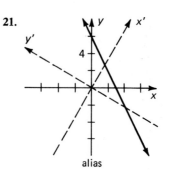

alias

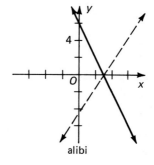

alibi

23.

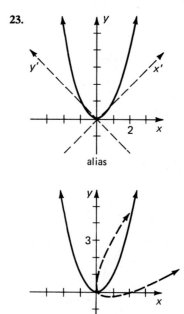

25.

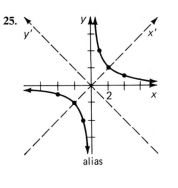

alias

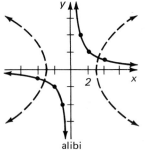

alibi

27. $x = \dfrac{5x' - y'}{\sqrt{26}}, \; y = \dfrac{x' + 5y'}{\sqrt{26}}$

29. $x = \dfrac{2x' + y'}{\sqrt{5}}, \; y = \dfrac{-x' + 2y'}{\sqrt{5}}$

31. $x = \dfrac{x' - y'}{\sqrt{2}}, \; y = \dfrac{x' + y'}{\sqrt{2}}$

33. $x = \dfrac{2x' + y'}{\sqrt{5}}, \; y = \dfrac{-x' + 2y'}{\sqrt{5}}$

Chapter 9, pages 324–326

1. (a)

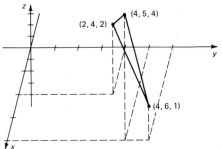

(b) sides are 3, 3, $\sqrt{10}$; Δ is isosceles
3. $x^2 + y^2 + z^2 - 9 = 0$
5. $x^2 + y^2 + z^2 - 4x - 4z - 17 = 0$
7. Center $(0, 0, 0)$, radius 4 **9.** Center
$(0, -1, 2)$, radius 2 **11. (a)** $2\mathbf{i} + 8\mathbf{j} - 6\mathbf{k}$,
$4\mathbf{i} - 14\mathbf{j} + 8\mathbf{k}$ **(b)** $\sqrt{6}, \; \sqrt{13}, \; \sqrt{3}, \; \sqrt{35}$
13. $61.9°$ **15.** $118.1°$

17. $\mathbf{v} = \dfrac{2}{9}\mathbf{a} + \dfrac{7}{9}\mathbf{b} - \dfrac{1}{9}\mathbf{c}$

19. $\mathbf{r} = (1 + t)\mathbf{i} + t\,\mathbf{j} + (1 - 2t)\mathbf{k}$
21. $\mathbf{r} = (1 + t)\mathbf{i} + t\,\mathbf{j} + (3 + t)\mathbf{k}$

23. $(1, 0, 1)$ **25.** $\dfrac{1}{3}\sqrt{222}$

27. $3x + y + 2z - 7 = 0$

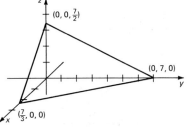

29. $\mathbf{r} = (2 + 3t)\mathbf{i} + (2 - t)\mathbf{j} + 2t\,\mathbf{k}$ **31.** $\dfrac{\sqrt{14}}{7}$
33. $x + 2y - 3z = 0$ **35. (a)** The
vector $2\mathbf{i} + \mathbf{j} - 3\mathbf{k}$ is perpendicular to both
planes, so they are parallel. **(b)** $\dfrac{5\sqrt{14}}{7}$

Chapter 10, pages 326–327 **1.** 2223 km
3. 13,420 km **5.** 16° **7.** $\angle A = 120°$,
$\angle B = 90°$, $\angle C = 90°$ **9.** $\angle A = 60°$,
$\angle B = 90°$, $c = 90°$ **11.** $a = 58°10'$,
$b = 114°0'$, $\angle B = 110°40'$ **13.** $a = 46°50'$,
$b = 107°20'$, $c = 101°50'$ **15.** $a = 36°50'$,
$b = 53°0'$, $c = 61°10'$ **17.** $c = 95.0°$,
$\angle A = 75.3°$; $\angle B = 108.6°$ **19.** $b = 83.8°$,
$c = 85.0°$, $\angle A = 35.6°$
21. $\sin(90 - B) = \tan a \tan(90 - c)$
$\sin(90 - c) = \cos a \cos b$
$\sin c = \cos(90 - A)\cos(90 - c)$

23. $\sin(90 - c) = \tan(90 - B)\tan(90 - A)$
$\sin(90 - A) = \tan(90 - c)\tan b$
$\sin a = \cos(90 - c)\cos(90 - A)$ **25.** no
solution **27.** $b = 86°10'$, $\angle A = 40°10'$,
$\angle C = 94°30'$ **29.** $\angle C = 107°50'$,
$\angle B = 158°20'$, $b = 157°10'$ or $\angle B = 21°40'$,
$b = 22°50'$, $\angle C = 72°10'$
31. $\angle A = 69°40'$ $\angle B = 110°20'$
$\angle C = 80°30'$ **33.** $a = 130°20'$ $\angle B = 45°30'$
$c = 121°50'$ **35.** $b = 85.6°$, $c = 46.6°$,
$\angle A = 126.6°$ **37.** $c = 33.7°$, $\angle A = 52.5°$,
$\angle B = 82.7°$ **39. (a)** 2932 **(b)** 2468 **(c)** 8332
41. $48°14'$ N, $98°7'$ W. **43.** $51°55'$ N,
$26°28'$ W

Chapter 11, page 328 **1. (a)** 0.868 **(b)** 0.710
(c) $\sin \dfrac{\pi}{3} = \dfrac{\sqrt{3}}{2} \approx 0.866$;

$\sin \dfrac{\pi}{4} = \dfrac{\sqrt{2}}{2} \approx .707$ **3. (a)** 0.805 **(b)** -0.808
(c) one is approximately the opposite of
the other, for $-.94 \approx 2.2 - \pi$ **5.** 1.543 $\cdots$

7. -1 **9.** $\dfrac{\sqrt{2}}{2} + \dfrac{\sqrt{2}}{2} i$

ALGEBRA AND GEOMETRY
REVIEW EXERCISES

Similar Triangles, page 329 **1.** $x = 10.5$
3. $x = 45$ **5.** $x = 34.2$ **7.** $x = 40$ **9.** $x = 6.6$

The Pythagorean Theorem, page 330
1. $c = 17.0$ **3.** $b = 36.5$ **5.** $c = 82.5$
7. $b = 77.9$ **9.** $b = 15.7$

Simplifying Radical Expressions, page 331
1. $\dfrac{3}{5}\sqrt{2}$ **3.** 2 **5.** $3\sqrt{2} + \sqrt{6}$ **7.** $\dfrac{4x^2\sqrt{3}}{3}$
9. $\dfrac{x\sqrt{3y}}{9y}$

Special Right Triangles, page 331
1. $x = 14$ **3.** $x = 68$ **5.** $x = \dfrac{17\sqrt{3}}{2}$
7. $x = 20\sqrt{3}$ **9.** $x = 2\sqrt{6}$

Slope of a Line, page 332 **1.** $m = 2$
3. $m = 0$ **5.** $m = 5$ **7.** m is undefined.

Parallel Lines, page 333 **1.** $\theta = 90°$
3. $\theta = 110°$ **5.** $\theta = 40°$ **7.** $\theta = 55°$
9. $\theta = 50°$

Graphing, page 334

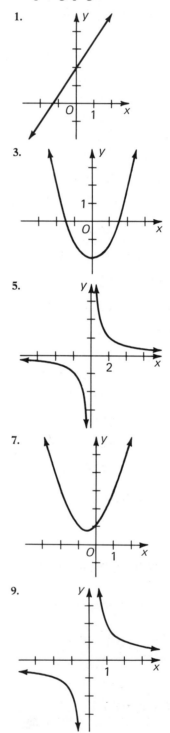

Simplifying Polynomial Expressions, page 335   **1.** $\dfrac{x}{x + 5}$ $(x \neq -2, x \neq -5)$ **3.** $\dfrac{2z - 3}{3}$ $(z \neq -1)$ **5.** $\dfrac{4x}{x + 1}$ $(x \neq 1, x \neq -1)$

7. $\dfrac{(x + 2)}{(x + 3)}$ $(x \neq -3, x \neq 1)$

9. $(a + 2)(a - 1)$ $(a \neq 2, a \neq -1)$

The Distance Formula, page 335   **1.** $\sqrt{85}$ **3.** $\sqrt{157}$ **5.** $2\sqrt{5}$ **7.** $\sqrt{122}$ **9.** $\sqrt{689}$

Solving Equations, page 336   **1.** $x = -1$ or $x = -4$ **3.** $x = \pm\dfrac{\sqrt{42}}{7}$ **5.** no solution

7. $x = 14$ or $x = -4$ **9.** $x = 4$

Area of a Polygon, page 337   **1.** 108

3. 169   **5.** 35   **7.** $\dfrac{363\sqrt{3}}{2}$ **9.** 1200

Completing the Square, page 338
1. $y = -5 \pm \sqrt{30}$ **3.** $y = -4 \pm \sqrt{22}$
5. $x = 3 \pm \sqrt{5}$ **7.** $x = -3$ or $x = -7$

9. $x = \dfrac{1}{2}$ or $x = -3$

Conic Sections, page 338
1. $(x - 1)^2 + (y - 2)^2 = 9$; circle   **3.** $xy = 3$; hyperbola   **5.** $\dfrac{x^2}{4} + y^2 = 1$; ellipse

7. $y = (x + 6)^2 - 32$; parabola

Systems of Equations, page 339   **1.** $\left(\dfrac{20}{3}, 6\right)$

3. $(1, 1)$ **5.** $(-4, -3)$ **7.** $\left(\dfrac{2}{5}, \dfrac{8}{5}, 0\right)$

9. $(3, -2, 0)$ **11.** $(4, 2, 3)$

APPENDIX. COMPUTATIONS WITH LOGARITHMS

Exercises A-1, pages 342–343   **1.** $y = 1$
3. $y = 0$ **5.** $y = -2$
7. $\log_{10} 127 + \log_{10} 42$ **9.** $15 \log_{10} 17$
11. $\log_{10} 59.7 - \log_{10} 23.2$
13. $\log_{10} 14.1 + 2\log_{10} 2.5 - \log_{10} 3.2$
15. $\dfrac{2}{3}\log_{10} 10.2 + \dfrac{1}{3}\log_{10} 5.1 - \log_{10} 4.1$
$- 3\log_{10} 6.7$ **17.** $\log_{10} 4.7 + 1$
19. $\log_{10} 7.36 + (-1)$ **21.** $\log_{10} 8.592 + 2$

Exercises A-2, page 346   **1.** 0.8331
3. 1.6021   **5.** 1.7875   **7.** 3.8745
9. 8.3617 − 10   **11.** 7.5933 − 10   **13.** 20.4
15. 0.556   **17.** 0.752   **19.** 0.955   **21.** 1544
23. 0.006003

Exercises A-3, pages 347–348
1. $\log_{10} 31.4 + \log_{10} 253$
3. $\log_{10} 517 - \log_{10} 37.1$
5. $2\log_{10} 22.4 + \log_{10} 12.4 - \log_{10} 8.54$
7. $\dfrac{1}{6}\log_{10} 0.0598$

9. $\dfrac{1}{3}\log_{10} 51.2 + \dfrac{1}{3}\log_{10} 7.81 - \dfrac{1}{3}\log_{10} 2.93$

11. $\dfrac{1}{5}\log_{10} 6 + \dfrac{1}{5}\log_{10} 3.91 - \dfrac{1}{5}\log_{10} 2$

$- \dfrac{1}{5}\log_{10} 3.02$ **13.** $\dfrac{1}{12}\log_{10} 16.4$

15. $\log_{10} 3.251 + \dfrac{1}{2}\log_{10} 64.9 + \dfrac{1}{2}\log_{10} 21.2$

$- \dfrac{1}{2}\log_{10} 15.7$ **17.** 7940   **19.** 13.9   **21.** 728

23. 0.625   **25.** 5.15   **27.** 1.31   **29.** 1.26
31. 30.4

Exercises A-4, page 349   **1.** $\angle B = 47°$, $a = 33.9$, $c = 49.8$ **3.** $\angle A = 44°00'$, $\angle B = 46°00'$, $b = 17.4$ **5.** $\angle A = 21°34'$, $\angle B = 68°26'$, $b = 36.00$ **7.** $\angle A = 63°$, $b = 5.21$, $c = 7.71$ **9.** $\angle B = 32°$, $a = 202$, $c = 208$

Mixed Review • Chapters 1–3, pages 382–383   **1.** $\sin 60° = \dfrac{\sqrt{3}}{2}$; $\cos 60° = \dfrac{1}{2}$; $\tan 60° = \sqrt{3}$; $\csc 60° = \dfrac{2\sqrt{3}}{3}$; $\sec 60° = 2$; $\cot 60° = \dfrac{\sqrt{3}}{3}$ **3.** $\cos(6.03) = 0.97$

7. $307°21'$ **9.** $\sin 285° = -\dfrac{\sqrt{2} - \sqrt{6}}{4}$; $\tan 285° = \dfrac{-\sqrt{3} - 3}{3 - \sqrt{3}}$ **11.** $(-2\sqrt{2}, 2\sqrt{2})$

15. $\dfrac{2}{\sin \theta}$ **17.** $\angle A = 20°$; $\angle B = 70°$; $b = 35.0$ **21.** $-\cos^2 \theta$

23. $\sin 75° = \dfrac{\sqrt{6} + \sqrt{2}}{4}$ **25.** $100°$ **27.** $\dfrac{1}{8}$

29. $\dfrac{\sqrt{14}}{4}$ **31.** 26.2

33.

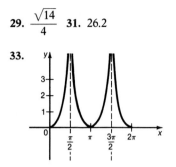

Mixed Review • Chapters 1–7, pages

384–387 1. (a) $\dfrac{5\pi}{6}$, **(b)** $-35°20'$

3. $y = 4\cos\left(\dfrac{2\pi}{3}t + \pi\right)$; $\dfrac{1}{3}$ **5.** 256

7. $167°54'46''$ **9.** 8.4   **13.** 1.025π **15.** 18.9

17. $\dfrac{-\sqrt{2} + \sqrt{3}}{2}$

19.

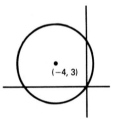

(−4, 3)

21. (a) $2\sqrt{2} - i$, **(b)** 3, **(c)** $\dfrac{2\sqrt{2}}{9} - \dfrac{1}{9}i$

23. $3\sqrt{5}(\cos 206.6° + i\sin 206.6°)$ **25.** 59.2

27. $s = 2$; $t = \dfrac{5}{3}$

29. (a) $\cos(-10°) + i\sin(-10°)$,
(b) $\cos(170°) + i\sin(170°)$

$\dfrac{z}{i}$ •

iz •

z •

31. $\left(0, -\dfrac{1}{2}\right)$ **33.** 0.8704; -0.4924; -1.767

35. $\csc^2\theta$ **37.** $\sin\theta = \dfrac{20}{29}$; $\cos\theta = -\dfrac{21}{29}$;

$\tan\theta = -\dfrac{20}{21}$; $\cot\theta = -\dfrac{21}{20}$; $\sec\theta = -\dfrac{29}{21}$

45.

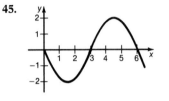

47. $0 + i$; $-\dfrac{\sqrt{3}}{2} + \dfrac{-1}{2}i$; $\dfrac{\sqrt{3}}{2} + \dfrac{-1}{2}i$

49. $\sin 255° = \dfrac{-\sqrt{2} - \sqrt{6}}{4}$;

$\cos 255° = \dfrac{-\sqrt{6} + \sqrt{2}}{4}$;

$\tan 255° = \dfrac{\sqrt{3} + 3}{3 - \sqrt{3}}$ **51.** There are no such

values.   **53.** Translate the graph of
$y = \cos x$ π units to the left. Then reflect
the result in the x-axis. Translate the result
1 unit up.

55.

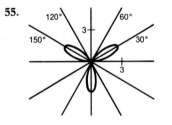

57. (a) i, **(b)** -1 **59.** $-53.7 - 34.9i$

63. $\dfrac{3\pi}{4} + 2n\pi$, $\dfrac{7\pi}{4} + 2n\pi$